Lecture Notes in Control and Information Sciences 426

For further volumes:
http://www.springer.com/series/642

Abdellah Benzaouia

Saturated Switching Systems

Springer

Author
Professor Abdellah Benzaouia
Department of Physics
Faculty of Science Semlalia
University of Cadi Ayyad
B.P. 2390, 40 000 Marrakech
Morocco
E-mail: benzaouia@uca.ma

ISSN 0170-8643 e-ISSN 1610-7411
ISBN 978-1-4471-2899-1 e-ISBN 978-1-4471-2900-4
DOI 10.1007/978-1-4471-2900-4
Springer London Heidelberg New York Dordrecht

Library of Congress Control Number: 2012931319

Printed on acid-free paper

Springer is part of Springer Science+Business Media (www.springer.com)

To

Laila, Soufiane, Rim,
Samy, Mohamed Walid.

To

All my family.

Contents

Acronyms

C. T. S	Continuous-time system
D. T. S	Discrete-time system
C. T. C	Continuous-time case
D. T. C	Discrete-time case
T-S	Takagi-Sugeno
SOFC	Static output feedbcak control
LMI	Linear matrix inequality
2 D	Two dimensional
OAE	Output algebraic equation
w.r.t.	with respect to
s.t.	such that
resp.	respectively
Eq.	Equation
Eqs.	Equations
co{.}	Convex hull of {.}
PDC	parallel distributed compensation
SISO	Single input single output
MIMO	Multiple input multiple output

Notations

- If x, y are vectors of $\mathbb{R}^n$, then $x \leq y$ stands componentwise.
- For a matrix $A \in \mathbb{R}^{n \times m}$, $|A|$ is the matrix formed by the absolute value of the components of A, while $\sigma(A)$ denotes its spectrum.
- For a vector $\theta_i \in \mathbb{R}^n$, θ_l^i indicates the lth component of the vector.
- $intD$ denotes the interior of the set D.
- For a square matrix $Q > 0, (Q \geq 0)$ if $Q \in \mathbb{R}^{n \times n}$ is positive definite (positive semi definite, respectively).
- $Q_j, j = 1, \ldots, n$, denotes the jth row of matrix Q.
- $A \succeq 0$ stands for a *positive* matrix A, that is, a matrix with nonnegative elements: $a_{ij} \geq 0$.
- $\rho(A)$ stands for the radius spectrum of matrix A.
- $\mathscr{I} := \{1, \ldots, N\}$, while $\mathscr{I}^2 = \mathscr{I} \times \mathscr{I}$.
- A matrix whose its off-diagonal entries are non positive is called Z-matrix.
- Scalar η denotes, $\eta = 2^m$.
- $\mathbb{I}$ denotes the identity of appropriate size.
- For a square matrix $H \in \mathbb{R}^{m \times m}$, $\tilde{H}_d, \widetilde{H}_c \in \mathbb{R}^{2m \times 2m}$ are defined as

$$\tilde{H}_d = \begin{bmatrix} H^+ & H^- \\ H^- & H^+ \end{bmatrix},$$
$$\widetilde{H}_c = \begin{bmatrix} H_1 & H_2 \\ H_2 & H_1 \end{bmatrix}$$

with

$$H^+(i,j) = h_{ij}^+ = \text{Sup}(h_{ij},\ 0),\ H^-(i,j) = h_{ij}^- = \text{Sup}(-h_{ij},\ 0),$$

for $i,\ j = 1,\ \ldots,\ n$, where h_{ij} denotes the matrix component $H(i,j)$ and

$$H_1(i,j) = \begin{cases} h_{ij} & if \quad i = j \\ h_{ij}^+ & if \quad i \neq j \end{cases} \qquad H_2(i,j) = \begin{cases} 0 & if \quad i = j \\ h_{ij}^- & if \quad i \neq j \end{cases}$$

- For a complex vector ξ, $\bar{\xi}$ stands for the conjugate vector while $\bar{\lambda}$ stands for the conjugate of the complex scalar λ.

Introduction and Preview

Introduction

Switched systems are a class of hybrid systems encountered in many practical situations which involve switching between several subsystems depending on various factors [43, 46, 54, 75, 118, 142]. Generally, a switching system consists of a family of continuous-time subsystems and a rule that supervises the switching between them. For example, many process in the chemical and pharmaceutical industries operate following batches, composed of different operations that are carried out in sequence. This changes discontinuously the dynamics of the operation [144]. In manufacturing, hybrid switched systems are found in steel rolling mills [112], used for producing thin metal sheets, following several steps based on pressing the metal strip with rolling cylinders: the dynamics are known to change at each pass due to the variation in thickness [50]. Many other examples can be found in the automotive industry, in aircraft and air traffic control, and many other fields. Some stabilizability problems for switching/switched systems are studied in [77, 120, 159, 160].

Two main problems are widely studied in the literature according to the classification given in [45]: The first one, which is the one solved in this book, looks for testable conditions that guarantee the asymptotic stability of a switched system under arbitrary switching rules, called in this book switching system, while the second is to determine a switching sequence that renders the switched system asymptotically stable (see [118]) and the reference therein).

A main problem which is always inherent to all dynamical systems is the presence of actuator saturations. Even for linear systems, this problem has been an active area of research for many years. Besides approaches using anti-windup techniques [131], model predictive controls [56] and asymmetric l_1 [136], two main approaches have been developed in the literature: The first is the so-called positive invariance approach which is based on the design of controllers which work inside a region of linear behavior where saturations do not occur (see [10, 15, 17, 44], and the references therein). This approach has already being applied to a class of hybrid systems involving jumping parameters [51]. It has also been used to design controllers for switching systems with constrained control under complete modelling taking into account reset functions at each switch and different system's dimension. The second approach, however allows saturations to take effect while guaranteeing asymptotic stability (see [26, 103, 104, 150] and the references therein). The main challenge in these two approaches is to obtain large domains of initial states which ensures asymptotic stability for the system despite the presence of saturations [17, 18, 94, 103].

The objective of this book is to present the available results in the literature for switching systems subject to actuator saturations. These results follow generally two ways: the first concerns the synthesis of non saturating controllers (controllers working inside a large region of linear behavior where the saturations do not occur), while the second extends the results obtained for unsaturated switching systems leading to saturating controllers (controllers tolerating saturations to take effect)[65, 66, 85]. The first approach was successfully used to study a general class of switching systems, with reset functions and different subsystem orders [27]. Necessary and sufficient condition of positive invariance is obtained. A design method based on the resolution of equation $XA + XBX = HX$ is proposed. A new topic, using the same approach, presented for the first time in this book, concerns the study of saturated singular switching systems. The controller design is developed upon the resolution of the algebraic equation $XA + XBXE = HE$ also presented in this book. The second method was firstly used in [22] with the use of a multiple Lyapunov function. However, only the intersection of all the corresponding level sets of the local functions was considered as a region of asymptotic stability of the switching system. This drawback is improved in [25, 35] by considering, for the first time, a large set of asymptotic stability composed by the union of all the level sets. In this context, different sufficient conditions of asymptotic stability were obtained for switching systems subject to actuator saturations. Furthermore, these conditions were presented in the form of LMIs for the state feedback control case.

The static output feedback control (SOFC) for dynamical systems plays a very important role in control theory and applications. The purpose is to design controllers such that the resulting closed-loop system is asymptotically stable without using any reconstruction method of the unavailable states. This type of control was already used for switching systems in [66] and extended to saturated switching systems in [25] and positive switching systems in [29]. In this book, a particular attention is given to the output feedback case which has an additive complexity due to the output equation. It is also shown, upon examples, that the LMIs obtained for computing controllers working inside a large region of linear behavior are not always more conservative. The obtained results are then extended to uncertain switching systems subject to actuator saturations as developed in [30, 31]. The uncertainty types considered in these two works are the polytopic one and the structured one. This second type of uncertainty was also studied, without saturation, in [98].

A frequent and inherent constraint in dynamical systems is the nonnegativity of the controls and/or the states. Systems with nonnegative states are important in practice because many physical and chemical processes involve quantities that have intrinsically a constant and nonnegative sign: temperatures, level of liquids, concentration of substances, etc, are of course positive or nonnegative. This problem is of relevance in many practical applications of switched systems, as the states frequently cannot take negative values for safety or production regulations. For example, control of the force applied in rolling mills must take into account that the width of the metal strip must always be positive. Unfortunately, in the presence of switches it is not simple to check that a given control technique will respect this nonnegativity, which prompts the development of specific control design techniques for switched systems that take into account this request.

In the literature, systems where the states are nonnegative whenever the initial conditions are nonnegative are referred to as being positive [84]. The design of controllers for these positive systems has been extensively studied even with constraints, see for example [1, 2, 99, 100] and references therein. To our knowledge, few works have directly considered positive switching linear systems. For example, the stability of switching positive systems composed of two subsystems has been studied by [96, 146]. In these works, the authors studied the difficulty of constructing a common Lyapunov function for a positive switching continuous-time system. The problem of switching positive systems was investigated in [29, 34] by using multiple Lyapunov functions but without saturation. The obtained results are recalled in this book and extended to saturated positive switching systems in discrete-time.

This book also studies the stabilization problem by state and output feedback control for linear two dimensional (2D) switching systems. In the last two decades, the 2D system theory has received a considerable attention by many researchers. The 2D linear models were introduced in the seventies [87, 88] and have found many applications, in digital data filtering, image processing, modeling of partial differential equations [127], among others. On the other hand, the stabilization problem is not fully investigated and still not completely solved. It has been shown that the stability of 2D systems can be reduced to checking the stability of 2D characteristic polynomial [4, 141, 157]. Further, in the literature, various types of easily checkable, but only sufficient conditions, for asymptotic stability and stabilization for 2D linear systems have been proposed [90, 91, 101, 116, 164]. However, in the best of our knowledge, no works on 2D switching systems were developed before [33, 36].

The last problem studied in this book concerns discrete-time switching nonlinear systems. Each subsystem is written as an equivalent T-S fuzzy model. Since the introduction of T-S fuzzy models by Takagi and Sugeno [147] in 1985, fuzzy model control has been extensively studied because T-S fuzzy models provide an effective representation of complex non linear systems [28, 38, 39, 132, 133, 135, 154]. In the literature, few works were interested to switching T-S fuzzy systems. One can consult [148, 166] where a detailed study of the problem of stability and controller switching for switching fuzzy systems is presented. In all the previous works, only a common Lyapuniv function was used for all the T-S fuzzy subsystems. Even for switching linear systems, it was proven that the use of such a common Lyapunov function leads to conservative results [54, 146]. A different approach using sampled-data fuzzy controller for nonlinear systems based on switching T-S fuzzy model is proposed in [115]. While robust control problem for uncertain switching fuzzy systems is studied in [121, 153].

In this book, sufficient conditions of stabilizability, by state feedback control, are obtained by using two types of multiple Lyapunov functions. These conditions are then worked out to be presented in a LMI form.

Preview of Chapters

This book is composed of ten chapters: The first and second chapters present the background of the approach called positive invariance which was developed in the eighteen and nineteen for linear systems to deal with saturations on the control of the system. This approach is known by designing controllers working in a linear behavior and does not allow saturation of the control. The use of this technique is extended to the class of switching systems in Chapter *IV* and Chapter V. The main idea of the technique of saturation is also recalled in this chapter. This technique presents conditions of building stabilizing controllers, for linear systems, allowing saturation to take effect on the control. These results are then extended to switching systems in Chapter VI and Chapter VII.

Chapter III introduces the class of switching systems by showing their particular comportment and the difficulties of studying the stability of this class. Some examples are presented to illustrate the different evolutions depending on the choice of the switching sequence which can stabilize a switching system composed of two unstable subsystems and destabilize a switching system composed of two stable subsystems. The main tools used to study the stability of switching systems are then recalled. A particular sight is given to multiple Lyapunov functions.

Chapter *IV* presents the first extension of the approach called positive invariance to the class of switching systems. The switching system here is taken in a general form with reset functions. In this chapter, necessary and sufficient condition of positive invariance is obtained. A design method based on the resolution of equation $XA + XBX = HX$ is proposed. A main result obtained for the first time with this approach is also presented. This result consists in proving that the union of all the sets of local positive invariance constitutes a set of positive invariance of the switching system. The developed controllers with this technique work only in regions of linear behavior.

Chapter *V* presents a different type of switching systems: systems where the sequence of switching is random and follows a Markovian law. This class is called Systems with Markovian Jumping Parameters. For this class of systems, the approach of positive invariance is again applied. Necessary and sufficient condition of positive invariance of a common set to all the subsystems is proposed. The cases of discrete-time and continuous-time systems are studied.

In Chapter VI, the technique of saturation is applied to switching systems allowing the design of stabilizing controllers tolerating saturations to take effect on the control. The obtained results are presented in LMI form. The objective of using this technique is to obtain larger domains of initial values. However, this objective is not theoretically guaranteed at all. As particular results, one find those obtained with positive invariance approach. A comparison between the two techniques is presented for some examples showing the advantage and the drawback of each technique. The problem of stabilizing switching systems by using output feedback is also studied.

Chapter *VII* extends the results obtained in Chapter *VI* to uncertain switching systems. Two types of uncertainties are considered: polytopic uncertainties and parametric uncertainties. Stabilizing controllers are developed while tolerating saturations despite the presence of uncertainties. Illustrative examples are presented for each type. While Chapter *VII* deals with saturated discrete-time uncertain switching systems with arbitrary switching sequences, is also concerned with saturated switched discrete-time uncertain systems where the switching sequence is considered as the control.

Chapter *VIII* deals with the stability and stabilizability analysis of positive switching systems with multiple Lyapunov functions. A necessary and sufficient condition for the positive switching system to admit a multiple Lyapunov function is proposed. A linear programming is presented to select positive diagonal matrices to construct this Lyapunov function. The obtained results are used to study the problem of stabilization of positive switching systems. Stabilizing controllers are designed by using state feedback and output feedback control. All the results are presented in LMIs form making easy their application. Illustrative examples are also studied. The extension to saturated positive switching systems is also developed.

Chapter *IX* studies the stabilization problem by sate feedback and output feedback control respectively for linear two dimensional (2D) switching systems. New conditions of stabilizability are obtained in this chapter by using common and multiple Lyapunov functions for both state feedback and output feedback controls. The case of saturated switching 2D systems is also studied. The unsaturating controllers ensuring asymptotic stability for this class of systems is then presented.

In Chapter *X*, sufficient conditions of stabilizability are obtained for switching nonlinear systems described under Takagi-Sugeno form. A switched Lyapunov function and state feedback control are used. The proposed Lyapunov function is multiple for both the T-S fuzzy subsystems and the switching modes. This choice allows a general approach which may be less conservative. The obtained conditions are then worked out to be presented in LMI form.

for some examples showing the advantages and the drawback of each technique. The problem of stabilizing switching systems by [illegible]

[illegible]

List of Figures

Chapter 1
Saturated Linear Systems: Analysis

1.1 Introduction

This chapter recalls the main results developed on saturated linear systems to be used for saturated switching systems. Namely, two approaches are recalled:

- The so-called positive invariance approach based on the idea of working inside linear behavior regions without tolerating the control to be saturated;
- The technique of saturation tolerating the control to be saturated while guaranteeing the asymptotic stability of the system.

This chapter is devoted to the study of saturated linear systems as presented in Fig. 1.1:

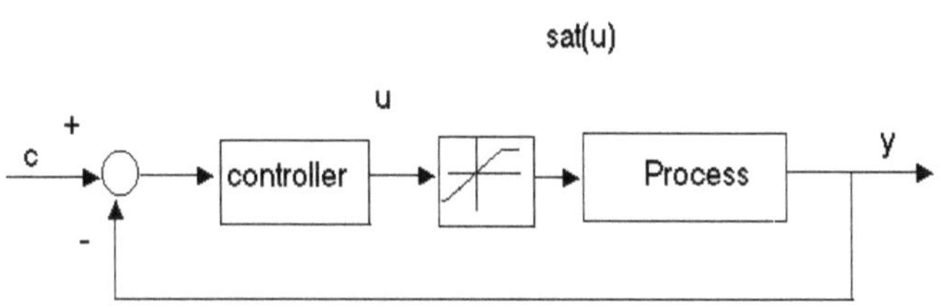

Fig. 1.1 Block schema of saturated systems

1.2 Discrete-Time Systems

In this section, we deal with linear discrete-time systems described by

$$x_{k+1} = Ax_k + Bu_k \tag{1.1}$$

with $k \in \mathbb{N}$, x is the state vector in $\mathbb{R}^n$ and u is the constrained control satisfying

$$u_k \in \Omega \subset \mathbb{R}^m, \; m \leq n. \tag{1.2}$$

A. Benzaouia: Saturated Switching Systems, LNCIS 426, pp. 1–35.
springerlink.com

Matrices A and B are constant and satisfy the assumption :

$$(A,B) \text{ is stabilizable.} \tag{1.3}$$

As it generally occurs in practical situations, the set of admissible controls Ω is an asymmetric polyhedral set defined as

$$\Omega = \{u \in \mathbb{R}^m : -q_2 \leq u \leq q_1,\ q_1,\ q_2 \in \mathbb{R}^m_+\}. \tag{1.4}$$

In order to study the problem control design under inequality constraints, we follow the approach adopted by [97]. Let us first consider the unconstrained case where the regulator problem for system (1.1) consists in the design of a feedback law as

$$u_k = F x_k, \text{ with } F \in \mathbb{R}^{mxn}. \tag{1.5}$$

Applying the control law as defined above, system (1.1) becomes

$$x_{k+1} = (A+BF)x_k = A_o\, x_k \tag{1.6}$$

Let $x(k;x_o)$ be the motion of system (1.6), at time k, starting at x_o.

Generally speaking, the matrix F is chosen in order to speed up the closed-loop system dynamics with (1.6) asymptotically stable, that is,

$$\rho(A+BF) < \rho(A) \text{ and } \rho(A+BF) < 1$$

and

$$\text{rank}(F) = m.$$

In the constrained case, the feedback law is defined as

$$u_k = sat(Fx_k) = \begin{cases} q_1 & \text{for } Fx_k > q_1 \\ Fx_k & \text{for } Fx_k \in \Omega \\ -q_2 & \text{for } Fx_k < -q_2, \end{cases}$$

This feedback law implies two possible models for the system in the closed loop :

(i) the linear model

$$x_{k+1} = (A+BF)x_k = A_o x_k, \qquad \text{for } Fx_k \in \Omega, \tag{1.7}$$

(ii) the non-linear model

$$x_{k+1} = Ax_k + Bsat(Fx_k), \qquad \text{for} Fx_k \notin \Omega. \tag{1.8}$$

Both representations are obtained in two different regions of the state space as indicated in Fig. 1.2.

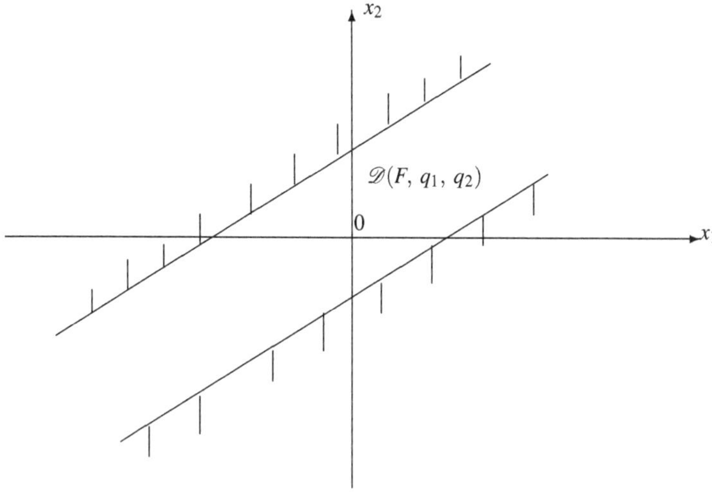

Fig. 1.2 Validity regions for the linear and the nonlinear model

The approach we deal with in this chapter consists in proceeding in such a way that the model (1.7) remains valid every time. This is only possible if the state is constrained to evolve in a specified region defined by : $F^{-1}\Omega = \mathscr{D}(F, q_1, q_2)$. where $F^{-1}\Omega$ stands for the inverse image of the Ω without the requirement of the invertibility of F. From (1.4), (1.7) and (1.7), the set of admissible states is defined as

$$\mathscr{D}(F, q_1, q_2) = \{x \in \mathbb{R}^n \;/\; -q_2 \le Fx \le q_1 ;\, q_1, q_2 \in \mathbb{R}^m_+\}. \tag{1.9}$$

Note that this domain is unbounded in the general case when $m < n$.

Clearly, if $x_k \in \mathscr{D}(F, q_1, q_2)$ we may get $x_{k+1} \notin \mathscr{D}(F, q_1, q_2)$.

Further, if a Lyapunov function $\mathscr{V}(x)$ is known for the system (1.7), then there always exists a scalar $c \in Int\mathbb{R}$ such that the set

$$\mathscr{D}_L = \{x \in \mathbb{R}^n \;/\; \mathscr{V}(x) \le c\}, \tag{1.10}$$

is a subset of $\mathscr{D}(F, q_1, q_2)$. Hence, for every $x_k \in \mathscr{D}_L \subset \mathscr{D}(F, q_1, q_2)$, we have $u_k \in \Omega$; consequently, the model (1.7) remains valid. Further, since $\mathscr{V}(x)$ is a Lyapunov function for the system (1.7), then for every $x \in \mathscr{D}_o \subset \mathscr{D}_L$, where $\mathscr{D}_o$ denotes the set of admissible initial states, we obtain $x(k,x_o) \in \mathscr{D}_L$, $\forall\, k \in \mathbb{N}$ and $x(k,x_o) \to 0$ as k goes to ∞.

It may be noted, from (1.4), that the set $\mathscr{D}(F, q_1, q_2)$ is generally of a polyhedral asymmetric nature. Thus, the largest domain of admissible initial values of system (1.1) is obtained if

$$\mathscr{D}_o = \mathscr{D}_L = \mathscr{D}(F, q_1, q_2). \tag{1.11}$$

The use of a quadratic Lyapunov function only allows one to obtain an ellipsoidal stability domain [97]. The idea of constructing the largest polyhedral stability domain $\mathscr{D}_L \subset \mathscr{D}(F, q_1, q_2)$ was put forward by [62] by using simplicial cones. Its formulation in the symmetrical case, was given by [47] who gives necessary and sufficient conditions for the set $\mathscr{D}(F, q_1, q_2)$ with $q_1 = q_2 > 0$ to be positively invariant w.r.t. the system.

Consider the following definitions:

Definition 1.1. *A subset $\mathscr{S}$ of $\mathbb{R}^n$ is said to be positively invariant with respect to (w.r.t.) the system (1.1)-(1.2), if for every initial state $x_o \in \mathscr{S}$, every admissible sequence*

$$\mathscr{U}_k = \{u_o, u_1, \ldots, u_{k-1};\, u_i \in \Omega\},$$

the motion $x(x_o, \mathscr{U}_k, k) \in \mathscr{S}$, for every $k \in \mathbb{N}$.

Definition 1.2. *A subset $\mathscr{S}$ of $\mathbb{R}^n$ is said to be*

- *contractive w.r.t. the system (1.1)-(1.2), if for every $x_k \in \partial(\tau_k \mathscr{S})$, there exists $\tau_{k+1} > 0$, satisfying $\tau_{k+1} < \tau_k$ such that $x_{k+1} \in \partial(\tau_{k+1}\mathscr{S})$, for every admissible control u_k and $k \in \mathbb{N}$. If $\tau_k < 1$ (resp. $\tau_k > 1$), we say that the set $\mathscr{S}$ is in-contractive (resp. out-contractive) w.r.t. the system.*
- *attractive for a subset $\mathscr{T}$ of $\mathbb{R}^n$ w.r.t. the system (1.1)-(1.2) if, for every $x_o \in \mathscr{T} \backslash \mathscr{S}$, there exists $k_o \in \mathbb{N}$ such that $x(x_o, \mathscr{U}_k, k) \in \mathscr{S}$, for every $k \geq k_o$, and every admissible sequence $\mathscr{U}_k$.*
- *globally attractive w.r.t. the system (1.1)-(1.2) if $\mathscr{T} = \mathbb{R}^n$.*

Note that the contractiveness property defined here is a one-step contractivity.

In the approach proposed by Gutman and Hagander [97], the necessity of the positive invariance property of domain $\mathscr{D}_L$ w.r.t. the system (1.7) (*i.e.*, $A_o\, \mathscr{D}_L \subset \mathscr{D}_L$), when we are interested in achieving (1.11), requires one to find conditions under which the set $\mathscr{D}(F, q_1, q_2)$ is positively invariant w.r.t. the system (1.7), the same system. This will be the main purpose of this chapter.

Hence, we will present in the subsequent paragraphs of this section, the necessary and sufficient conditions allowing the design of a regulator for linear discrete-time systems with symmetric and asymmetric constrained control.

Let us now define the null space $\mathscr{K}er(F)$ of F as follows :

$$\mathscr{K}er(F) = \{x \in \mathbb{R}^n \;/\; Fx = 0, F \in \mathbb{R}^{mxn}\}. \tag{1.12}$$

Consider the following state transformation

$$z_k = F\, x_k, \qquad F \in \mathbb{R}^{mxn}, \tag{1.13}$$

then, from (1.7), one can obtain

$$z_{k+1} = F(A + BF)x_k. \tag{1.14}$$

If a matrix $H \in \mathbb{R}^{mxm}$ exists such that,

$$FA_o = HF, \tag{1.15}$$

or equivalently,

$$FA + FBF = HF,$$

the n-order dynamical system (1.7) can be transformed to an m-order dynamical system given by

$$z_{k+1} = Hz_k, \;\; z_k \in \mathbb{R}^m, \tag{1.16}$$

and domain (1.9) becomes

$$\mathscr{D}(\mathbb{I}, q_1, q_2) = \{z \in \mathbb{R}^m \;/ -q_2 \le z \le q_1, q_1, q_2 \in \mathbb{R}^m_+\}. \tag{1.17}$$

The invariance positivity of (1.17) implies necessarily the stability of H. Further, comparing (1.9) with (1.17) leads to

$$x_k \in \mathscr{D}(F, q_1, q_2) \text{ iff } z_k \in \mathscr{D}(\mathbb{I}, q_1, q_2) \;\forall\, k \in N. \tag{1.18}$$

It is obvious, that in this case, the domain (1.9) is positively invariant (resp. positively invariant and contractive) w.r.t. the system (1.7) if and only if domain (1.17) is positively invariant (resp. positively invariant and contractive) w.r.t. the system (1.16).

In this approach, matrices A, B and F are given while matrix H is obtained as a solution to Equation (1.15). This approach is known as the direct approach and is based on the state transformation (1.13), Equation (1.15) and the transformed dynamical system (1.16) with domain (1.17) and property (1.18).

The first result presented in this section, concerns a necessary and sufficient condition for domain $\mathscr{D}(\mathbb{I}, q_1, q_2)$ to be positively invariant w.r.t. the system (1.16).

Theorem 1.1. *The subset $\mathscr{D}(\mathbb{I}, q_1, q_2)$ of $\mathbb{R}^n$ defined by (1.17) is positively invariant (resp. positively invariant and contractive) w.r.t. the system (1.16) if and only if*

$$\tilde{H}_d q \le q \;\; (resp.\;\; \tilde{H}_d q < q), \tag{1.19}$$

with

$$\tilde{H}_d = \begin{bmatrix} H^+ & H^- \\ H^- & H^+ \end{bmatrix}, \tag{1.20}$$

$$q = \begin{bmatrix} q_1 \\ q_2 \end{bmatrix}, \tag{1.21}$$

and

$$H^+_{ij} = sup(0, H_{ij}), \; H^-_{ij} = sup(0, -H_{ij}), \text{ for } i,\; j = 1 \ldots n. \tag{1.22}$$

Proof:

(Sufficiency): Let $z_k \in \mathscr{D}(\mathbb{I}, q_1, q_2)$, that is,

$$-q_2 \leq z_k \leq q_1. \tag{1.23}$$

Decompose H as follows, $H = H^+ - H^-$ with H^+ and H^- both non-negative matrices. By pre-multiplying inequality (1.23) by H^+ (resp. by $-H^-$, one obtains

$$-H^+ q_2 \leq H^+ z_k \leq H^+ q_1 \text{ (resp. } -H^- q_1 \leq -H^- z_k \leq H^- q_2). \tag{1.24}$$

Summing up both inequalities yields

$$-H^+ q_2 - H^- q_1 \leq H z_k \leq H^+ q_1 + H^- q_2. \tag{1.25}$$

Taking account of (1.19), one can deduce that $z_{k+1} \in \mathscr{D}(\mathbb{I}, q_1, q_2)$, for every $k \in \mathbb{N}$. Note that the contractivity property is guaranteed with $\tilde{H}_d q < q$.

(Necessity:) Assume that the set $\mathscr{D}(\mathbb{I}, q_1, q_2)$ is positively invariant w.r.t. the system (1.16) and condition (1.19) is violated. Thus, without loss of generality, assume there exists an integer $k \in \mathbb{N}$ and a subscript $i \in [1, \ldots, n]$ such that,

$$H_i^+ q_1 + H_i^- q_2 > (q_1)_i, \tag{1.26}$$

where H_i^+ is the ith row vector of the matrix H^+, the same definition applies for H_i^-. The ith component of q_1 is denoted as $(q_1)_i$. Consider the following state χ_k given by

$$(\chi_k)_j = \begin{cases} (q_1)_j \text{ if } H_{ij} > 0, \\ 0 \text{ if } H_{ij} = 0, \\ -(q_2)_j \text{ if } H_{ij} < 0, \end{cases} \qquad j = 1 \ldots n.$$

Note that $\chi_k \in \mathscr{D}(\mathbb{I}, q_1, q_2)$. The state $\chi_{k+1} = H\chi_k$ has the ith component

$$(\chi_{k+1})_i = H_i \chi_k \tag{1.27}$$

$$= H_i^+ q_1 + H_i^- q_2$$

which, by virtue of (1.26), leads to $(\chi_{k+1})_i > (q_1)_i$. The consequence is that $\chi_{k+1} \notin \mathscr{D}(\mathbb{I}, q_1, q_2)$ for $\chi_k \in \mathscr{D}(\mathbb{I}, q_1, q_2)$ which contradicts our initial assumption. □

1.2.1 Analysis of Equation FA + FBF = HF

This section deals with the analysis of Equation (1.15) by relating the existence of the matrix H to $\mathscr{K}er(F)$, matrix A and matrix F. This equation represents the pivot of this approach like the Riccati equation in the linear optimal control. The following results can be found in [10, 13].

Lemma 1.1. *The set $\mathscr{K}er(F)$ is positively invariant with respect to the motion of the system (1.7) if and only if there exists a matrix $H \in \mathbb{R}^{mxm}$ satisfying (1.15).*

Proof:

(Sufficiency): Assume that there exists a matrix H satisfying (1.15) and let $x \in \mathscr{K}er(F)$, that is $F\,x = 0$. Then, $F\,A_o x = H\,F\,x = 0$, *i.e.*, $A_o x \in \mathscr{K}er(F)$.

(Necessity:) Let $w \in \mathscr{K}er(F)$ and assume that $\mathscr{K}er(F)$ is positively invariant w.r.t. (1.7), then it follows that $F\,A_o w = 0$. Besides, without loss of generality, let F, $F\,A_o$ and w be partitioned as :

$$F = [F_1\ F_2],\ \text{with}\ F_1 \in \mathbb{R}^{mxm},\ F_2 \in \mathbb{R}^{mx(n-m)},$$

$$\text{rank}F_1 = m, \text{and}$$

$$F\,A_o = [M_1\ \ M_2],\ \text{with}\ M_1 \in \mathbb{R}^{mxm}\ \text{and}\ M_2 \in \mathbb{R}^{mx(n-m)},$$

and

$$w = [w_1^t\ \ w_2^t]^T\ ,\ \text{with}\ w_1 \in \mathbb{R}^m\ \text{and}\ w_2 \in \mathbb{R}^{n-m}.$$

Hence, $F\,w = 0$ can be written as follows :

$$F_1\,w_1 + F_2\,w_2 = 0,$$

which implies $w_1 = -\,F_1^{-1}\,F_2\,w_2$. Let us underline that for every $w_2 \in \mathbb{R}^{n-m}$, we have

$$\begin{bmatrix} -\,F_1^{-1}\,F_2\,w_2 \\ w_2 \end{bmatrix} \in \mathscr{K}er(F).$$

By substituting the latter equality into $F\,A_o\,w = 0$, it follows that for every $w_2 \in \mathbb{R}^{n-m}$ we have $M_1\,F_1^{-1}\,F_2\,w_2 = M_2\,w_2$, which can also be written as $H\,F_2 = M_2$, where the matrix H is taken to be

$$H = M_1 F_1^{-1} \in \mathbb{R}^{mxm} \tag{1.28}$$

or equivalently, satisfying the linear equation $HF_1 = M_1$, then we get finally $HF = [M_1\ M_2] = FA_o$. □

Lemma 1.2. *If the set $\mathscr{D}(F,\ q_1,\ q_2)$ is positively invariant w.r.t. the system (1.7), then, the set $\mathscr{K}er(F)$ is also positively invariant w.r.t. the system.*

Proof: Let $w \in \mathscr{K}er(F)$, then it is clear that $w \in \mathscr{D}(F,\ q_1,\ q_2)$. Set $F\,A_o\,w = r$. From the development in the proof of Lemma 1.1, it can be easily seen that $FA_o\,w = r$ can be written, with the same notations as above, as $Gw_2 = r$, with $G = M_2 - M_1\,F_1^{-1}\,F_2 \in \mathbb{R}^{mx(n-m)}$.

Recall that for every $w_2 \in \mathbb{R}^{n-m}$, the vector $w = [w_1^t\ \ w_2^t]^T \in \mathscr{K}er(F)$, with $w_1 = -\,F_1^{-1}\,F_2\,w_2$.

Assume there exists, at least, one nonzero component of G, for instance g_{il} . For w_2 chosen such that,

$$(w_2)_l = \alpha \frac{(q_1)_i}{g_{il}}, \text{ with } \alpha > 1, \text{ and } (w_2)_j = 0, \text{ for } j \neq l = 1, \dots, n-m.$$

Further, $Gw_2 = r$ yields

$$r_s = \alpha\, (q_1)_i \frac{g_{sl}}{g_{il}} \text{ for } s = 1, \dots, m.$$

In particular we have for the ith component, $r_i = \alpha\,(q_1)_i$ which yields, as $\alpha > 1$, $(F\,A_o\,w)_i = (Gw_2)_i = r_i > (q_1)_i$, or in other words

$$FA_o w \notin \mathscr{D}(F, q_1, q_2),$$

for $w \in \mathscr{D}(F, q_1, q_2)$. Hence, the vector r will belong to the domain $\mathscr{D}(F, q_1, q_2)$ for every $w_2 \in \mathbb{R}^{n-m}$ only if $G = 0$, which yields $r = F\,A_o\,w = 0$. □

Lemma 1.3. *If a stable matrix $H \in \mathbb{R}^{mxm}$ satisfying (1.15) exists, then the spectrum of the matrix A contain a set of $n-m$ stable eigenvalues, closed under complex conjugacy, corresponding to $n-m$ common eigenvectors to matrices A and A_o, belonging to $\mathscr{K}er(F)$ and parallel to $\partial\mathscr{D}(F, q_1, q_2)$.*

Proof: Let H be a matrix in $\mathbb{R}^{mxm}$ satisfying (1.15), and consider an eigenvector v of A_o associated to the eigenvalue λ (real or complex). Then Equation (1.15) allows us to write $F\,A_o\,v = H\,F\,v$. Since $A_o\,v = \lambda\,v$, it follows that $H(Fv) = \lambda\,(Fv)$. Hence Fv is an eigenvector of H corresponding to the same eigenvalue λ. The matrix $H \in \mathbb{R}^{mxm}$ could admit only m eigenvalues from the spectrum of the matrix A_o. Let $\sigma(A_o)$ be the spectrum of A_o given by $\sigma(A_o) = \Lambda = \Lambda_1 \cup \Lambda_2$, with $\sigma(H) = \Lambda_1 \subset \mathbb{C}^m$ and $\Lambda_2 \subset \mathbb{C}^{n-m}$.

Let α be an eigenvector of A_o associated to $\sigma \in \Lambda_2$, then

$$FA_o\alpha = HF\alpha = \sigma F\alpha,$$

which leads to $F\alpha = 0$ since σ is not assumed to belong to $\sigma(H)$. Further, $F\alpha = 0$ implies that $A_o\alpha = A\alpha = \sigma\alpha$, or in other words, that σ is an eigenvalue of A which means that $\Lambda_2 \subset \sigma(A)$.

Further, condition (1.7) ensures that the spectrum of A_o is stable *i.e.*, $|\lambda_i(A_o)| < 1$, for $i = 1, \dots n$. Consequently, Λ_2 contains $n - m$ stable eigenvalues corresponding to $n-m$ common eigenvectors to matrices A and A_o and belonging to $\mathscr{K}er(F)$. Further, $F\chi = r$, with $r = q_1$ or $r = -q_2$, represents the equations of m boundary hyper plans of the set $\mathscr{D}(F, q_1, q_2)$ and are obviously parallel to $\mathscr{K}er(F)$.

Then, these $n-m$ vectors are parallel to $\partial\mathscr{D}(F, q_1, q_2)$. Finally, since the matrix A_o is real, its spectrum Λ is obviously closed under complex conjugacy. The required matrix H is also real, then its spectrum Λ_1 is also closed under complex conjugacy. All this implies that the set Λ_2 is necessarily closed under complex conjugacy. □

Equation (1.15) was also studied by [139]. The following lemma is the direct application of Porter's result. We first write Equation (1.15) in the following equivalent form:

$$FA = H_o F, \text{ with } H_o = H - FB. \tag{1.29}$$

Lemma 1.4. *[139] There exists a solution $H \in \mathbb{R}^{mxm}$ of Equation (1.29), where $F \in \mathbb{R}^{mxn}$ and $rank(F) = m$, with $m \leq n$, if and only if :*

$$rank(S) = m, \text{ with } S = \begin{bmatrix} FA \\ F \end{bmatrix}. \tag{1.30}$$

Remark 1.1.

(*i*) *When condition (1.30) is satisfied, matrix H can be computed from (1.28). Lemma 1.3 allows us to relate the resolution of Equation (1.15) to an eigenvalue assignment problem. According to this Lemma, it is clear that, if A does not possess $n-m$ stable eigenvalues which are closed under complex conjugacy, the required matrix H, solution to Equation (1.15) or (1.29), will not exist for every given matrix F satisfying (1.7). However, the necessary and sufficient condition given by Lemma 1.4 has to be satisfied, after Lemma 1.3, for a given matrix F satisfying (1.7).*

(*ii*) *If A is non-singular, i.e. $rank(A) = n$, it follows that $N = FA$ and F have the same rank which yields $rank(S) \geq m$. Consequently, a solution H to Equation (1.29) may not always exist.*

1.2.2 Non-symmetrical Saturations

In this section, we extend some results given previously to the asymmetric case. A necessary and sufficient condition for domain (1.9) to be positively invariant and contractive w.r.t. the system (1.7) is given by the result below [10, 13].

Theorem 1.2. *The subset $\mathscr{D}(F, q_1, q_2)$ of $\mathbb{R}^n$ given by (1.9), with $rank(F) = m$, is positively invariant (resp. positively invariant and contractive) with respect to the motions of system (1.7) if and only if the conditions below hold :*

(*i*) *There exists a matrix $H \in \mathbb{R}^{mxm}$ satisfying (1.15), i.e.,*

$$FA + FBF = HF, \tag{1.31}$$

(*ii*) *for q_1 and $q_2 \in \mathbb{R}^m_+$, we have*

$$(\mathbb{I} - \tilde{H}_d)q \geq 0 \ (resp. \ (\mathbb{I} - \tilde{H}_d)q > 0 \text{ and } q \in Int\mathbb{R}^{2m}_+), \tag{1.32}$$

where

$$\tilde{H}_d = \begin{bmatrix} H^+ & H^- \\ H^- & H^+ \end{bmatrix}, \; q = \begin{bmatrix} q_1 \\ q_2 \end{bmatrix}. \tag{1.33}$$

Proof: (Sufficiency): Assume that there exists a matrix H satisfying conditions (1.31)-(1.32). Applying the transformation $z_k = Fx_k$ to the closed-loop system (1.7), leads to the set $\mathscr{D}(\mathbb{I}, q_1, q_2)$ and the reduced order system $z_{k+1} = Hz_k$. Using Theorem 1.1, condition (1.32) implies that the set $\mathscr{D}(\mathbb{I}, q_1, q_2)$ is positively invariant (resp. positively invariant and contractive) with respect to the reduced order system $z_{k+1} = Hz_k$. This means that the control $u_k = Fx_k$ is always admissible which implies that the set $\mathscr{D}(F, q_1, q_2)$ given by (1.9) is positively invariant (resp. positively invariant and contractive) with respect to the system (1.7).

(Necessity): Assume that the set $\mathscr{D}(F, q_1, q_2)$ given by (1.9), with $rank(F) = m$, is positively invariant (resp. positively invariant and contractive) with respect to the system (1.7). According to Lemma 1.2, the set $\mathscr{K}er(F)$ is also positively invariant (resp. positively invariant and contractive) with respect to the system (1.7). By virtue of Lemma 1.1, there exists a matrix H solution of Equation (1.31). Applying again the transformation $z_k = Fx_k$ to the closed-loop system (1.7), leads to the set $\mathscr{D}(\mathbb{I}, q_1, q_2)$ and the reduced order system $z_{k+1} = Hz_k$. By using Theorem 1.1, condition (1.32) is obtained. □

Remark 1.2.

(*i*) *Note that conditions (1.32) can be rewritten as follows*

$$\tilde{H}_d q \leq q \; (resp. \;\; \tilde{H}_d q < q), \tag{1.34}$$

with $\tilde{H}_d \in \mathbb{R}^{2mx2m}$ *and* $q \in \mathbb{R}_+^{2m}$.

Furthermore, it is worth noting that the asymmetric case requires more computation time.

(*ii*) *Condition (1.32) of Theorem 1.2 contains conditions of the symmetrical case. To show this, let* $q_1 = q_2 = \rho$, *then* $\tilde{H}_d q \leq q$ *is equivalent to*

$$H^+\rho + H^-\rho \leq \rho \text{ and } H^-\rho + H^+\rho \leq \rho$$

which leads to $(H^+ + H^-)\rho = |H|\rho \leq \rho$.

Two particular interesting cases of Theorem 1.2 can now be stated by means of the following result.

Corollary 1.1.

(*i*) *Domain* $\mathscr{D}(F, q_1, q_2)$ *is positively invariant (resp. positively invariant and contractive) w.r.t. the system (1.7) if and only if domain* $\mathscr{D}(F, q_2, q_1)$ *is positively invariant (resp. positively invariant and contractive) w.r.t. the system.*

(ii) If domain $\mathscr{D}(F, q_1, q_2)$ is positively invariant (resp. positively invariant and contractive) w.r.t. the system (1.7), then the set

$$\mathscr{D}(F, q_1, q_2) \cup \mathscr{D}(F, q_2, q_1)$$

is positively invariant (resp. positively invariant and contractive) w.r.t. the system.

Proof: Obvious. □

When we are interested in a control vector with non-negative components, condition (1.32) allows this possibility only in the case of positive invariance property of domain $\mathscr{D}(F, q_1, q_2)$. If the control vector is non-negative (resp. non-positive), *i.e.*, $0 \leq u \leq \rho$ (resp. $-\rho \leq u \leq 0$), we need the following result :

Corollary 1.2. *The subset $\mathscr{D}(F, \rho, 0)$, (resp. $\mathscr{D}(F, 0, \rho)$), of $\mathbb{R}^n$ given by (1.9), with rank$(F) = m$, is positively invariant w.r.t. the motion of the system (1.7) if and only if there exists a non-negative matrix $H \in \mathbb{R}^{mxm}$*
(i) solution to the equation :

$$FA + FBF = HF,$$

(ii) and satisfying

$$H\rho \leq \rho.$$

Proof: Follows readily by replacing $q_1 = \rho, q_2 = 0$ in condition (1.32). □

Theorem 1.3. *The domain*

$$\mathscr{D}(F, \rho, \infty) = \{x \in \mathbb{R}^n \, / \, Fx \leq \rho, \rho \in Int\mathbb{R}^m_+\},$$

with rank$(F) = m$, is positively invariant (resp. positively invariant and contractive) w.r.t. the motion of system (1.7) if and only if there exists a non-negative matrix $H \in \mathbb{R}^{mxm}$
(i) solution to the equation

$$FA + FBF = HF,$$

(ii) and satisfying

$$H\rho \leq \rho \quad (resp. \;\; H\rho < \rho).$$

Proof: Follows readily by following the same reasoning used to obtain condition (1.32). The same conditions hold for the set $\mathscr{D}(F, \infty, \rho)$. The same result is obtained in [47]. □

Example 1.1. *Consider the linear discrete-time system described by the following matrices :*

$$A = \begin{bmatrix} -0.5 & -1.5 \\ 1.8 & 3.2 \end{bmatrix}, \qquad B = \begin{bmatrix} -1 \\ 2 \end{bmatrix}.$$

The matrix A is unstable, with $\lambda_1(A) = 2.2$ and $\lambda_2(A) = 0.5$, then it contains $n-m=1$ stable real eigenvalue. The set of admissible controls Ω is given by

$$\Omega = \{u \in \mathbb{R}^m : -\delta \leq u \leq 10, \qquad \delta \geq 0\}.$$

In this case, we choose the following spectrum Λ and matrix F:

$$\Lambda = \{\ 0.5\ ,\ 0.2\},$$

$$F = [\ -1 \quad -1.5\ ] \qquad \text{with} \qquad rank(F) = rank(S) = 1.$$

Hence,

$$A_o = A + BF = \begin{bmatrix} 0.5 & 0 \\ -0.2 & 0.2 \end{bmatrix}.$$

It is obvious that, equation (1.15) yields $H = 0.2$. Then, for $\delta = 1$, we have

$$(\mathbb{I} - \tilde{H}_d)q = [8 \qquad 0.8]^t > 0,$$

and for $\delta = 0$, we have

$$(\mathbb{I} - \tilde{H}_d)q = [0 \qquad 0.8]^t \geq 0,$$

that is, condition (1.32) is satisfied in both cases of $\delta = 1$ and $\delta = 0$.

Consequently, the set $\mathscr{D}(F, q_1, q_2)$ is positively invariant with respect to the motion of the system (1.7).

1.2.3 Asymptotic Stability

In this section, we link both Lemma 1.4 and Theorem 1.2 to the asymptotic stability of a linear discrete-time system with constrained control (1.1)-(1.5).

It is now obvious that the positive invariance property of domain $\mathscr{D}(F, q_1, q_2)$ w.r.t. the system (1.7) guarantees that every motion emanating from this domain can not leave it. This property ensures that the linear model (1.7) remains valid despite the existence of input saturations (1.4). Besides, the contractiveness property of the same domain w.r.t. the system (1.7) implies the asymptotic stability of system (1.1)-(1.5) for every $x_o \in \mathscr{D}(F, q_1, q_2)$. In addition, the positive invariance property guarantees also the asymptotic stability of system (1.1)-(1.5).

Theorem 1.4. *If there exists a matrix $F \in \mathbb{R}^{mxn}$ such that:*

(i) $\text{rank}(F) = m$,

(ii) $\rho(A + BF) < min(1, \rho(A))$,

(iii) $\text{rank}\begin{bmatrix} FA \\ F \end{bmatrix} = m$,

(vi) the solution $H \in \mathbb{R}^{mxm}$ to equation $FA + FBF = HF$ satisfies $\tilde{H}_d q \leq q$, *then, the system (1.1)-(1.5) with (1.7) is asymptotically stable and the set* $\mathscr{D}(F, q_1, q_2)$ *is positively invariant w.r.t. the closed-loop system.*

Proof: Obvious. □

Let us underline that in Example (1.1), the system is also asymptotically stable.

1.2.4 System Input Augmentation

In this section, we present a technique allowing to avoid the conservatism of Lemma 1.3 by augmenting the system entries such that the required number $n-m$ stable eigenvalues of matrix A becomes equal to zero. The price of this operation will be the augmentation of the computation time and the restriction of domain $\mathscr{D}(F, q_1, q_2)$, which is unbounded, to a bounded set.

Let us rewrite Equation (1.1) as follows:

$$x_{k+1} = Ax_k + \begin{bmatrix} B & 0 \end{bmatrix} \begin{bmatrix} u_k \\ v_k \end{bmatrix} \tag{1.35}$$

with v_k a fictitious input of the system. The set of admissible control Ω given by (1.2) remains unchanged when the vector v_k satisfies

$$v_k \in \Omega_f = \left\{ v \in \mathbb{R}^{n-m} \,/\, -r_2 \leq v \leq r_1;\, r_1, r_2 \in \mathbb{R}_+^{n-m} \right\}, \tag{1.36}$$

where vectors r_1 and r_2 stand for fictitious saturations. Let

$$g_1 = \begin{bmatrix} q_1 \\ r_1 \end{bmatrix};\; g_2 = \begin{bmatrix} q_2 \\ r_2 \end{bmatrix};\; g = \begin{bmatrix} g_1 \\ g_2 \end{bmatrix}.$$

The purpose remains unchanged and the feedback law becomes

$$w_k = \begin{bmatrix} u_k \\ v_k \end{bmatrix} = \begin{bmatrix} F \\ E \end{bmatrix} x_k = G\,x_k, \text{ with } G \in \mathbb{R}^{nxn}. \tag{1.37}$$

In this case, domain (1.9) takes the form

$$\mathscr{D}(G, g_1, g_2) = \mathscr{D}(F, q_1, q_2) \cap \mathscr{D}(E, r_1, r_2). \tag{1.38}$$

This technique obviously implies a limitation of domain $\mathscr{D}(F, q_1, q_2)$. The existence of a matrix $H \in \mathbb{R}^{nxn}$ solution to Equation (1.15) is now guaranteed if the matrix G is invertible which implies that matrix E and vectors r_1 and r_2 are chosen such that:

$$\begin{aligned} &(i)\;\; rank(G) = n \\ &(ii)\;\; \mathscr{D}_o \subset \mathscr{D}(G, g_1, g_2). \end{aligned} \tag{1.39}$$

Recall that the set $\mathscr{D}_o$ is the set of initial values and note that the domain $\mathscr{D}(G, g_1, g_2)$ is a bounded polyhedral set. In this case, the necessary and sufficient condition of Theorem 1.2 is expressed in $\mathbb{R}^{2nx2n}$ instead of $\mathbb{R}^{2mx2m}$.

Example 1.2. *Here, we consider the same example as the one given in [97]*

$$A = \begin{bmatrix} 1 & 1 \\ 0 & 1 \end{bmatrix}, B = \begin{bmatrix} 0.5 \\ 1 \end{bmatrix},$$

with a different set of admissible controls

$$\Omega = \{u \in R / -\delta \leq u \leq 10\},$$

instead of $|u| \leq 1$ *as used in [97].*

In this case, the matrix A does not contain $n - m = 1$ *stable eigenvalue. Thus, we proceed according to the previous technique. Let us take any matrix F of full rank such that* $A_o = A + BF$ *is stable, in particular the one computed in [97], i.e.,*

$$F = [-1 \qquad -1.5].$$

This leads to

$$A_o = \begin{bmatrix} 0.5 & 0.25 \\ -1 & -0.5 \end{bmatrix}.$$

We now choose a matrix E such that

$$rank \begin{bmatrix} F \\ E \end{bmatrix} = 2,$$

for instance

$$E = [-1 \qquad -0.5].$$

Moreover the fictitious constraints r_1, r_2 *are chosen as* $r_1 = \delta$ *and* $r_2 = 10$*; then, one gets* $g = [10 \ \delta \ \delta \ 10]^T$*. Further, the matrix H is given by*

$$H = \begin{bmatrix} 0 & -1 \\ 0 & 0 \end{bmatrix},$$

which yields $\tilde{H}_d g = [10 \ 0 \ \delta \ 10]^T \leq g$*, for every* $\delta \geq 0$*. If the domain* $\mathscr{D}_o$ *is known, many choices of E,* r_1 *and* r_2 *are possible. Thus the set* $\mathscr{D}(G, g_1, g_2)$ *is positively invariant w.r.t. the closed-loop system for every* $x_o \in \mathscr{D}(G, g_1, g_2)$ *(cf. Fig. 1.3).*

In Figure 1.3, we show domain $\mathscr{D}(G, g_1, g_2)$ *for* $\delta = 1$ *as a positively invariant and asymptotic stability set w.r.t. the system (1.7) with two motions : one of them is starting inside the set* $\mathscr{D}(G, g_1, g_2)$ *belonging to* $\mathscr{D}(G, g_1, g_2)$ *whereas the second, starting outside, leaving the set* $\mathscr{D}(G, g_1, g_2)$*.*

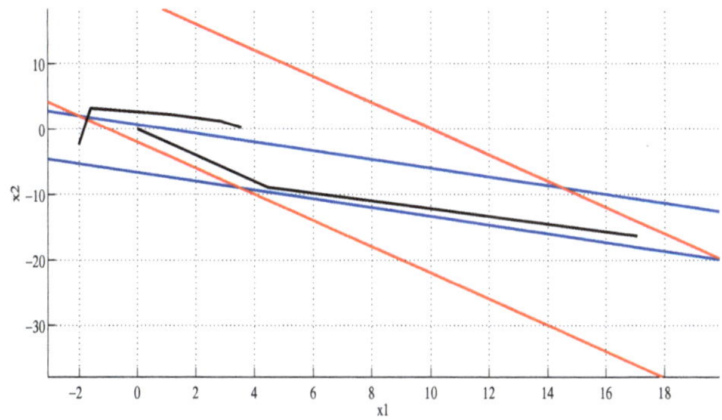

Fig. 1.3 Non-symmetrical domain as a positively invariant set

1.2.4.1 Algorithm

In order to recapitulate the steps required to satisfy our purpose, the following algorithm is presented taking into account the assumption (1.3) of stabilizability of the pair (A, B).

Algorithm 1 Computation of matrices H and F from the data A and B

Step 1 : If the matrix A does not possess a set of $n-m$ non null and stable eigenvalues, closed under complex conjugacy for $m < n$, proceed according to paragraph 1.2.4.

Step 2 : Take a set $\Lambda = \{\lambda_1\ \lambda_2\ \dots\ \lambda_n\}$ (subject to complex conjugacy) containing $n-m$ non null and stable eigenvalues of A closed under complex conjugacy, such as $\rho(\Lambda) < min(1,\ \rho(A))$.

Step 3 : Compute F such as $\sigma(A+BF) = \Lambda$. Recall that, according to assumption (1.3), the matrix F always exists.

Step 4 : If $rank(F) = rank(S) = m$, go to step 5, else go to step 1 and change the set Λ.

Step 5 : Compute H by means of Equation (1.28).

Step 6 : If $\tilde{H}_d q \leq q$, stop, else go to step 1.

In this section, necessary and sufficient conditions, given in [10, 13], for an asymmetric bounded polyhedral set to be positively invariant and contractive w.r.t. the system (1.7) are successfully used. The application of these conditions to the regulator problem of linear discrete-time system with asymmetric constrained control enables us to specify conditions under which, system (1.7) admits a largest positively invariant and contractive asymmetric polyhedral subset of $\mathscr{D}(F,\ q_1,\ q_2)$.

1.3 Continuous-Time Systems

This section is devoted to the study of linear continuous-time systems described by Equation (1.40)

$$\dot{x} = Ax + Bu, \tag{1.40}$$

where x is the state vector in $\mathbb{R}^n$, and u is the constrained control, satisfying

$$u \in \Omega \subset \mathbb{R}^m. \tag{1.41}$$

Matrices A and B are constant, of appropriate size and satisfy the following assumption.

$$(A,B) \text{ is stabilizable.} \tag{1.42}$$

Ω is the set of admissible controls defined as

$$\Omega = \{u \in \mathbb{R}^m / -q_2 \leq u \leq q_1;\, q_1,\, q_2 \in \mathbb{R}^m_+ - \{0\}\}. \tag{1.43}$$

This is an asymmetric polyhedral set as is generally the case in practical situations.

Let us first consider the unconstrained case where the regulator problem for system (1.40) consists in realizing a feedback law as

$$u = Fx,\ F \in \mathbb{R}^{mxn} \text{ with } rank(F) = m. \tag{1.44}$$

In such a case, system (1.40) becomes

$$\dot{x} = (A+BF)x = A_o x, \tag{1.45}$$

where F is generally chosen in such a way that an increase of system dynamics is obtained with the asymptotic stability of the closed-loop system (1.45), or equivalently

$$Re(\lambda_i(A+BF)) < 0;\ i = 1,\ \ldots,\ n. \tag{1.46}$$

In the constrained case, we follow the approach proposed in Section 1.2. Recall that this approach consists in giving conditions allowing the choice of a stabilizing controller (1.44) such that the model (1.45) remains valid every time. This is only possible if the state is constrained to evolve in a specified region defined by

$$\mathscr{D}(F,\, q_1,\, q_2) = \{x \in \mathbb{R}^n \,/ -q_2 \leq Fx \leq q_1;\, q_1,\, q_2 \in \mathbb{R}^m_+ - \{0\}\}. \tag{1.47}$$

Note that this domain is unbounded when $m < n$. In addition, if

$$x(t) \in \mathscr{D}(F,\, q_1,\, q_2),$$

one would get

$$x(t+\tau) \notin \mathscr{D}(F,\, q_1,\, q_2),\ \forall \tau \geq 0.$$

Recall that $x(t+\tau) = e^{A_o \tau} x(t)$, with

$$e^{A_o t} = \sum_{k=o}^{\infty} \frac{1}{k!} t^k A_o^k.$$

This matrix makes the problem very difficult to deal with compared to the case of the discrete-time systems (*cf.* 1.2). However, an attempt will be given in the following sections.

The present problem has been studied by many authors. For instance, [48, 49] give a solution to the regulator problem for linear continuous-time systems with constrained control. Their approach consists in constructing a positively invariant domain w.r.t. the system (1.45) included in the set (1.47) and it allows to construct an admissible control law with the asymptotic stability for the closed-loop.

Bitsoris [48] presents necessary and sufficient conditions for domain (1.47) to be positively invariant w.r.t. the system (1.45) derived by means of an indirect proof.

While in [149, 55] the domains used are given by

$$\mathscr{D}_{[a,b]}\{\mathscr{K}\} = (\mathscr{K}_+ - b) \cap (\mathscr{K}_- - a),$$

with $\mathscr{K} = \mathscr{K}_+ = -\mathscr{K}_-$ is a proper cone, and

$$\mathscr{K}_+ - a = \{x \in \mathbb{R}^n, \,/\, x = y + a, \, y \in \mathscr{K}_+\}.$$

They give necessary and sufficient conditions for such domains to be positively invariant w.r.t. a large class of continuous-time linear systems given by

$$\dot{x} = A_o x + c(t), \text{ with } e^{A_o t} \mathscr{K}_+ \subset \mathscr{K}_+,$$

and $c(t)$ defined in a compact set. Their results can be extended to systems described by (1.45), whose state belongs to the set (1.47), with the restrictive assumption that

$$e^{A_o t} \mathscr{K}_+ \subset \mathscr{K}_+.$$

Benzaouia and Hmamed [14, 15] present necessary and sufficient conditions for domain (1.47) to be positively invariant w.r.t. the system (1.45) based on a new and direct proof using asymmetric Lyapunov function given for the first time in [11, 12].

In particular, domain $\mathscr{D}(\mathbb{I}, q_1, q_2)$ defined by

$$\mathscr{D}(\mathbb{I}, q_1, q_2) = \{z \in \mathbb{R}^m \,/ -q_2 \leq z \leq q_1\}, \qquad (1.48)$$

which is described by the function

$$\vartheta(z) = \underset{i}{Max}\ max \left\{ \frac{z_i^+}{q_1^i}, \frac{z_i^-}{q_2^i} \right\}, \text{ when } q_1 > 0 \text{ and } q_2 > 0 \qquad (1.49)$$

i.e.,

$$\mathscr{D}(\mathbb{I}, q_1, q_2) = \{z \in \mathbb{R}^m \,/ \vartheta(z) \leq 1\}.$$

It follows from above that the main purpose of this chapter is to present necessary and sufficient conditions under which the polyhedral asymmetric domain $\mathscr{D}(F, q_1, q_2)$ is positively invariant w.r.t. the system (1.45).

1.3.1 Preliminary Results

In this section, we present some useful definitions and results. Consider a continuous-time nonlinear system

$$\dot{z}(t) = f(z(t)),\ z \in \mathbb{R}^m,\ f(0) = 0. \tag{1.50}$$

Consider a function $\varphi : \mathbb{R}^m \to \mathbb{R}_+$ with $\varphi(0) = 0$ and assume that φ is directionally differentiable at each point in each direction and define $\dot{\varphi}(z)$ by

$$\begin{aligned} \dot{\varphi}(z(t)) &= \tfrac{d^+}{dt}\left[\varphi(z(t))\right] = \tfrac{\partial \varphi}{\partial z}^T f(z(t)), \\ &= \lim_{\varepsilon \to 0^+} \tfrac{\varphi(z+\varepsilon f(z)) - \varphi(z)}{\varepsilon}, \end{aligned} \tag{1.51}$$

Definition 1.3. *[151] $\dot{\varphi}(z)$ is the directional derivative of φ at z in the direction $f(z)$, with $f(0) = 0$ and $\dot{z}(t) = f(z(t))$.*

Theorem 1.5. *[151]*

(*i*) *If a function φ is positive definite on a set $\mathscr{R} \subseteq \mathbb{R}^m$, φ is a Lyapunov function of system (1.50) on $\mathscr{R}$ if and only if $\dot{\varphi}(z)$ is negative semi-definite along any motion of the system (1.50) starting at $z(t_o)$, for every $z(t) \in \mathscr{R}$ and $\forall t > t_o$, i.e., $\dot{\varphi}(z) < 0$, $\forall\, z(t) \in \mathscr{R}$ and $\forall t > t_o$.*

(*ii*) *If a function φ is a Lyapunov function of system (1.50) on a subset $\mathscr{R}$ of $\mathbb{R}^m$ then the set*

$$\mathscr{D}(\varphi, c) = \{z \in \mathscr{R} \,/\, \varphi(z) \leq c,\ c > 0\}, \tag{1.52}$$

is a stability domain of the system.

Consider the following linear continuous-time stationary system,

$$\dot{z}(t) = Hz(t),\ z \in \mathscr{R} \subseteq \mathbb{R}^m \text{ and } 0 \in Int\mathscr{R}. \tag{1.53}$$

Lemma 1.5. *Domain $\mathscr{D}(\varphi, 1)$ given by (1.52) with $\varphi(z)$ a positive definite function satisfying*

$$\varphi(\alpha z) = \alpha^\sigma \varphi(z);\ \alpha > 0,\ \sigma > 0 \tag{1.54}$$

is positively invariant w.r.t. the system (1.53) if and only if $\varphi(z)$ is a Lyapunov function of the system.

Proof:

(Sufficiency): Let $\varphi(z)$ be a Lyapunov function of the system, then domain $\mathscr{D}(\varphi,1)$ is a stability domain , which is obviously positively invariant w.r.t. the system.

(Necessity:) Let domain $\mathscr{D}(\varphi,1)$ be positively invariant w.r.t. the system, that is,

$$\varphi(z(t)) \leq 1 \Longrightarrow \varphi(z(t+\tau)) \leq 1, \forall\tau > 0.$$

In this case, we have to show that domain $\mathscr{D}(\varphi,c)$ is also positively invariant w.r.t. the system for every positive scalar c. Let $z(t) \in \mathscr{D}(\varphi,c)$, that is, $\varphi(z(t)) < c$. Since function $\varphi(z(t))$ satisfies (1.54), we should have $c^{-1}\varphi(z(t)) = \varphi(c^{-\chi}z(t)) \leq 1$, with $\chi = \sigma^{-1}$. It follows that $c^{-\chi}z(t) \in \mathscr{D}(\varphi,1)$, this implies that $c^{-\chi}z(t+\tau) \in \mathscr{D}(\varphi,1)$. Using relation (1.54) a second time, we can write equivalently $\varphi(z(t+\tau)) \leq c$, for every positive scalar c. In particular, consider the case when $c = \varphi(z(t))$, this allow us to state that

$$\varphi(z(t+\tau)) \leq \varphi(z(t)), \ \forall\, \tau \geq 0,$$

and

$$\lim_{\tau\to 0^+} \frac{\varphi(z(t+\tau)) - \varphi(z(t))}{\tau} \leq 0.$$

Consequently, the positive definite function $\varphi(z(t))$ is a Lyapunov function of the system.

(Necessity bis:) Let domain $\mathscr{D}(\varphi,1)$ be positively invariant w.r.t. the system $\dot{z}(t) = f(z(t))$, or explicitly

$$\varphi(z(t)) \leq 1 \text{ implies } \varphi(z(t+\tau)) \leq 1, \forall\tau > 0.$$

Besides, without loss of generality, take a vector z in $\mathscr{D}(\varphi,1)$ such as $\varphi(z(t)) = c^{-\sigma}$, with $c > 0$. It follows that $c^{\sigma}\varphi(z(t)) = \varphi(cz(t)) = 1$, which means that for every $\tau \geq 0$, $\varphi(cz(t+\tau))$ is in $\mathscr{D}(\varphi,1)$ since the latter set is assumed to be positively invariant. Further, using (1.54) one can deduce easily that

$$\varphi(z(t+\tau)) \leq \varphi(z(t)),$$

for every $\tau \geq 0$, or explicitly that $\varphi(\cdot)$ is a Lyapunov function. □

This lemma enables us to give a necessary and sufficient condition for domain $\mathscr{D}(\mathbb{I}, q_1, q_2)$ defined by (1.48) to be a positively invariant set w.r.t. the system (1.53) by using the associated asymmetric function.

Theorem 1.6. *[15] Domain $\mathscr{D}(\mathbb{I}, q_1, q_2) \subseteq \mathscr{R}$ given by (1.48) with $q_1,q_2 > 0$ is positively invariant w.r.t. the system (1.53) if and only if*

$$\tilde{H}_c q \leq 0, \tag{1.55}$$

with

$$\tilde{H}_c = \begin{bmatrix} H_1 & H_2 \\ H_2 & H_1 \end{bmatrix}; \qquad q = \begin{bmatrix} q_1 \\ q_2 \end{bmatrix}$$
$$H_1 = \begin{cases} h_{ii} & \text{for } i = j \\ h_{ij}^+ & \text{for } i \neq j \end{cases}; \quad H_2 = \begin{cases} 0 & \text{for } i = j \\ h_{ij}^- & \text{for } i \neq j \end{cases}. \tag{1.56}$$

Proof:

(Sufficiency): Consider the set $\mathscr{D}(\varphi, 1)$ defined by (1.52) with function $\varphi(z)$ given by (1.49). This function is continuous, positive definite [11, 12] and satisfies relation (1.54) with $\sigma = 1$. According to Lemma 1.5, domain $\mathscr{D}(\varphi, 1)$ is positively invariant w.r.t. the system (1.53) if and only if $\varphi(z)$ is a Lyapunov function for the system. It is now obvious that $\mathscr{D}(\varphi, 1) = \mathscr{D}(\mathbb{I}, q_1, q_2)$. Let us compute the directional derivative of the function $\varphi(\cdot)$ at z in the direction $f(z) = Hz$. For this, recall that φ is a continuous positive definite function. By virtue of Definition 1.3

$$\dot{\varphi}(z) = \lim_{\varepsilon \to 0^+} \frac{\varphi(z + \varepsilon Hz) - \varphi(z)}{\varepsilon}. \tag{1.57}$$

Denote $C = \mathbb{I} + \varepsilon H$ and compute the rate of increase $\Delta\varphi(z) = \varphi(Cz) - \varphi(z)$. For this, we use the result of [11, 12].

$$\Delta\varphi(z) \leq \underset{i}{Max}\ max \left\{ \left[\frac{(C^+ q_1)_i + (C^- q_2)_i}{q_1^i} - 1 \right] \varphi(z); \right.$$
$$\left. \left[\frac{(C^- q_1)_i + (C^+ q_2)_i}{q_2^i} - 1 \right] \varphi(z) \right\} \tag{1.58}$$

It is worth noting that in the vicinity of zero, ε can be chosen sufficiently small, for instance, such as the term $1 + \varepsilon h_{ii} \geq 0$ or in other words

$$\frac{1}{\varepsilon} > \underset{i}{Max} |h_{ii}|.$$

The matrices C^+ and C^- when $\varepsilon \to 0^+$ are given by

$$C^+ = (\mathbb{I} + \varepsilon H)^+ = \begin{cases} \varepsilon h_{ij}^+ & \text{for } i \neq j \\ 1 + \varepsilon h_{ii} & \text{for } i = j \end{cases}$$
$$= \mathbb{I} + \varepsilon H_1,$$
$$C^- = (\mathbb{I} + \varepsilon H)^- = \begin{cases} \varepsilon h_{ij}^- & \text{for } i \neq j \\ 0 & \text{for } i = j \end{cases}$$
$$= \varepsilon H_2.$$

By substituting matrices C^- and C^+ into (1.58), we obtain :

$$\Delta\varphi(z) \le \underset{i}{Max}\ max \left\{ \left[\varepsilon \frac{(H_1q_1+H_2q_2)_i}{q_1^i}\right] \varphi(z); \left[\varepsilon \frac{(H_2q_1+H_1q_2)_i}{q_2^i}\right] \varphi(z) \right\}$$

then,

$$\dot{\varphi}(z) \le \underset{i}{Max}\ max \left\{ \left[\frac{(H_1q_1+H_2q_2)_i}{q_1^i}\right] \varphi(z); \left[\frac{(H_2q_1+H_1q_2)_i}{q_2^i}\right] \varphi(z) \right\}$$

From condition (1.55), we conclude that

$$\dot{\varphi}(z) \le 0, \forall z \in \mathscr{R} \subseteq \mathbb{R}^m.$$

That is, $\varphi(z)$ is a Lyapunov function of the system on the set $\mathscr{R}$. Consequently, domain $\mathscr{D}(\mathbb{I}, q_1, q_2)$ is positively invariant w.r.t. the system (1.53).

(Necessity:) Assume that there exists a subscript k in the set $[1, \ldots, m]$ such that

$$\underset{i}{Max}\{(\tilde{H}_cq)_i\} = (\tilde{H}_cq)_k > 0.$$

Moreover, without loss of generality, assume that $k \le m$ and define the vector χ as follows :

$$\chi_j = \begin{cases} (q_1)_j & \text{for } j = k, \\ (q_1)_j & \text{for } h_{kj} \ge 0, \\ -(q_2)_j & \text{for } h_{kj} < 0, \end{cases}$$

for $j = 1 \ldots m$, or in other words $h_{kj}\chi_j \ge 0$ for every j in the set $[1, \ldots, m] - \{k\}$ and $\chi_k = (q_1)_k$.

It is obvious that for every $k \in [1, \ldots, m]$, we have

$$\begin{aligned}(\tilde{H}_cq)_k &= (H_1q_1 + H_2q_2)_k, \\ &= h_{kk}(q_1)_k + \sum_{j=1,\ j\ne k}^{m} \left\{h_{kj}^+(q_1)_j + h_{kj}^-(q_2)_j\right\}, \\ &= (H\chi)_k,\end{aligned}$$

which implies that $(H\chi)_k > 0$ and consequently that $(C\chi)_k > 0$.

Note also that

$$(C\chi)_k = \chi_k + \varepsilon(H\chi)_k = (q_1)_k + \varepsilon(H\chi)_k = (C\chi)_k^+ > 0,$$

from which we deduce that $\frac{(C\chi)_k^+}{(q_1)_k} > 1$ or equivalently that $\varphi(C\chi) > 1$. Moreover, as it was assumed that $\varphi(\chi) = 1$, it follows that $\Delta\varphi(\chi) \ge 0$ which leads to

$$\dot{\varphi}(\chi) = \lim_{\varepsilon \to 0^+} \frac{\varphi(\chi + \varepsilon H\chi) - \varphi(\chi)}{\varepsilon} \ge 0,$$

or similarly, that φ is not a Lyapunov function and this contradicts surely the result of Lemma 1.5. □

It is possible to give the same result by using directly the development of the matrix e^{Ht}.

Theorem 1.7. *The domain $\mathscr{D}(\mathbb{I},\ q_1,\ q_2)$ given by (1.48), with $q \geq 0$, is positively invariant w.r.t. the system (1.53) if and only if (1.55) and (1.56) are satisfied.*

Proof:

(Sufficiency): Consider the solution to (1.53) given by $z(t+\varepsilon) = e^{H\varepsilon}z(t)$ which can be developed as follows:

$$z(t+\varepsilon) = z(t) + \varepsilon Hz(t) + \varepsilon^2\left\{\sum_{k=2}^{\infty}\frac{1}{k!}\varepsilon^{k-2}H^k z(t)\right\}.$$

Take the matrix Γ as follows :

$$\Gamma = \mathbb{I} + \varepsilon H + \varepsilon^2\Gamma_o,$$

with

$$\Gamma_o = \sum_{k=2}^{\infty}\frac{1}{k!}\varepsilon^{k-2}H^k.$$

In this case, the system (1.53) can be equivalently described as

$$z(t+\varepsilon) = \Gamma z(t), \tag{1.59}$$

for every $\varepsilon > 0$ as small as possible.

Let $z(t)$ be in the set $\mathscr{D}(\mathbb{I},\ q_1,\ q_2)$, with $q \geq 0$, *i.e.*,

$$-q_2 \leq z(t) \leq q_1.$$

The same arguments used in the proof of Theorem 1.2 for the discrete-time systems lead to

$$z(t+\varepsilon) \in \mathscr{D}(\mathbb{I},\ q_1,\ q_2),$$

with $q \geq 0$, if

$$\tilde{\Gamma}q \leq q,$$

or equivalently

$$\begin{cases}\Gamma^+q_1 + \Gamma^-q_2 \leq q_1,\\ \Gamma^-q_1 + \Gamma^+q_2 \leq q_2.\end{cases} \tag{1.60}$$

Further, compute Γ^+ and Γ^- for $\varepsilon > 0$ as small as possible, as follows:

$$\begin{aligned}\Gamma^+ &= \mathbb{I} + \varepsilon H_1 + \varepsilon^2\Gamma_o^+,\\ \Gamma^- &= \varepsilon H_2 + \varepsilon^2\Gamma_o^-,\end{aligned}$$

and by substituting the latter into (1.60), one can deduce, for $\varepsilon > 0$,

$$\begin{cases} H_1 q_1 + H_2 q_2 + \varepsilon(\Gamma_o^+ q_1 + \Gamma_o^- q_2) \leq 0, \\ H_2 q_1 + H_1 q_2 + \varepsilon(\Gamma_o^- q_1 + \Gamma_o^+ q_2) \leq 0. \end{cases} \tag{1.61}$$

Taking into account that $\tilde{H}_c q \leq 0$ and ε is as small as possible, one can always satisfy (1.61).

(Necessity:) Let the domain $\mathscr{D}(\mathbb{I}, q_1, q_2)$ with $q \geq 0$ be positively invariant w.r.t. the system (1.53) and condition $\tilde{H}_c q \leq 0$ is violated. For that, let $i \in [1, m]$ exists in such a way that equation

$$h_{ii}(q_1)_i + \sum_{j=1,\, j\neq i}^{m} \left\{ h_{ij}^+(q_1)_j + h_{ij}^-(q_2)_j \right\} > 0, \tag{1.62}$$

is satisfied.

Besides, consider the vector $\xi(t) \in \mathscr{D}(\mathbb{I}, q_1, q_2)$ defined as

$$\xi(t) = \begin{cases} (q_1)_j & \text{if } j = i, \\ (q_1)_j & \text{if } h_{ij} > 0,\ j \neq i, \\ -(q_2)_j & \text{if } h_{ij} < 0,\ j \neq i, \\ 0 & \text{if } h_{ij} = 0,\ j \neq i. \end{cases} \quad j = 1, \ldots m, \tag{1.63}$$

Note that we can also compute $\xi(t+\varepsilon)$ by using the solution to (1.53), given by (1.59).

It follows that

$$\xi(t+\varepsilon) - \xi(t) = \varepsilon H \xi(t) + \varepsilon^2 \Gamma_o \xi(t).$$

This implies, for a small $\varepsilon > 0$, that we can write,

$$\xi_i(t+\varepsilon) - \xi_i(t) = \varepsilon (H\xi(t))_i + o(\varepsilon^2).$$

Moreover, taking into account Equations (1.62) and(1.63), one can conclude that,

$$\xi_i(t+\varepsilon) - \xi_i(t) > 0,$$

or in other words that

$$\xi_i(t+\varepsilon) > (q_1)_i, \text{ i.e., } \xi(t+\varepsilon) \notin \mathscr{D}(\mathbb{I}, q_1, q_2) \text{ for } \xi(t) \in \mathscr{D}(\mathbb{I}, q_1, q_2)$$

which means that the motion of the system leaves the domain $\mathscr{D}(\mathbb{I}, q_1, q_2)$. This contradicts our previous assumption. Consequently, $\tilde{H}_c q \leq 0$. □

The symmetrical case is directly obtained by the corollary below.

Corollary 1.3. *The domain*

$$\mathscr{D}(\mathbb{I}, \rho) = \left\{ z \in \mathbb{R}^m \,/\, -\rho \leq z \leq \rho,\, \rho \in \mathbb{R}_+^m \right\},$$

is positively invariant w.r.t. the system (1.53) if and only if

$$\hat{H}\rho \leq 0 \text{ with } \hat{H} = H_1 + H_2 = \begin{cases} h_{ii}, & \text{if } i = j, \\ |h_{ij}| & \text{otherwise.} \end{cases} \tag{1.64}$$

Proof: Follows readily from Theorems 1.6 and 1.7. Note that in the case when $\rho > 0$,

$$\vartheta(z) = \underset{i}{Max} \frac{|z_i|}{\rho_i},$$

is a Lyapunov function of the system. □

Based on the arguments above, we are able to extend the results of Theorem 1.7 as follows.

Corollary 1.4.
(i) The domain $\mathscr{D}(\mathbb{I}, q_1, q_2)$ with $q \geq 0$ is positively invariant w.r.t. the system (1.53) if and only if $\mathscr{D}(\mathbb{I}, q_2, q_1)$ is positively invariant w.r.t. the system.
(ii) If the domain $\mathscr{D}(\mathbb{I}, q_1, q_2)$ with $q \geq 0$ is positively invariant w.r.t. the system (1.53) then

$$\mathscr{D}(\mathbb{I}, q_1, q_2) \cup \mathscr{D}(\mathbb{I}, q_2, q_1),$$

is positively invariant w.r.t. the system.

Proof: Follows the same arguments as in the previous section. □

1.3.2 System with State Matrix of Metzler Type

Consider the continuous-time system

$$\dot{\zeta}(t) = L\zeta(t), \ \zeta \in \mathbb{R}^m, \tag{1.65}$$

where L is a Metzler matrix, *i.e.*, $L_{ij} \geq 0$, for $i \neq j$.

Corollary 1.5. *The Domain $\mathscr{D}(\mathbb{I}, \rho, 0)$ with $\rho \geq 0$ is positively invariant w.r.t. the system (1.65) if and only if*

$$L\rho \leq 0. \tag{1.66}$$

Proof: Consider the same development given in Theorem 1.7, that is,

$$z(t+\varepsilon) = Mz(t), \tag{1.67}$$

with,

$$M = \mathbb{I} + \varepsilon L + \varepsilon^2 M_o$$

$$M_o = \sum_{k=2}^{\infty} \frac{1}{k!} \varepsilon^{k-2} L^k.$$

Note that,

$$M_{ii} = 1 + \varepsilon L_{ii} + \varepsilon^2 (M_o)_{ii}$$
$$M_{ij} = \varepsilon L_{ij} + \varepsilon^2 (M_o)_{ij},\ for\ i \neq j.$$

For ε as small as possible, one can always obtain M a non-negative matrix. The use of Theorem 1.2 enables one to state that a necessary and sufficient condition for domain $\mathscr{D}(\mathbb{I}, \rho, 0)$ with $\rho \geq 0$ to be positively invariant w.r.t. the system (1.67) is given for ε as small as possible and $\varepsilon > 0$ by, $M\rho \leq \rho$, or equivalently, $L\rho \leq 0$. □.

Corollary 1.6. *The Domain $\mathscr{D}(\mathbb{I}, \rho, \infty)$ with $\rho \geq 0$ is positively invariant w.r.t. the system (1.65) if and only if condition (1.66) holds.*

Proof: Follows readily by the use of the same arguments as in the proof of Corollary 1.5 and Theorem 1.2. □

In addition, the following lemma states the stability of the matrix H, as follows.

Lemma 1.6. *If H satisfies $\tilde{H}_c q < 0$, with $q > 0$, then H is Hurwitz.*

Proof: Let H be a matrix satisfying $\tilde{H}_c q < 0$. Then from (1.56), one can deduce easily that $\hat{H} w < 0$, with $w = q_1 + q_2 > 0$ and $\hat{H}$ given by (1.64) which means that H is Hurwitz. □

1.3.3 Controller Design by Direct Procedure

1.3.3.1 Analysis of Equation $FA + FBF = HF$

In this section, we apply the results of Theorem 1.7 to the problem of the constrained regulator described in Section 1.3.

Consider the system (1.40)-(1.43) with the feedback law (1.44) and (1.46). The closed-loop system is then given by (1.45). Let us make the transformation

$$z = Fx,\ F \in \mathbb{R}^{mxn}, \tag{1.68}$$

with the matrix F given by (1.44) and (1.46). It follows that $\dot{z} = F(A+BF)x$. If there exists a matrix $H \in \mathbb{R}^{mxm}$ such that

$$FA + FBF = HF \tag{1.69}$$

then, the coordinate transformation (1.68) allows one to obtain the dynamical system (1.53) from (1.45). The controller design for the system (1.45) with $x \in \mathscr{D}(F, q_1, q_2)$ defined by (1.47), becomes possible by the use of (1.53) and Theorem 1.7, or Theorem 1.6, with $z \in \mathscr{R} = \mathscr{D}(\mathbb{I}, q_1, q_2)$.

Before going further, let us present some useful lemmas.

Lemma 1.7. *The set $\mathcal{K}erF$ is positively invariant w.r.t. the system (1.45) if and only if there exists a matrix $H \in \mathbb{R}^{mxm}$ such that $FA_o = HF$.*

Proof:

(Sufficiency): Let $x(t) \in \mathcal{K}er(F)$, that is $Fx(t) = 0$. For every $\tau \in \mathbb{R}$, $x(t+\tau) = e^{A_o\tau}x(t)$. It follows that $Fx(t+\tau) = Fe^{A_o\tau}x(t)$. Equation $FA_o = HF$ implies obviously that $FA_o^k = H^kF$, for every $k \in \mathbb{N}$. Thus, $Fx(t+\tau) = e^{H\tau}Fx(t)$. Consequently, $x(t+\tau) \in \mathcal{K}er(F)$, for every $\tau \in \mathbb{R}$.

(Necessity:) Let the kernel $\mathcal{K}er(F)$ be positively invariant w.r.t. the system (1.45). Then, stating that for $w(t)$, $Fw(t) = 0$ implies necessarily $Fw(t+\tau) = 0$, $\forall \tau \in \mathbb{R}$. It is clear that in this situation the first derivative is surely zero, that is,

$$\frac{d}{dt}(Fw(t)) = 0,$$

which obviously leads to

$$F\dot{w}(t) = FA_o w(t) = 0.$$

The same decomposition method used in the proof of Lemma 1.1 leads to the existence of a matrix H solution to $FA_o = HF$. □

Lemma 1.8. *If domain $\mathscr{D}(F, q_1, q_2)$ is positively invariant w.r.t. the system (1.45), then $\mathcal{K}er(F)$ is also positively invariant w.r.t. the system.*

Proof: The proof is similar to that of Lemma 1.2 of Section 1.3, except that we change in the development, the matrix A_o by $e^{A_o\tau}$. □

Lemma 1.9. *If there exists a stable matrix $H \in \mathbb{R}^{mxm}$ satisfying (1.69), the spectrum of A_o must contain a set of $n-m$ stable eigenvalues, closed under complex conjugacy, corresponding to $n-m$ common eigenvectors to both matrices A and A_o belonging to $\mathcal{K}er(F)$ and parallel to $\partial\mathscr{D}(F, q_1, q_2)$.*

Proof: It follows the same arguments as in the proof of Lemma 1.3. □

1.3.3.2 Application to the Regulator Problem with Saturated Control

We are now able to give the main result of this section which is a necessary and sufficient condition for domain $\mathscr{D}(F, q_1, q_2)$ to be positively invariant w.r.t. the motion of system (1.45).

This result was presented by [48, 15, 59] for the case $q > 0$.

Theorem 1.8. *Domain $\mathscr{D}(F, q_1, q_2)$ is positively invariant w.r.t. the system (1.45), if and only if there exists a matrix $H \in \mathbb{R}^{mxm}$, solution to*

$$FA + FBF = HF, \tag{1.70}$$

and satisfying

$$\tilde{H}_c q \leq 0, \tag{1.71}$$

where $\tilde{H}_c$ and q are defined by (1.56).

Proof: The proof given here is taken from [15] for $q > 0$.

(Sufficiency): Consider the coordinates transformation (1.68). Condition (1.70) implies that (1.45) can be transformed to (1.53) and domain $\mathscr{D}(F,\ q_1,\ q_2)$ with $q > 0$ to domain $\mathscr{R} = \mathscr{D}(\mathbb{I},\ q_1,\ q_2)$.
Theorem 1.6, means that condition (1.71) ensures the positive invariance property of $\mathscr{D}(\mathbb{I},\ q_1,\ q_2)$ w.r.t. the system (1.53).
As a consequence, it follows that domain $\mathscr{D}(F,\ q_1,\ q_2)$ is positively invariant w.r.t. the system (1.45).

(Necessity:) Let domain $\mathscr{D}(F,\ q_1,\ q_2)$ with $q > 0$ be positively invariant w.r.t. the system (1.45). By virtue of Lemma 1.8, $\mathscr{K}er(F)$ is also positively invariant w.r.t. (1.45). According to Lemma 1.7, there exists a matrix $H \in \mathbb{R}^{mxm}$ satisfying Equation (1.69). The same coordinates transformation (1.68) allows one to obtain (1.53) from (1.45) and domain $\mathscr{D}(\mathbb{I},\ q_1,\ q_2)$ from $\mathscr{D}(F,\ q_1,\ q_2)$, underlying that the former is also positively invariant w.r.t. the system (1.53). Taking account of Theorem 1.6, condition (1.71) holds. □

The symmetrical case is obtained directly by the result below.

Corollary 1.7. *If $q_1 = q_2 = \rho$, domain $\mathscr{D}(F,\ \rho)$ is positively invariant w.r.t. the system (1.45) if and only if there exists a matrix $H \in \mathbb{R}^{mxm}$, such that :*

(i) $FA + FBF = HF$,

(ii) $\hat{H}\rho \leq 0$,

where $\hat{H}$ is given as in Corollary 1.3.

Proof: Follows readily from Theorem 1.8. □

Recall that the result of this theorem is based on the existence of a matrix $H \in \mathbb{R}^{mxm}$ satisfying (1.71). A necessary and sufficient condition of the existence of a matrix H is given by [139] that is,

$$\text{rank}\begin{bmatrix} FA \\ F \end{bmatrix} = m. \tag{1.72}$$

Comment 1.1.

- The use of Theorem 1.7 leads to the same result as that of Theorem 1.8 with $q \geq 0$. In this case, the design approach becomes more interesting as it was the case for discrete-time systems (cf. Section 1.2).

- Note that the conditions presented by [59], are slightly different when compared to the present conditions. More precisely, conditions (1.71) are to be replaced by

(i) *There exist a positive scalar s_o and a matrix $P \in \mathbb{R}^{mxm}$ such that*

$$FA + FBF = (-s_o\mathbb{I} + P)F$$

(ii) *and*

$$\left(-s_o\mathbb{I}+\begin{bmatrix} P^+ & P^- \\ P^- & P^+ \end{bmatrix}\right) q \leq 0.$$

These conditions are obviously equivalent to (1.71) by taking $s_0 > \underset{i}{Max}\{p_{ii}^+\}$ *and* $p_{ii}^- = 0$ *and setting* $H = -s_o\mathbb{I}+P$ *which implies that*

$$\tilde{H}_c = -s_o\mathbb{I}+\tilde{P},$$

with

$$(H_1)_{ij} = \begin{cases} -s_o + p_{ii}^+, & \text{for } i=j \\ p_{ij}^+ & \text{for } i \neq j, \end{cases} \quad \text{and } (H_2)_{ij} = \begin{cases} 0 & \text{for } i=j, \\ p_{ij}^-, & \text{for } i \neq j. \end{cases}$$

- *Conditions (1.70) and (1.71) guarantee that domain* $\mathscr{D}(F,\, q_1,\, q_2)$ *defined by (1.47) is positively invariant w.r.t. the system (1.40)-(1.45), despite the existence of asymmetric constraints on the control. The choice of* $F \in \mathbb{R}^{mxn}$ *satisfying (1.46), (1.72) and (1.70)-(1.71) allows the system (1.40)-(1.46) to be asymptotically stable, even when* $\tilde{H}_c q = 0$. *Further, Lemma 1.9 gives a necessary condition on A for the existence of H solution to Equation (1.69), i.e., the matrix A must possess* $n-m$ *stable eigenvalues. When it is not the case, we proceed by augmenting the control vector dimension without losing the assumption (1.42) by writing the system (1.40) into the equivalent representation.*

$$\dot{x} = Ax + [B \;\; 0]\begin{bmatrix} u \\ v \end{bmatrix}.$$

This technique is presented in Section 1.3.

The result below can be easily deduced from Lemmas 1.6 and 1.9.

Theorem 1.9. *If there exists a matrix* $H \in \mathbb{R}^{mxm}$ *satisfying*

$$(i) \;\; FA + FBF = HF,$$
$$(ii) \;\; \tilde{H}_c q \leq 0, [-2mm]$$

then system (1.40)-(1.46) is asymptotically stable for every $x_o \in \mathscr{D}(F,\, q_1,\, q_2)$.

Corollary 1.8.
(i) *The domain* $\mathscr{D}(F,\, q_1,\, q_2)$ *with* $q \geq 0$ *is positively invariant w.r.t. the system (1.45) if and only if* $\mathscr{D}(F,\, q_2,\, q_1)$ *is positively invariant w.r.t. the system.*
(ii) *If the domain* $\mathscr{D}(F,\, q_1,\, q_2)$ *with* $q \geq 0$ *is positively invariant w.r.t. the system (1.45) then*

$$\mathscr{D}(F,\, q_1,\, q_2) \cup \mathscr{D}(F,\, q_2,\, q_1),$$

is positively invariant w.r.t. the system.

Proof: Follows readily from Corollary 1.4. □

Corollary 1.9. *The domain $\mathscr{D}(F, \rho, 0)$ with $\rho \geq 0$ is positively invariant w.r.t. the system (1.45) if and only if there exists a matrix $H \in \mathbb{R}^{mxm}$ with $h_{ij} \geq 0$ for $i \neq j$, such that :*

$$\begin{array}{ll} (i) & FA + FBF = HF, \\ (ii) & H\rho \leq 0. \end{array} \tag{1.73}$$

Proof: Follows readily from Corollary 1.5 . □

Corollary 1.10. *The domain $\mathscr{D}(F, \rho, \infty)$ with $\rho \geq 0$ is positively invariant w.r.t. the system (1.45) if and only if conditions (1.73) hold.*

Proof: Obtained by following the same reasoning as Corollary 1.6. This result is also given by [48]. □

Example 1.3. *Consider the following double integrator system with matrices A and B given by*

$$A = \begin{bmatrix} 0 & 1 \\ 0 & 0 \end{bmatrix} \text{ and } B = \begin{bmatrix} 0 \\ 1 \end{bmatrix},$$

and let $-5 \leq u \leq 10$. According to (1.44) and (1.46), we have to find a gain matrix $F \in \mathbb{R}^{mxn}$ such that $A + BF$ is stable. A solution to this is given by

$$F = [\,-2 \;\; -3\,].$$

In this case we have

$$A_o = A + BF = \begin{bmatrix} 0 & 1 \\ -2 & -3 \end{bmatrix}.$$

Note that matrix A is unstable. The eigenvalue $\lambda = 0$ has a degree of multiplicity equal to 2.

Further, this matrix does not possess $n - m$ stable eigenvalues. For this, we proceed as indicated in Section 1.2. That is,

$$\begin{array}{l} \dot{x} = Ax + [B \;\; 0] \begin{bmatrix} u \\ v \end{bmatrix}, \text{ with } v \in \mathbb{R}^{n-m} \\ -r_2 \leq v \leq r_1, \end{array}$$

where r_1 and r_2 are some fictitious constraints on the control law v. The whole control law is given by $\begin{bmatrix} u \\ v \end{bmatrix} = \begin{bmatrix} F \\ E \end{bmatrix} x$, where the matrix E is chosen in such a way that the matrix $G = \begin{bmatrix} F \\ E \end{bmatrix}$ is invertible.

Take for instance $E = [\,-1\;\;0\,]$, *then* H *is given by*

$$H = GA_oG^{-1} = \begin{bmatrix} -2.3333 & -1.3333 \\ 0.3333 & -0.6667 \end{bmatrix}.$$

The choice of the fictitious constraints is given such that domain $\mathscr{D}(G,\, g_1,\, g_2)$ *becomes the largest possible domain with conditions* $\tilde{H}_c g \leq 0$ *standing with*

$$g = \begin{bmatrix} g_1 \\ g_2 \end{bmatrix},\; g_1 = \begin{bmatrix} q_1 \\ r_1 \end{bmatrix} \text{ and } g_2 = \begin{bmatrix} q_2 \\ r_2 \end{bmatrix}.$$

We find out that $r_1 = 5$ *and* $r_2 = 11$ *which means that they satisfy condition (1.71).*

We also obtain $H_1 = \begin{bmatrix} -2.3333 & 0 \\ 0.3333 & -0.6667 \end{bmatrix}$ *and* $H_2 = \begin{bmatrix} 0 & 1.3333 \\ 0 & 0 \end{bmatrix}$.

Thus,

$$\tilde{H}_c g = [-8.67\;\; 0\;\; -5\;\; -0.61]^t \leq 0.$$

Consequently, $\mathscr{D}(G,\, g_1,\, g_2)$ *is a positively invariant set w.r.t. the system in closed-loop.*

In this section, the regulator problem of linear continuous-time systems with asymmetric constrained control is studied. Necessary and sufficient conditions for domain $\mathscr{D}(F,\, q_1,\, q_2)$, which generates admissible control by feedback law, to be a positively invariant set w.r.t. the system (1.45), are given. These conditions guarantee that system (1.40)-(1.46) is asymptotically stable for every motion emanating from domain $\mathscr{D}(F,\, q_1,\, q_2)$.

1.4 Saturated Singular Systems

This section studies the class of saturated singular systems by using the concept of positive invariance.

Singular systems have been of great interest in the control literature since they can model many systems in electrical circuits networks, robotic and economics [67]. Some problems of observers and synthesis of stabilizing controllers for linear singular systems can be cited in this category [67, 69] and the references therein. Among the subjects of continuous interest, the pole assignment problem takes a remarkable place due to the large amount of works on this area [63, 67, 68, 79, 86, 108, 137, 167] and the references therein. Two main algebraic approaches are usually followed, the first consists in transforming the singular system into a slow and fast subsystem. The pole assignment is then formed indirectly [67]. The second approach deals with the solution of a generalized Sylvester equation as used in [68, 167] and the references therein. A different way was, however, followed by [86] and [108], where the pole assignment is deeply studied, giving the necessary and sufficient conditions of existence of a state feedback allowing an arbitrary finite set of self-conjugate eigenvalues to be placed. The expression of this controller is also given.

In this section, new, necessary and sufficient conditions of positive invariance are presented for continuous-time singular system with constraints on the control. These conditions are obtained directly and without use of any transformation of the initial system, as used in [89, 140, 161], by using state feedback control. These results can be found in [42].

1.4.1 Problem Formulation

In this section, we give the problem formulation related to singular linear systems studied in this section.

Consider the following singular linear system described by

$$E\delta x(t) = Ax(t) + Bu(t) \qquad (1.74)$$
$$x(0) = x_0$$

where $x \in \mathbb{R}^n$ is the state, $u \in \mathbb{R}^m$ is the control. Matrices A, B, E are real of appropriate size with E a square matrix such that $Rank(E) = r \leq n$. $\delta x(t)$ denotes $\dot{x}(t)$ for continuous-time singular systems while $\delta x(t)$ denotes $x(t+1)$ for discrete-time singular systems. The control is assumed here to be constrained as follows:

$$u \in \Omega = \{u \in \mathbb{R}^m | -u_{min} \leq u \leq u_{max}; u_{max}, u_{min} \geq 0\}; g = \begin{bmatrix} u_{max} \\ u_{min} \end{bmatrix}. \qquad (1.75)$$

Definition 1.4. *[67]*

- *The pair (E,A) is said to be regular if $det(sE - A)$ is not identically zero.*
- *The pair (E,A) is said to be impulse free if $deg(det(sE - A)) = rank(E)$.*

In order to present some useful results, matrices E and B are decomposed as follows:

$$E = [R \;\; 0][S_0 \;\; S_\infty]^T \qquad (1.76)$$
$$B = U[Z_B^T \;\; 0]^T, U = [U_0 \;\; U_1] \qquad (1.77)$$

where matrix $[S_0 \;\; S_\infty]$ is orthogonal, $R \in \mathbb{R}^{n \times r}$ is of full column rank and matrix U is orthogonal with $rankZ_B = rankB$. These orthogonal matrices always exist and can be obtained by using the singular value decomposition.

Lemma 1.10. *[108]*
The pencil (E,A) is regular and has r finite eigenvalues if and only if

$$rank[E + AS_\infty S_\infty^T] = n. \qquad (1.78)$$

Note that the pencil (E,A) has r finite eigenvalues if $deg(det(sE - A)) = rankE = r$, according to Definition 1.4, the pencil is said impulse free.

We assume that,

H1) The pencil (E,A) is regular and impulse free, *i.e.*, $rank[E+AS_\infty S_\infty^T]=n$.
H2) The singular system (1.74) is controllable, *i.e.*, $rank[B,(\lambda E-A)]=n, \forall\lambda\in\mathbb{C}$.

Consider a state feedback given by $u(t)=Fx(t)$, the closed-loop singular system is then obtained as

$$E\delta x(t)=(A+BF)x(t)=A_cx(t). \tag{1.79}$$

The problem of pole assignment can be stated as follows: given real matrices E,A,B where $rankE=r\leq n$ and a set of r finite self-conjugate complex numbers $\mathscr{L}=\{\lambda_1\dots\lambda_r\}$, find a matrix $F\in\mathbb{R}^{m\times n}$ such that

$$det(A+BF-\lambda E)=0, \forall\lambda, |\lambda|<\infty, \ \in\mathscr{L} \tag{1.80}$$
$$det(A+BF-\lambda E)\neq 0, \forall\lambda, |\lambda|<\infty, \notin\mathscr{L}. \tag{1.81}$$

In this case, one should have

$$(A+BF)V_r=EV_r\Lambda \tag{1.82}$$

where, $\Lambda=diag\{\lambda_1\dots\lambda_r\}$ and $V_r\in\mathbb{C}^{n\times r}$ a given matrix. Equations (1.80)-(1.81) mean that only the r elements of the set $\mathscr{L}$ are the generalized eigenvalues of the pencil $(A+BF,E)$. One can notice that only the r finite eigenvalues are placed while the infinite eigenvalues remain unchanged. Besides, since the open-loop system is impulse free and the closed-loop system must be kept impulse free by state feedback control, these eigenvalues have no impulsive impact in closed-loop. According to [108], this problem has a solution if and only if the singular system is controllable and condition (1.78) holds. This justifies the assumptions H1) and H2).

We recall hereafter an interesting result of pole assignment.

Theorem 1.10. *[108]*
For matrix B of full rank, given Λ and a matrix $V_r\in\mathbb{C}^{n\times r}$ such that $[V_r \ S_\infty]$ is non-singular, then there exists F satisfying (1.82) and such that the pencil $(A+BF,E)$ is regular if and only if

$$U_1^T(AV_r-EV_r\Lambda)=0 \tag{1.83}$$
$$U_1^T(E+AS_\infty S_\infty^T) \ \textit{has full rank}. \tag{1.84}$$

Then matrix F is given by

$$F=\left[Z_B^{-1}U_0^T(EV_r\Lambda-AV_r) \ Z_B^{-1}W\right]\left[V_r \ S_\infty\right]^{-1} \tag{1.85}$$

where W is any matrix such that

$$rank[E+AS_\infty S_\infty^T+U_0WS_\infty]=n. \tag{1.86}$$

1.4.2 Conditions of Positive Invariance for Singular Systems

The objective of this subsection is to design a stabilizing controller by state feedback for the continuous-time singular system while respecting the constraints on the control. The unsaturated singular system in closed-loop is then given by (1.79). Assume that F is stabilizing system (1.79), which means that this system is admissible (regular and impulse free). The induced constraint set on the state is given by,

$$\mathscr{D} = \{x \in \mathbb{R}^n | -u_{min} \leq Fx \leq u_{max}\} \tag{1.87}$$

The main idea of the approach of positive invariance is to impose to the system trajectories to evolve only inside the region of linear behavior defined by (1.87) to have Equation (1.79) valid. This representation of the singular system is used directly to design the stabilizing controller in the presence of constraints. Assume that matrix F can be decomposed as follows: $F = \Gamma E$, where matrix $\Gamma \in \mathbb{R}^{m \times n}$.

Now, we are able to state the main result of this subsection.

Theorem 1.11. *The set $\mathscr{D}$ given by (1.87) is positively invariant with respect to (w.r.t) the continuous-time system (1.79) if and only if there exists a matrix $H \in \mathbb{R}^{m \times m}$ such that*

$$\Gamma A + \Gamma B \Gamma E = H \Gamma E \tag{1.88}$$

$$\tilde{H}_c g \leq 0. \tag{1.89}$$

Proof: (Sufficiency):
Consider the change of variables $z(t) = Fx(t)$, with this new variable, the set $\mathscr{D}$ becomes

$$\mathscr{D}_z = \{z \in \mathbb{R}^m | -u_{min} \leq z \leq u_{max}\}. \tag{1.90}$$

Compute now the new dynamical system obtained with this change of variables by using $F = \Gamma E$.

$$\dot{z}(t) = F\dot{x}(t) = \Gamma E\dot{x}(t) = \Gamma(A + BF)x(t) \tag{1.91}$$

According to Equation (1.88), one obtains

$$\dot{z}(t) = Hz(t). \tag{1.92}$$

The proof is similar to the one of Theorem 1.6 given in Section 1.3 applied to the set (1.90) w.r.t the system (1.92) by means of condition (1.89). It is also obvious that this implies that the set $\mathscr{D}$ given by (1.87) is positively invariant w.r.t system (1.79).

(Necessity):
Assume that the set $\mathscr{D}$ is positively invariant w.r.t the singular system in closed-loop (1.79) and $rank(F) = m$. Consider a vector $\omega \in Ker(F)$, that is $F\omega(t) = 0$. Decompose without loss of generality ω, F and ΓA_c as follows:

$$\omega = \begin{bmatrix} \omega_1 \\ \omega_2 \end{bmatrix}; F = [F_1 \;\; F_2]; \Gamma A_c = [M_1 \;\; M_2] \tag{1.93}$$

with $\omega_1 \in \mathbb{R}^m$, $F_1 \in \mathbb{R}^{m\times m}$ and $rank(F_1) = m$. Then, $F\omega(t) = F_1\omega_1(t) + F_2\omega_2(t) = 0$, and $\omega_1(t) = -F_1^{-1}F_2\omega_2(t)$. That means that with such $\omega_1(t)$, for any $\omega_2(t) \in \mathbb{R}^{n-m}$, $\omega(t) \in Ker(F)$. Since by definition,

$$Lim_{\tau\to 0}\frac{F\omega(t+\tau) - F\omega(t)}{\tau} = F\dot{\omega}(t),$$

one can write, $F\omega(t+\tau) = \tau F\dot{\omega}(t) = \tau\Gamma A_c\omega(t)$, where τ is as small as possible but non null. Using the decomposition obtained before leads to, $F\omega(t+\tau) = \tau\Gamma A_c\omega(t) = \tau(M_2 - M_1F_1^{-1}F_2)\omega_2(t) = G\omega_2(t) = \eta$. Assume that matrix G is non null, that is, it contains at least one component $g_{ij} \neq 0$. Since ω_2 is arbitrary chosen, take a vector with only one non null component $\omega_2^j = \alpha u^i_{max}/g_{ij}, \alpha > 1$. It follows obviously that $G\omega_2 = \alpha u^i_{max} > u^i_{max}$, consequently, $[F\omega(t+\tau)]_i > u^i_{max}$ which means that the set $\mathscr{D}$ is not positively invariant w.r.t the system. This contradicts the assumption. Besides, $G = 0$ implies by noting $H = M_1F_1^{-1}$, $M_2 = HF_2$ and $M_1 = HF_1$ or in a condensed form, $\Gamma A_c = [M_1 \;\; M_2] = H[F_1 \;\; F_2] = HF$. The second part of the proof is then easily derived: the existence of matrix H solution of Equation (1.88) enables one to use the same transformation as in the sufficiency to obtain the reduced order dynamic system (1.92) which is a standard system. Applying the result of Theorem 1.6 given in Section 1.3, condition (1.89) follows. □

Remark 1.3. *The proof of the necessity of Theorem 1.11 is based on the full row rankness of matrix F :, however, by virtue of Remark 2.2, $rank(F) = r$ if and only if $r \leq m$. If $r > m$, $rank(F) = m$ as shown in Example 2.5.2.1 case* 3. *That means that if $m > r$, conditions (1.88)-(1.89) are only sufficient.*

Corollary 1.11. *The singular system (1.79) is admissible (stable and impulse free) $\forall x(t) \in \mathscr{D}$, if there exists a matrix $H \in \mathbb{R}^{m\times m}$ satisfying (1.88) such that*

$$\tilde{H}_c g < 0. \tag{1.94}$$

Proof: It is worth noting that the stability of the singular system is also realized since there exists a matrix H satisfying (1.88), enabling one to transform the initial system into the reduced order system (1.92). Besides, condition (1.94) guarantees that the set (1.90) is positively invariant and contractive w.r.t the system (1.92). That is, the set (1.87) is also positively invariant w.r.t the singular system in closed-loop (1.79) which always has a linear comportment inside this region. According to the assumptions on matrix F, this singular linear system is admissible $\forall x(t) \in \mathscr{D}$. □

In this section, conditions of positive invariance for saturated singular linear systems are developed. These conditions are obtained without using any one of the classical transformations.

1.5 Conclusion

In this chapter, the regulator problem of linear discrete-time and continuous-time systems with asymmetric constrained control is studied. Necessary and sufficient conditions for domain $\mathscr{D}(F, q_1, q_2)$, which generates admissible control by feedback law, to be a positively invariant set w.r.t. the system in closed-loop, are given. These conditions guarantee that the system in closed-loop is asymptotically stable for every motion emanating from domain $\mathscr{D}(F, q_1, q_2)$. A spectral analysis of equation $FA + FBF = HF$ is also given together with the existence conditions of the matrix H. Necessary condition of the main result is established by using an important property of the kernel of the matrix F, that is, when domain $\mathscr{D}(F, q_1, q_2)$ is positively invariant w.r.t. the system in closed-loop, $\mathscr{K}erF$ is also positively invariant w.r.t. the system.

The case of symmetrical constrained control is obtained as a special case by taking $q_1 = q_2 = \rho$ together with some particular sets as $\mathscr{D}(F, \rho, 0)$, $\mathscr{D}(F, 0, \rho)$, $\mathscr{D}(F, \rho, \infty)$ and $\mathscr{D}(F, \infty, \rho)$. An extension to the output feedback case is then developed. Besides, conditions of positive invariance for saturated singular linear systems are also developed. These conditions are obtained without using any one of the classical transformations.

Chapter 2
Saturated Linear Systems: Controller Design

2.1 Introduction

In this chapter, a design method based on the resolution of equation $XA + XBX = HX$ is proposed. Controllers working in a linear region of behavior are obtained. This region, which is asymmetric and polyhedral, is the largest one. A new topic, using the same approach, presented for the first time in this book, concerns the study of saturated singular linear systems. The controller design is developed upon the resolution of the algebraic equation $XA + XBXE = HE$ also presented in this book. Two other techniques, allowing the design of unsaturating and saturating controllers by means of LMIs, are proposed. The corresponding regions of stability are only symmetric and of ellipsoid form.

2.2 Resolution of the Algebraic Equation $XA + XBX = HX$

The analysis of the problem of the regulator for linear discrete-time (resp. continuous-time) systems with constrained control is presented in the previous chapter. The approach presented in this chapter and called direct approach is mainly based on the positive invariance concept.

Implementation of this approach seems to be very simple with acceptable computation time. However, this direct approach still needs improvement both in the design and in the implementation of such a regulator.

The basic idea of this approach is to find, for a given stabilizing matrix F, a matrix H solution to $FA + FBF = HF$ satisfying in the same time the condition $\tilde{H}_d q \leq q$ (resp. $\tilde{H}_c q \leq 0$).

The approach presented in this chapter will be referred to as the *inverse procedure*. This approach consists in choosing a matrix H satisfying $\tilde{H}_d q < q$ (resp. $\tilde{H}_c q < 0$) and computing a matrix F solution to $XA + XBX = HX$, when such a matrix exists.

This problem is similar to the pole assignment problem. The literature on this subject is a prolific one since the famous Wonham's fundamental result [156].

A. Benzaouia: Saturated Switching Systems, LNCIS 426, pp. 37–75.
springerlink.com

2.2.1 Pole Assignment Problem

In this section, we recall the necessary and sufficient conditions of the existence of a matrix K which assigns a given symmetric set of eigenvalues.

Before going further, we first present the relevant basic results on the subject.

Consider the following linear system:

$$\delta x(t) = Ax(t) + Bu(t) \tag{2.1}$$

$$y(t) = Cx(t) \tag{2.2}$$

where $x \in \mathbb{R}^n$ is the state vector, $u \in \mathbb{R}^m$ is the control vector and $y \in \mathbb{R}^p$ is the output vector.

The operator denoted here as δ is defined as follows :

$$\delta x(t) = \begin{cases} \dot{x}(t), & \text{for continuous} - \text{time systems, with } t \in \mathbb{R} \\ x(t+1), & \text{for discrete} - \text{time systems, with } t \in \mathbb{N}. \end{cases}$$

Besides, we assume that

$$\begin{array}{ll} A1: & (A,B) \text{ is controllable} \\ A2: & (C,A) \text{ is observable} \\ A3: & \text{matrices } B \text{ and } C \text{ are of full rank} \\ A4: & m+p>n \end{array} \tag{2.3}$$

We also consider a feedback law given by,

$$u(t) = Fx(t) = Ky(t) \tag{2.4}$$

with

$$KC = F,\ F \in \mathbb{R}^{mxn} \text{ and } K \in \mathbb{R}^{mxp}. \tag{2.5}$$

The pole assignment problem consists in giving a symmetric set Λ (every element is accompanied by its conjugate) of stable eigenvalues and computing a matrix K which ensures,

$$\sigma(A+BKC) = \Lambda \tag{2.6}$$

According to Wonham's result [156], condition $A1$ of (2.3) guarantees the existence of a matrix $F = KC$ achieving (2.6).

Definition 2.1. *A symmetric set Λ is said to be assignable to the system (2.1)-(2.2) if there exists a real constant matrix K in $\mathbb{R}^{mxp}$ such as*

$$\sigma(A+BKC) = \Lambda.$$

We define two subspaces $\mathscr{F}(\lambda)$ and $\mathscr{W}(\lambda)$ which will play a fundamental role in the following development.

Definition 2.2. *For any complex number λ, the subspace $\mathscr{F}(\lambda)$ is defined as,*

$$\mathscr{F}(\lambda) = \{\xi \in \mathbb{C}^n : (\lambda\mathbb{I} - A)\xi \in \mathscr{R}ange(B)\}, \tag{2.7}$$

where $\mathscr{R}ange(B)$ denotes the range space of the matrix B.

Definition 2.3. *For any complex number λ, the subspace $\mathscr{W}(\lambda)$ in $\mathbb{C}^n$ is defined as,*

$$\mathscr{W}(\lambda) = \left\{\omega \in \mathbb{C}^n : (\lambda\mathbb{I} - A)^T\omega \in \mathscr{R}ange(C^T)\right\} \tag{2.8}$$

Obviously, $\mathscr{W}(\lambda)$ is the dual of $\mathscr{F}(\lambda)$.

Definition 2.4. *A symmetric set Λ is said to be p-decomposable if there exists a non-overlapping decomposition $\Lambda = \Lambda_1 \cup \Lambda_2$ such that $\mathscr{C}ard\{\Lambda_1\} = p$ and Λ_1 symmetric. Such a decomposition is called a p-decomposition.*

The case when Λ is not p-decomposable occurs only when Λ does not contain real values (which implies that n is even and p is odd). Therefore, the requirement that Λ is p-decomposable is by no means a strong restriction from a practical point of view.

The main result guaranteeing the existence of a matrix K is as follow.

Theorem 2.1. *[113, 114] A symmetric set*

$$\Lambda = \{\lambda_1, ..., \lambda_n\},$$

is assignable to the system (2.1)-(2.2) if and only if there exist $\xi_i \in \mathscr{F}(\lambda_i)$ and $\omega_i \in \mathscr{W}(\lambda_i)$ for $i = 1, ..., n$, such that:

$$\begin{aligned} &(i) \quad \{\xi_i,\ i = 1, ..., n\} \text{ are linearly independent in } \mathbb{C}^n \\ &\qquad \text{and if } \lambda_i = \bar{\lambda}_j \text{ then } \xi_i = \bar{\xi}_j, \end{aligned} \tag{2.9}$$

$$\begin{aligned} &(ii) \quad \{\omega_i,\ i = 1, ..., n\} \text{ are linearly independent in } \mathbb{C}^n \\ &\qquad \text{and if } \lambda_i = \bar{\lambda}_j \text{ then } \omega_i = \bar{\omega}_j, \end{aligned} \tag{2.10}$$

$$(iii) \quad \omega_j^T\xi_i = 0 \text{ for any } i \neq j. \tag{2.11}$$

From the proof of this theorem, one can deduce the following:

$$(\lambda_i\mathbb{I} - A)\xi_i = B\theta_i, \text{ for } i = 1, ..., n, \tag{2.12}$$

$$\omega_j^T(\lambda_j\mathbb{I} - A) = \varphi_j^T C, \text{ for } j = 1, ..., n, \tag{2.13}$$

where θ_i's $\in \mathbb{C}^n$ represent the directions of the pole assignment and φ_i's $\in \mathbb{C}^n$. When conditions (2.9)-(2.11) hold, the feedback gain F is then given by

$$F = [\theta_1\ \theta_2\ ...\ \theta_n][\xi_1\ \xi_2\ ...\ \xi_n]^{-1}. \tag{2.14}$$

From condition (2.9) and relation (2.12), we can state that

$$\text{if } \xi_i = \bar{\xi}_j, \text{ then } \theta_i = \bar{\theta}_j,$$

which means that F is a real matrix. By reporting the equality $\theta_i = F\xi_i$ into (2.12), one can easily show that $\sigma(A+BF) = \Lambda$.

Hence, conditions (2.10)-(2.11) ensure the existence of a matrix K solution to Equation (2.5). Note that a necessary and sufficient condition for the existence of a matrix K solution to (2.5) was given by [139], *i.e.*,

$$\text{rank}\begin{bmatrix} F \\ C \end{bmatrix} = p.$$

In some special cases, one can obtain the output matrix K directly, with a significant reduction in computation time.

Theorem 2.2. *[114] A p-decomposable set Λ is assignable to the system (2.1)-(2.2) if and only if a p-decomposition $\Lambda = \Lambda_1 \cup \Lambda_2$ exists such that there exist $\xi_i \in \mathscr{F}(\lambda_i)$ and $\omega_j \in \mathscr{W}(\lambda_j)$, respectively satisfying :*

$$\begin{aligned} (i) \quad & \{C\xi_i,\ i = 1,\ ...,\ p\} \text{ are linearly independent in } \mathbb{C}^p && (2.15)\\ & \text{and if } \lambda_i = \bar{\lambda}_j \text{ then } \xi_i = \bar{\xi}_j,\ j,\ i = 1,\ ...,\ p, \\ (ii) \quad & \{\omega_i,\ i = p+1,\ ...,\ n\} \text{ are linearly independent in } \mathbb{C}^n && (2.16)\\ & \text{and if } \lambda_i = \bar{\lambda}_j \text{ then } \omega_i = \bar{\omega}_j,\ j,\ i = p+1,\ ...,\ n, \\ (iii) \quad & \omega_i^T \xi_j = 0 \text{ for } i = 1,\ ...,\ p \text{ and } j = p+1,\ ...,\ n, && (2.17) \end{aligned}$$

with, $\Lambda_1 = \{\lambda_1\ ...\ \lambda_p\}$ and $\Lambda_2 = \{\lambda_{p+1}\ ...\ \lambda_n\}$. In this case, the gain matrix K is given by,

$$K = [\theta_1\ ...\ \theta_p]\,[C\xi_1\ ...\ C\xi_p]^{-1} \tag{2.18}$$

Other new techniques of pole assignment based on LMIs and algebraic equations are given in [3] and [7].

In order to give an inverse procedure to the application of Theorem 1.2 (resp. Theorem 1.8), we first look for a matrix $H \in \mathbb{R}^{mxm}$ satisfying the necessary requirements and then carry out the resolution of equation

$$XA + XBX = HX. \tag{2.19}$$

This procedure is described in the following section. Equation (2.19) will be referred to as the *Algebraic Equation* or *(AE)*. The same resolution procedure of (2.19) is extended to the case of output feedback control where the algebraic equation takes the form

$$XCA + XCBXC = HXC. \tag{2.20}$$

Equation (2.20) will be referred to as the *Output Algebraic Equation* or *(OAE)*.

2.2.2 *Resolution of the Algebraic Equation*

2.2.2.1 Resolution of the *AE*

In this section, we present the resolution of the algebraic equation (AE). It is now known that matrix H exists only if the spectrum of A contains a symmetric set Λ_2 of $n-m$ stable eigenvalues (*cf.* Lemma 1.3). For this, assume that

$$\sigma(A) \supset \Lambda_2 = \{\lambda \in \mathbb{C} \,/\, \lambda_j \neq 0,\, \lambda_j = \bar{\lambda}_i; j,i = m+1, \ldots n\}. \tag{2.21}$$

In this case, define the corresponding eigenvectors,

$$A\zeta_j = \lambda_j \zeta_j, \text{ for } j = m+1, \ldots n \tag{2.22}$$

with

$$\{\zeta_{m+1}, \ldots, \zeta_n\} \text{ linearly independent.} \tag{2.23}$$

It is worth noting that this assumption was completed by [17].

Let $H \in \mathbb{R}^{mxm}$ be a given diagonalizable matrix satisfying :

$$\Lambda_1 = \sigma(H) = \{\lambda_i \in \mathbb{C} \,/\, \lambda_i \neq 0 \text{ for } i = 1, \ldots, m\}, \tag{2.24}$$

$$H\theta_i = \lambda_i \theta_i; \quad B\theta_i \neq 0 \text{ for } i = 1, \ldots, m, \tag{2.25}$$

$$\{\theta_1\ \theta_2\ \ldots\ \theta_m\} \text{ are linearly independent.} \tag{2.26}$$

Assumption (2.25), *i.e.*, $B\theta_i \neq 0$, is easily satisfied by rank$(B) = m$ as in Theorem 2.1 or by an adequate choice of H. It is obvious that the assumption rank$(B) = m$ presents a great restriction as shown in Section 2.2.3. Moreover, (2.26) restricts the class of solutions to equation (2.19).

More precisely, this means that the solution to *the AE* is presented for a matrix H even with repeated eigenvalues, save that assumption (2.26) holds, *i.e.*,

$$\dim(\mathscr{K}er(\lambda_i \mathbb{I} - H)) = n_i,$$

where n_i stands for the multiplicity of λ_i for $i = 1, \ldots, m$.

Further, we assume that the given matrix H satisfies

$$\Lambda_1 \cap \sigma(A) = \emptyset. \tag{2.27}$$

The assigned spectrum to the closed-loop matrix $A_o = A + BF$ is then given by

$$\Lambda = \Lambda_1 \cup \Lambda_2, \tag{2.28}$$

$$\text{with } A_o \xi_i = \lambda_i \xi_i \text{ for } i = 1, \ldots, n. \tag{2.29}$$

Now we can give a necessary and sufficient condition for the existence of a matrix F solution to (2.19) expressed by the following result given by [13].

Theorem 2.3. *For a given matrix $H \in \mathbb{R}^{mxm}$ satisfying (2.24) to (2.27), there exists a real full rank matrix $F \in \mathbb{R}^{mxn}$ given by*

$$F = [\theta_1 \dots \theta_m \, 0 \dots 0] \, [\xi_1 \dots \xi_m \, \xi_{m+1} \dots \xi_n]^{-1}, \tag{2.30}$$

which represents the unique solution to (2.19) if and only if,

$$\{\xi_1 \dots \xi_m \, \xi_{m+1} \dots \xi_n \} \tag{2.31}$$

are linearly independent.

Proof:

(Necessity:) Assume that there exists a real matrix $F \in \mathbb{R}^{mxn}$ which represents the unique solution to the *Algebraic Equation* (2.19).

- Consider $\lambda_i \in \Lambda_1$, with the associated eigenvectors θ_i and ξ_i such that

  $$H\theta_i = \lambda_i \theta_i \text{ and } A_o \xi_i = \lambda_i \xi_i \text{ for } i = 1, \dots, m.$$

  In this case, equation (2.19) gives :

  $$FA_o\xi_i = HF\xi_i \text{ for } i = 1, \dots, m,$$

  which yields,

  $$H(F\xi_i) = \lambda_i (F\xi_i) \text{ for } i = 1, \dots, m.$$

  Since λ_i is also an eigenvalue of H, we have necessarily

  $$F\xi_i = \theta_i \text{ for } i = 1, \dots, m.$$

  However, ξ_i is unknown at this step since A_o is not yet known. To compute ξ_i, remember that $A_o\xi_i = \lambda_i\xi_i$, which is the same as,

  $$(A+BF)\xi_i = \lambda_i\xi_i \text{ or } A\xi_i + BF\xi_i = \lambda_i\xi_i$$

  Besides, using the fact that $F\xi_i = \theta_i$ and assumption (2.27) (*i.e.* $\lambda_i \notin \sigma(A)$), it follows that

  $$\xi_i = (\lambda_i \mathbb{I} - A)^{-1} B\theta_i \quad \text{for } i = 1, \dots, m \tag{2.32}$$

  It is now obvious that if $B\theta_i = 0$ with $\theta_i \neq 0$, we should have an indetermination, since $\xi_i \neq 0$, and $\lambda_i \, (\neq 0) \notin \sigma(A)$ (see assumption (2.25)).
- Now, consider an eigenvalue $\lambda_j \in \Lambda_2$, and its associated eigenvector ξ_j of A_o. Thus, $FA_o\xi_j = HF\xi_j$, for $j = m+1, \dots, n$ which is equivalent to $H(F\xi_j) = \lambda_j(F\xi_j)$. Since λ_j is not an eigenvalue of H and H nonsingular, it follows then that

  $$F\xi_j = 0, \text{ for } j = m+1, \dots, n, \; (i.e., \; \xi_j \in \mathscr{K}er(F)). \tag{2.33}$$

Further, $A_o\xi_j = \lambda_j\xi_j$ can be expanded as,

$$A\xi_j + BF\xi_j = \lambda_j\xi_j.$$

Since, $F\xi_j = 0$, we should have

$$A\xi_j = \lambda_j\xi_j,$$

or similarly that the eigenvectors ξ_j of A_o are also eigenvectors of A, *i.e.*,

$$\xi_j = \zeta_j, \ j = m+1, \ ..., \ n. \tag{2.34}$$

From above, it becomes clear that

$$F\xi_j = \theta_j, \ \text{for } j = 1, \ ..., \ m,$$

and (2.35)

$$F\xi_j = 0, \ \text{for } j = m+1, \ ..., \ n,$$

which can be written in a condensed form as

$$F\left[\xi_1 \ ... \ \xi_m \ \xi_{m+1} \ ... \ \xi_n\right] = \left[\theta_1 \ ... \ \theta_m \ 0 \cdots 0\right]. \tag{2.36}$$

Since F is real and is the unique solution to (2.19), the n vectors

$$\{\xi_1 \ ... \ \xi_n\},$$

should be linearly independent and in this case, the matrix F is given by (2.30).
It is obvious that condition (2.31) and relations (2.29) imply that the matrix A_o is diagonalizable as in Theorem 2.1.

Note also that assumption (2.26) ensures that F is of full rank which implies that A needs to have only $n - m$ non-null eigenvalues as assumed by (2.21).

(Sufficiency): Let condition (2.31) hold and recall that Λ_1 and Λ_2 are symmetric by construction. Since both matrices H and A are real, we can write

$$\begin{aligned} &\text{if } \lambda_i = \bar{\lambda}_j \ \text{ then } \bar{\theta}_i = \theta_j, \ i, \ j = 1, \ ..., \ m, \\ &\text{if } \lambda_i = \bar{\lambda}_j \ \text{ then } \bar{\zeta}_i = \zeta_j, \ i, \ j = m+1, \ ..., \ n. \end{aligned} \tag{2.37}$$

Condition (2.31) ensures the existence of a matrix F given by (2.30). Let us underline that the latter is equivalent to (2.36).

In addition, it is obvious that we can conclude that (2.35) is also satisfied. The computation of the eigenvectors ξ_i of $A_o = A + BF$ leads to the following equation,

$$A\xi_i + BF\xi_i = \lambda_i\xi_i, \ i = 1, \ ..., \ n.$$

By reporting relations (2.35) into the equation above, we finally obtain (2.32) and (2.34). These relations and (2.37) allow us to obtain,

$$\text{if } \lambda_i = \bar{\lambda}_j \text{ then } \xi_i = \bar{\xi}_j,\ i,\ j = 1,\ ...,\ n, \tag{2.38}$$

Then F given by (2.30) exists, and is a full rank real matrix. However, one wonders if such a matrix is the unique solution to equation (2.19). To prove this, let us note that equation (2.30) is equivalent to (2.36). The post-multiplication of (2.36) by

$$\text{diag}\left[\lambda_1 \ ...\ \lambda_m\ \lambda_{m+1}\ ...\ \lambda_n\right],$$

which is nonsingular by virtue of (2.21) and (2.24), leads to

$$\begin{aligned} F\left[\lambda_1\xi_1\ \ ...\ \ \lambda_m\xi_m\ \ \lambda_{m+1}\xi_{m+1}\ \ ...\ \ \lambda_n\xi_n\right] = \\ \left[\lambda_1\theta_1\ \ ...\ \ \lambda_m\theta_m\ \ 0\ \ ...\ \ 0\right]. \end{aligned} \tag{2.39}$$

Moreover, relations

$$A_o\xi_i = \lambda_i\xi_i,\ \lambda_i \in \Lambda,\ \text{ for } i = 1,\ ...,\ n,$$

and

$$H\theta_i = \lambda_i\theta_i,\ \lambda_i \in \Lambda_1,\ \text{ for } i = 1,\ ...,\ m.$$

enable us to rewrite (2.39) as follows:

$$FA_o\left[\xi_1\ ...\ \xi_m\ \xi_{m+1}\ ...\ \xi_n\right] = H\left[\theta_1\ ...\ \theta_m\ \ 0\ ...\ 0\right].$$

Again, the use of (2.36) leads to

$$FA_o\left[\xi_1\ ...\ \xi_m\ \xi_{m+1}\ ...\ \xi_n\right] = HF\left[\xi_1\ ...\ \xi_m\ \xi_{m+1}\ ...\ \xi_n\right],$$

which yields

$$FA_o = FA + FBF = HF,$$

since the matrix $\left[\xi_1\ ...\ \xi_m\ \xi_{m+1}\ ...\ \xi_n\right]$ is invertible. Consequently, the matrix F is a solution to equation (2.19).

To show that F is the unique solution, let F^* be a second solution to equation (2.19). Following the same arguments as those in the necessary condition, we deduce equation (2.36) for both matrices F and F^* with

$$\left[\theta_1\ ...\ \theta_m\ 0\ ...\ 0\right],$$

as a common part.

Finally, condition (2.31) enables us to conclude that $F = F^*$, or explicitly that F given by (2.30) is the unique solution to equation (2.19). □

Example 2.1. *Consider the following discrete-time system described by*

$$A = \begin{bmatrix} -1 & 0.5 & 0 \\ 0.5 & -0.5 & 0 \\ 0 & 4.5 & -1.5 \end{bmatrix}, \; B = \begin{bmatrix} 1.5 & 0 \\ 0 & 1 \\ 0 & 0 \end{bmatrix},$$

the spectrum of A is then

$$\sigma(A) = \{-1.5, \; -1.309, \; -0.19098\}$$

In this case, the matrix A has $n > n - m$ *non-null eigenvalue. Take* $\Lambda_2 = \{-0.19098\}$. *The corresponding eigenvector is given by*

$$\xi_3 = \zeta_3 = [\,0.5257 \; 0.8506 \; 2.9242\,]^T.$$

Let us choose a matrix H satisfying (2.24)-(2.27), for instance

$$H = \begin{bmatrix} -0.5 & 0.01 \\ 0 & -0.4 \end{bmatrix}.$$

Thus, $\Lambda_1 = \{\,-0.5, \,-0.4\,\}$ *and the associated eigenvectors are :*

$$\theta_1 = \begin{bmatrix} 1 \\ 0 \end{bmatrix}, \; \theta_2 = \begin{bmatrix} 1 \\ 10 \end{bmatrix}$$

$$B\theta_1 = [1.5\,0\,0\,]^T, \; B\theta_2 = [1.5\;10\,0\,]^T.$$

The assigned spectrum is then as follows :

$$\Lambda = \{\,-0.191, \,-0.5, \,-0.4\,\},$$

with $\xi_1 = [\,0 \;-3\; -13.5\,]^T$, $\xi_2 = [\,-27.105 \;-35.526\; -145.334\,]^T$.

The solution to equation (2.19) gives

$$F = \begin{bmatrix} 0.3368 & 0.1969 & -0.1178 \\ -0.5362 & 1.4038 & -0.3119 \end{bmatrix}.$$

The closed-loop matrix A_o *is then given by*

$$A_o = \begin{bmatrix} -0.4948 & 0.7953 & -0.1767 \\ -0.0360 & 0.9037 & -0.3119 \\ 0.0000 & 4.5000 & -1.5000 \end{bmatrix}.$$

In order to check that the resolution of equation (2.19) is a partially pole assignment technique, we compute the spectrum of the closed-loop matrix A_o, *which is given by*

$$\sigma(A_o) = \{\,-0.4, \,-0.191, \,-0.5\,\},$$

which is the same as Λ.

The resolution of equation (2.19) shows that this is only a particular case of the general pole assignment problem. In our case, a part of the directions of the pole assignment problem together with the assigned spectrum are given by means of a given matrix H satisfying assumptions (2.24)-(2.27).

It is clear that the global assigned spectrum is completed by a part of the spectrum of the open-loop matrix A, which exists according to assumption (2.21). This hypothesis is not in any case a restriction to this approach, to avoid it, we can always proceed by augmenting the system entries as developed in Section 1.3, *i.e.*, $m = n$. In this case, the augmented system is still controllable since the *rank* of matrix $[B\,AB\,\ldots\,A^{n-1}B]$ is not affected by adding zero diagonals.

If the matrix A does not satisfy assumption (2.21) or when we are interested in assigning a global spectrum Λ which does not satisfy (2.28) even when A satisfies (2.21), then, we take an $n \times n$ real matrix H such that,

$$\begin{aligned} \Lambda &= \sigma(H) \subset \mathbb{C} - \{0\} \text{ with } \mathrm{Card}(\Lambda) = n, \\ H\theta_i &= \lambda_i \theta_i,\ B\theta_i \neq 0, \text{ for } i = 1, \ldots, n. \end{aligned} \tag{2.40}$$

We further assume that the given matrix H satisfies

$$\Lambda \cap \sigma(A) = \emptyset. \tag{2.41}$$

The assigned spectrum to the matrix $A_o = A + BF$ is the symmetric set Λ with the same eigenvectors defined by (2.29). Now we can give a necessary and sufficient condition stating the existence of the solution to equation (2.19), which is denoted in this case as $G \in \mathbb{R}^{nxn}$ for $H \in \mathbb{R}^{nxn}$, by the following result:

Corollary 2.1. *For a given matrix $H \in \mathbb{R}^{nxn}$ satisfying (2.40) and (2.41), there exists a unique real matrix $G \in \mathbb{R}^{nxn}$ solution to (2.19) given by,*

$$G = [\theta_1 \ldots \theta_m\ \theta_{m+1} \ldots \theta_n\,]\,[\xi_1 \ldots \xi_m\ \xi_{m+1} \ldots \xi_n\,]^{-1}, \tag{2.42}$$

if and only if,

$$\{\xi_1, \ldots \xi_m, \xi_{m+1}, \ldots \xi_n\,\} \text{ are linearly independent.} \tag{2.43}$$

Proof: The proof follows directly from Theorem 2.3. □

Example 2.2. *Consider the same matrices A and B as given in Example 2.1, and let the matrix H be chosen as follows:*

$$H = \begin{bmatrix} -0.50 & 0.01 & 0.10 \\ 0.00 & -0.40 & 0.00 \\ 0.00 & 0.00 & -0.20 \end{bmatrix}$$

Thus, $\sigma(H) = \{-0.5, -0.4, -0.2\}$ *and the associated eigenvectors are:*

$$\theta_1 = [\,1\ \ 0\ \ 0\,]^T,\ \theta_2 = [\,1\ \ 10\ \ 0\,]^T,\ \theta_3 = [\,1\ \ 0\ \ 3\,]^T.$$

The assigned spectrum is then the set $\Lambda = \sigma(H)$, *with*

$$\xi_1 = [\,0\ -3\ -13.5\,]^T,\ \xi_2 = [\,-2.697\ -3.5349\ -14.46\,]^T,$$

$$\xi_3 = [\,-14.229\ -23.715\ -82.09\,]^T.$$

The solution to equation (2.19) gives

$$G = \begin{bmatrix} 0.3400 & 0.1689 & -0.1160 \\ -0.5345 & 1.3898 & -0.3088 \\ 0.0290 & -0.2512 & -0.0558 \end{bmatrix}.$$

The matrix A_o *is then given by*

$$A_o = \begin{bmatrix} -0.489 & 0.7534 & -0.167 \\ -0.0345 & 0.8898 & -0.3 \\ 0 & 4.5 & -1.5 \end{bmatrix}.$$

In order to check that the resolution of equation (2.19) is a pole assignment problem technique, we compute the spectrum of the closed-loop matrix A_o *as*

$$\sigma(A_o) = \{\,-0.5,\ \ -0.4,\ \ -0.2\,\},$$

which is the same as Λ.

Note that for the actual implementation, the feedback gain matrix F *is extracted from* G *and although the assumption (2.21) is satisfied for this example, one can assign a set* Λ *such that*

$$\Lambda \cap \sigma(A) = \emptyset,$$

that is, a completely different set of eigenvalues.

2.2.2.2 Resolution of the *OAE* for Square Systems

In the case of square systems, that is $m = p$, an interesting application of the resolution of the algebraic equation is given in the case of output feedback by partitioning the matrix C. The algebraic equation arising in output feedback control is of the type (2.20).

Let the matrix C be partitioned as follows:

$$C = C_l\bar{C}C_r = C_l[\,\mathbb{I}\ \ 0\,]C_r,\ \text{with}\ C_l \in \mathbb{R}^{pxp}\ \text{and}\ C_r \in \mathbb{R}^{nxn}, \tag{2.44}$$

where C_l and C_r always exist and are non-singular matrices. Consider the following transformations

$$\bar{A} = C_rAC_r^{-1} = \begin{bmatrix} \bar{A}_{11} & \bar{A}_{12} \\ \bar{A}_{21} & \bar{A}_{22} \end{bmatrix}, \tag{2.45}$$

$$\bar{B} = \begin{bmatrix} \bar{B}_1 \\ \bar{B}_2 \end{bmatrix} = C_rB,\ \bar{K} = KC_l. \tag{2.46}$$

Let ξ_i be the eigenvector of A_o associated to the same eigenvalue λ_i of H, that is, for $i = 1, \ldots, m$,

$$\begin{array}{rcl} A_o\xi_i &=& \lambda_i\xi_i = C_r^{-1}\bar{A}_oC_r\xi_i, \\ H\theta_i &=& \lambda_i\theta_i, \end{array} \tag{2.47}$$

this implies that,

$$\bar{A}_o(C_r\xi_i) = \lambda_i(C_r\xi_i) = \lambda_i \begin{bmatrix} \gamma_i \\ \beta_i \end{bmatrix}, \text{ with } \gamma \in \mathbb{R}^p. \tag{2.48}$$

The vector $C\xi_i$ can be obtained by using the partitioning of C as given above, that is,

$$C\xi_i = C_l[\,\mathbb{I}\ \ 0\,]C_r\xi_i = C_l\gamma_i. \tag{2.49}$$

The resolution of equation (2.20) is stated by the following result.

Theorem 2.4. *For a given matrix $H \in \mathbb{R}^{mxm}$ satisfying (2.24) to (2.27), there exists a real matrix $K \in \mathbb{R}^{pxp}$ given by,*

$$K = [\theta_1 \ldots \theta_m]\,[C\xi_1 \ldots C\xi_m\,]^{-1}, \tag{2.50}$$

which represents the unique solution to (2.20) with $m = p$ if and only if,

(i) the m vectors

$$\{C\xi_1 \ldots C\xi_m\,\}, \tag{2.51}$$

are linearly independent,

(ii) and the 12-block of $\bar{A}$ satisfies

$$\bar{A}_{12} = 0. \tag{2.52}$$

Proof:

(Necessity:) Assume that there exists a real full rank matrix K the unique solution to equation (2.20). Rewriting (2.20) leads to

$$KC_l[\,\mathbb{I}\ \ 0\,]C_rA + KC_l[\,\mathbb{I}\ \ 0\,]C_rBKC_l[\,\mathbb{I}\ \ 0\,]C_r = HKC_l[\,\mathbb{I}\ \ 0\,]C_r,$$

and taking account of (2.45)-(2.46), one can easily deduce that

$$\bar{K}[\,\mathbb{I}\ \ 0\,]\bar{A} + \bar{K}[\,\mathbb{I}\ \ 0\,]\bar{B}\bar{K}[\,\mathbb{I}\ \ 0\,] = H\bar{K}[\,\mathbb{I}\ \ 0\,].$$

It follows then that $\bar{K}$ should satisfy

$$\bar{K}(\bar{A}_{11} + \bar{B}_1\bar{K}) = H\bar{K}, \tag{2.53}$$

and

$$\bar{K}\bar{A}_{12} = 0.$$

Recalling that K is assumed to be of full rank, so is $\bar{K}$, hence from the equation above it follows that (2.52) holds. Further, since we assume that K is given by (2.50), it is obvious that the m–vectors $\{C\xi_1 \ldots C\xi_m\}$ are linearly independent.

(Sufficiency): Assume that $\bar{A}_{12} = 0$ and $\{C\xi_1 \ldots C\xi_m\}$ are linearly independent with the vector ξ_i's given by (2.32). It follows that

$$\begin{aligned} C\xi_i &= C(\lambda_i\mathbb{I} - A)^{-1}B\theta_i \quad \text{for } i = 1, \ldots, m \\ &= C_l[\mathbb{I}\ 0]C_r(\lambda_i\mathbb{I} - A)^{-1}C_r^{-1}\begin{bmatrix}\bar{B}_1\\ \bar{B}_2\end{bmatrix}\theta_i, \\ &= C_l[\mathbb{I}\ 0](\lambda_i\mathbb{I} - \bar{A})^{-1}\begin{bmatrix}\bar{B}_1\\ \bar{B}_2\end{bmatrix}\theta_i. \end{aligned}$$

Since we assume that $\bar{A}_{12} = 0$, we have

$$(\lambda_i\mathbb{I} - \bar{A})^{-1} = \begin{bmatrix}(\lambda_i\mathbb{I} - \bar{A}_{11})^{-1} & 0\\ \bullet & \bullet\end{bmatrix},$$

where $\bullet$ stands for terms not needed in the next developments. This leads to

$$\begin{aligned} C\xi_i &= C_l(\lambda_i\mathbb{I} - \bar{A}_{11})^{-1}\bar{B}_1\theta_i \quad \text{for } i = 1, \ldots, m, \\ &= C_l\gamma_i. \end{aligned} \tag{2.54}$$

As we assume that the m–vectors $\{C\xi_i\}_{1\le i\le m}$ are linearly independent, so are the m–vectors $\{\gamma_i\}_{1\le i\le m}$. Hence, according to Theorem 2.3, there exists a unique full rank matrix $\bar{K}$ solution to (2.53) and given by

$$\bar{K} = [\theta_1 \ldots \theta_m]\,[\gamma_1 \ \cdots\ \gamma_m]^{-1}.$$

with γ_i's given by (2.54), for $i = 1 \ \ldots\ m$. From the expression of $\bar{K}$, we deduce that

$$\begin{aligned} \bar{K}C_l &= [\theta_1 \ldots \theta_m]\,[C\xi_1 \ \cdots\ C\xi_m]^{-1}, \\ &= K. \end{aligned}$$

Hence, K is unique and one can easily check that it is the solution to (2.20). □

Note that for systems given by an input/output representation as

$$G(\delta) = \mathscr{B}(\delta)\mathscr{A}^{-1}(\delta) = C(\delta\mathbb{I} - A)^{-1}B,$$

with $\delta = z$ for the discrete-time systems and $\delta = s$ for the continuous-time systems, vectors $C\xi_i$ can be obtained directly from $G(\delta)$ by the following equations, $C\xi_i = G(\lambda_i)\theta_i, i = 1, \ldots, m$. Matrix K is then given by

$$K = [\theta_1 \ldots \theta_m]\,[G(\lambda_1)\theta_1 \ldots G(\lambda_m)\theta_m]^{-1}. \tag{2.55}$$

This technique allows one to compute matrix K directly from an estimation or an identification algorithm in the case of a system with slowly varying parameters leading to obtain some heuristic adaptive controller acting in the presence of constraints.

2.2.2.3 A Pole Assignment Problem Approach

In this section, we relate the resolution to equation (2.19) to a pole assignment problem and show that Theorem 2.3 is a particular case of Theorem 2.1.

Corollary 2.2. *The set Λ given by (2.28), with (2.24)-(2.27) is assignable to the system (2.1)-(2.4) with (2.21)-(2.23) and $C = \mathbb{I}$ if and only if the n vectors*

$$\xi_i \text{ for } i = 1, ..., n$$

given by (2.32) and (2.34) are linearly independent with the feedback gain F given by (2.30).

Proof: According to Theorem 2.1, necessary and sufficient conditions for this pole assignment are given by (2.9), (2.10) and (2.11). Following the proof of Theorem 2.3, condition (2.31), with the existence of $\xi_i \in \mathscr{F}(\lambda_i)$ given by (2.32)-(2.34), is equivalent to condition (2.9). Note that, by construction, the set Λ is symmetric and H is a real matrix, which ensures that F is also a real matrix (see details in the proof of Theorem 2.3).

Let us now compute the vectors w_i, for $i = 1, ..., n$, given by (2.10) and (2.11). Since the output matrix is taken as $C = \mathbb{I}$, one should look for vectors w_i's satisfying,

$$(\lambda_i \mathbb{I} - A)^T w_i \in \mathscr{R}\text{ange}(\mathbb{I}). \tag{2.56}$$

As $\mathscr{R}\text{ange}(\mathbb{I}) = \mathbb{R}^{nxn}$, it is obvious that for every vector $w_i \in \mathbb{R}^n$, condition (2.56) is satisfied.

Consequently, we can always select a symmetric set of n linearly independent vectors $w_i \in \mathbb{R}^n$ satisfying the orthogonality condition (2.11). The gain matrix F is then given by (2.30). □

The result of Corollary 2.2 can be expressed in another equivalent form as follows:

Corollary 2.3. *The set Λ defined by (2.28), with (2.24)-(2.27) is assignable to the system (2.1)-(2.4) with (2.21)-(2.23) and $C = \mathbb{I}$, if and only if there exists a unique full rank real matrix $F \in \mathbb{R}^{mxn}$ solution to equation (2.19).*

Proof: The proof follows from Theorem 2.3 and Corollary 2.2.

This technique remains valid even when condition (2.21) is not fulfilled or when we are not interested in completing the set Λ among the spectrum of matrix A which satisfies condition (2.21).

Consequently, the result of Corollary 2.2 shows that the resolution of equation (2.19), with H given according to Theorem 2.3 or Corollary 2.1, can be used as a method of a multivariable pole assignment technique. Thus, the solution $F \in \mathbb{R}^{mxn}$ satisfies $\sigma(A+BF) = \Lambda$, with Λ the assigned spectrum in the direction $\theta_1, ..., \theta_m$, where the θ_i's are the eigenvectors of H. Nevertheless, the solution is easily obtained and the assigned eigenvalues are nonzero. Then, this method ensures a pole assignment in a specified region. Note that Theorem 2.4 may also be related to Theorem 2.2 by establishing the equivalence between condition (2.52) and equations (2.16)-(2.17).

2.2.3 Controller Design by Inverse Procedure

In this section, we propose an inverse procedure to simplify the application of the approach presented in Sections 1.2 and 1.3. This method uses the resolution of the algebraic equation (2.19) developed in Section 2.2 with H a given matrix satisfying (2.24)-(2.27). The objective of this technique is to simplify the design and the implementation of the direct approach which may require excessive computation time.

2.2.3.1 Case When A Has n-m Stable Eigenvalues

Consider the following linear system:

$$\delta x(t) = Ax(t) + Bu(t), \tag{2.57}$$

$$(A,B) \quad \text{is stabilizable} \tag{2.58}$$

$$\sigma(A) \supset \Lambda_2 = \{\lambda \in \mathbb{C} \,/\, \lambda_j \neq 0 \text{ is stable}, \ \lambda_j = \bar{\lambda}_i, \ \text{for } j,\, i = m+1, \ldots, n\}. \tag{2.59}$$

It is also assumed that the eigenvalues in Λ_2 are associated to distinct eigenvectors.

Note that $\lambda \in \mathbb{C}$ is said to be stable if :

$$0 \leq |\lambda| < 1, \text{ in the discrete} - \text{time case}$$
$$\text{Real}(\lambda) \leq 0, \text{ in the continuous} - \text{time case}$$
$$u \in \Omega = \left\{ u \in \mathbb{R}^m \,/ -q_2 \leq u \leq q_1,\, q_1,\, q_2 \in \mathbb{R}^m_+ \right\}. \tag{2.60}$$

A matrix H is said to be stable, if all its eigenvalues are stable. We first consider a state feedback law before extending the obtained result to the output feedback law as used in the pole assignment problem. That is,

$$u(t) = sat(Fx(t)), \tag{2.61}$$

which induces a linear behavior for the system in the region given by,

$$\mathcal{D}(F,\, q_1,\, q_2) = \left\{ x \in \mathbb{R}^n \,/ -q_2 \leq Fx \leq q_1,\, q_1,\, q_2 \in \mathbb{R}^m_+ \right\}, \tag{2.62}$$

where the closed-loop system is given by

$$\delta x(t) = A_o x(t), \tag{2.63}$$

with $A_o = A + BF$.

The objective remains unchanged as developed in the previous sections. That is, to find a couple of matrices $(F,\ H)$ such that,

$$\begin{aligned} \rho(A+BF) &< 1 \text{ (resp. } \mathrm{Real}(\lambda_i(A+BF)) < 0), \\ FA+FBF &= HF, \\ \text{and } \tilde{H}_\delta q &\leq 0, \end{aligned}$$

with

$$\tilde{H}_\delta = \begin{cases} \tilde{H}_d - \mathbb{I}, \text{ for discrete} - \text{time systems}, \\ \tilde{H}_c, \text{ for continuous} - \text{time systems}, \end{cases}$$

and

$$\tilde{H}_d = \begin{bmatrix} H^+ & H^- \\ H^- & H^+ \end{bmatrix},\ q = \begin{bmatrix} q_1 \\ q_2 \end{bmatrix},$$

$$h_{ij}^+ = \mathrm{Sup}(h_{ij},\ 0),\ h_{ij}^- = \mathrm{Sup}(-h_{ij},\ 0)$$

$$\tilde{H}_c = \begin{bmatrix} H_1 & H_2 \\ H_2 & H_1 \end{bmatrix},$$

where

$$H_1 = \begin{cases} h_{ij}^+ \text{ for } i \neq j \\ h_{ii} \text{ otherwise} \end{cases},\ H_2 = \begin{cases} h_{ij}^- \text{ for } i \neq j \\ 0 \text{ otherwise} \end{cases}.$$

for $i,\ j = 1,\ \ldots,\ n$.

Recall that condition $\tilde{H}_\delta q \leq 0$ is a necessary and sufficient condition for domain $\mathscr{D}(F,\ q_1,\ q_2)$ to be positively invariant w.r.t. the system (2.63) where F is assumed to be a stabilizing matrix.

It follows from above that our actual problem is to find a matrix H, in $\mathbb{R}^{mxm}$, satisfying all the necessary requirements and compute a feedback gain $F \in \mathbb{R}^{mxn}$ which guarantees the asymptotic stability of system (2.63) with the saturated inputs (2.61). The following result, based on Theorem 2.3, gives a solution to this problem [16].

Theorem 2.5. *For a given stable matrix H in R^{mxm} satisfying (2.24) to (2.27), (2.31) and*

$$\tilde{H}_\delta q \leq 0, \tag{2.64}$$

the system (2.57)-(2.61) with the matrix F given by

$$F = [\theta_1 \ldots \theta_m\ 0 \ldots\ 0\,][\xi_1 \ldots \xi_m\ \xi_{m+1} \ldots \xi_n\,]^{-1},$$

is asymptotically stable for every $x_o \in \mathscr{D}(F,\ q_1,\ q_2)$.

Proof: According to Theorem 2.3, condition (2.31) guarantees the existence of a unique mxn real full rank matrix F solution to (2.19), whenever H satisfies conditions (2.24)-(2.27) and A satisfies (2.59). This solution is given by (2.30) and ensures the desired pole assignment expressed by (2.28) and (2.29). The proof will be complete if we show that this regulator stabilizes the system (2.63), in the presence of input saturations given by (2.60) and (2.61).

Recall that H is also a stable matrix, then, $\rho(H) < 1$ (resp. $\text{Real}(\lambda_i(H)) < 0$). Taking account of (2.59), we should obtain $\rho(A+BF) < 1$ (resp. $\text{Real}(\lambda(A+BF)) < 0$).

Consequently, condition (2.64) ensures that the polyhedral domain $\mathscr{D}(F, q_1, q_2)$ is positively invariant w.r.t. the motion of the system (2.63), for every $x_o \in \mathscr{D}(F, q_1, q_2)$. The system is then asymptotically stable. Note that if $q > 0$, condition $\tilde{H}_\delta q < 0$ implies that matrix H is stable (see Lemma 1.6).

Example 2.3. *Consider the discrete-time system,*

$$x_{k+1} = Ax_k + Bu_k$$

with

$$A = \begin{bmatrix} -1. & 0.5 & 0. \\ 0.5 & -0.5 & 0. \\ 0. & 4.5 & -1.5 \end{bmatrix}, \; B = \begin{bmatrix} 1.5 & 0 \\ 0 & 1 \\ 0 & 0 \end{bmatrix},$$

and

$$-\begin{bmatrix} 3 \\ 2 \end{bmatrix} \le u \le \begin{bmatrix} 5 \\ 1 \end{bmatrix}.$$

The pole assignment problem by the resolution of equation (2.19) for this system was considered in Example 2.1, with H satisfying all the requirements of Theorem 2.3, that is

$$\tilde{H}_d q = [\, 1.51 \;\; 0.8 \;\; 2.52 \;\; 0.4 \,]^T \le q.$$

The solution to equation (2.19) given in Example 2.1 is recalled :

$$F = \begin{bmatrix} 0.3368 & 0.1969 & -0.1178 \\ -0.5362 & 1.4038 & -0.3119 \end{bmatrix}.$$

Remark that the asymptotic stability of the system is guaranteed inside the polyhedral domain $\mathscr{D}(F, q_1, q_2)$, despite the presence of input saturations.

When the spectrum of the matrix A does not contain a symmetric set of $n-m$ stable and non-null eigenvalues, we can proceed as described in the following section.

2.2.3.2 System Entries Augmentation

Let us rewrite equation (2.57) as follows:

$$\delta x(t) = Ax(t) + [\, B \; 0 \,] \begin{bmatrix} u(t) \\ v(t) \end{bmatrix}, \tag{2.65}$$

$$(A, B) \text{ is controllable.} \tag{2.66}$$

Note that in this case the matrix $\begin{bmatrix} B & 0 \end{bmatrix}$ is not of full rank, nevertheless, we can find a matrix H whose eigenvectors satisfy $B\theta_i \neq 0$ for $i = 1, ..., n$. Besides, the augmented system is still controllable since the *rank* of matrix $[B\,AB\,\ldots\,A^{n-1}B]$ is not affected by addition of zero diagonals. The input vector v is a fictitious control of the system. The set of admissible control Ω given by (2.60) remains unchanged. Thus,

$$u \in \Omega, \tag{2.67}$$

$$v \in \Omega_f = \left\{ v \in \mathbb{R}^{n-m} \,/ -r_2 \leq v \leq r_1,\, r_1,\, r_2 \in \mathbb{R}_+^{n-m} \right\}, \tag{2.68}$$

where r_1 and r_2 are some fictitious constraints. Denote

$$g_1 = \begin{bmatrix} q_1 \\ r_1 \end{bmatrix},\, g_2 = \begin{bmatrix} q_2 \\ r_2 \end{bmatrix},\, g = \begin{bmatrix} g_1 \\ g_2 \end{bmatrix}. \tag{2.69}$$

The purpose remains unchanged and the feedback law becomes

$$\begin{aligned} w(t) &= \begin{bmatrix} u(t) \\ v(t) \end{bmatrix} \\ &= \mathrm{sat}\left(\begin{bmatrix} F \\ E \end{bmatrix} x(t) \right) = \mathrm{sat}(Gx(t)), \text{ with } G \in \mathbb{R}^{nxn}. \end{aligned} \tag{2.70}$$

In this case, domain (2.62) takes the form

$$\mathscr{D}(G,\, g_1,\, g_2) = \mathscr{D}(F,\, q_1,\, q_2) \cap \mathscr{D}(E,\, r_1,\, r_2) \tag{2.71}$$

Relation (2.71) means that the new domain is bounded and the fictitious constraints add some limitations on the domain $\mathscr{D}(F,\, q_1,\, q_2)$. The resolution of equation $GA + GBG = HG$, where $H \in \mathbb{R}^{nxn}$ satisfies conditions (2.40)-(2.41) and (2.43), is given by Corollary 2.1. Once G is computed with relation (2.42), one can extract the required gain matrix $F \in \mathbb{R}^{mxn}$ by using (2.70). Further, if the given matrix H is stable and satisfies

$$\tilde{H}_\delta g \leq 0, \tag{2.72}$$

the asymptotic stability of system (2.65)-(2.70) is guaranteed for every $x_o \in \mathscr{D}(G,\, g_1,\, g_2)$. The fictitious constraints are chosen such that (2.72) holds with $\mathscr{D}_o \subset \mathscr{D}(G,\, g_1,\, g_2)$ where $\mathscr{D}_o$ is the set of initial states.

Corollary 2.4. *For a given nxn stable matrix H satisfying (2.40)-(2.41), (2.43) and (2.72), the system (2.65)-(2.70) with the gain matrix G given by*

$$G = [\theta_1 \ldots \theta_m\;\; \theta_{m+1} \ldots\; \theta_n\,][\xi_1 \ldots \xi_m\; \xi_{m+1} \ldots \xi_n\,]^{-1},$$

is asymptotically stable, for every $x_o \in \mathscr{D}(G,\, g_1,\, g_2)$.

Proof: The proof follows obviously from above. □△△

Example 2.4. *Consider the discrete-time double integrator studied in Example 1.2 described by :*

$$A = \begin{bmatrix} 1 & 1 \\ 0 & 1 \end{bmatrix}, \; B = \begin{bmatrix} 0.5 \\ 1 \end{bmatrix},$$

and the constraints

$$-\delta \leq u \leq 10, \text{ with } \delta \geq 0.$$

It is obvious that this system is unstable and the spectrum of the matrix A does not contain a symmetric set of $n-m$ non-null and stable eigenvalues. In this case, we proceed as described above by introducing a fictitious input v with the virtual constraint

$$-r_2 \leq v_k \leq r_1,$$

where r_1 and r_2 are chosen arbitrarily such that $r_1 + r_2 \neq 0$. The matrix

$$H = \begin{bmatrix} 0.5 & -1 \\ 0 & 0.4 \end{bmatrix},$$

satisfies all the requirements of Corollary 2.4. Its eigenvectors are given by

$$\theta_1 = \begin{bmatrix} 1 \\ 0 \end{bmatrix}, \; \theta_2 = \begin{bmatrix} 10 \\ 1 \end{bmatrix},$$

and they satisfy

$$\bar{B}\theta_1 \neq 0 \text{ and } \bar{B}\theta_2 \neq 0$$

with $\bar{B} = [B \;\; 0]$.

The fictitious constraints are chosen in such a way that condition (2.72) holds. Let r_1 and r_2 be chosen as follows:

$$r_1 = 0.4\delta \quad \text{and} \quad r_2 = 4,$$

which leads to

$$\tilde{H}_d g = [9 \;\; 0.16\delta \;\; 0.9\delta \;\; 1.6]^T \leq g, \forall\, \delta \geq 0.$$

The vectors ξ_1 and ξ_2 are computed from

$$\xi_i = (\lambda_i \mathbb{I} - A)^{-1} \bar{B}\theta_i, \text{ for } i = 1, 2,$$

which gives

$$\xi_1 = \begin{bmatrix} 3 \\ -2 \end{bmatrix} \text{ and } \xi_2 = \begin{bmatrix} 350/18 \\ -50/3 \end{bmatrix}.$$

From (2.42) we get

$$G = \begin{bmatrix} -0.3 & -0.95 \\ -0.18 & -0.27 \end{bmatrix}.$$

It follows that the feedback gain matrix F is given by

$$F = [-0.3 \ -0.95].$$

Hence, the closed-loop transition matrix A_o is then given by

$$A_o = \begin{bmatrix} 0.85 & 0.525 \\ -0.3 & 0.05 \end{bmatrix}, \text{ with } \sigma(A_o) = \{0.5, 0.4\}.$$

As expected, the obtained regulator is asymptotically stabilizing in the presence of the constraints, for every $x_o \in \mathscr{D}(G, g_1, g_2)$.

Example 2.5. *Consider the continuous-time double integrator studied in Example 1.3 described by*

$$A = \begin{bmatrix} 0 & 1 \\ 0 & 0 \end{bmatrix} \text{ and } B = \begin{bmatrix} 0 \\ 1 \end{bmatrix},$$

with

$$-\delta \le u(t) \le 10 \text{ and } \delta \ge 0.$$

It is obvious that this system is unstable and the spectrum of matrix A does not contain a symmetric set of $n-m$ non-null and stable eigenvalues. For these reasons, we will proceed according to the development in the subsection above by introducing the fictitious constraints

$$-r_2 \le v(t) \le r_1,$$

where $v(t)$ stands for the fictitious input. The matrix

$$H = \begin{bmatrix} -0.5 & 1 \\ 0 & -0.4 \end{bmatrix},$$

satisfies all the requirements of Corollary 2.4. Here again, the fictitious constraints are chosen such as condition (2.72) holds, which is satisfied if

$$r_1 \le 5 \text{ and } r_2 \le 0.5\delta.$$

Let $r_1 = 4.5$ and $r_2 = 0.4\delta$, which leads to

$$\tilde{H}_c g = [-0.5 \ -1.8 \ -0.1\delta \ -0.16\delta]^T \le 0, \ \forall \delta \ge 0.$$

The use of (2.32) gives

$$\xi_1 = \begin{bmatrix} 4 \\ -2 \end{bmatrix} \text{ and } \xi_2 = \begin{bmatrix} 6.25 \\ -2.25 \end{bmatrix}.$$

And from (2.42) we get

$$G = \begin{bmatrix} -0.2 & -0.9 \\ 8 & 16 \end{bmatrix}.$$

The obtained feedback gain matrix F is then given by

$$F = [-0.2 \ -0.9],$$

which yields

$$A_o = \begin{bmatrix} 0 & 1 \\ -0.2 & -0.9 \end{bmatrix}, \text{ with } \sigma(A_o) = \{-0.5, -0.4\}.$$

From the above, we note that the obtained regulator ensures the asymptotic stability of the system in the presence of the constraints, for every $x_o \in \mathscr{D}(G, g_1, g_2)$.

Comment 2.1. *With this design, we note that it is possible to stabilize a completely unstable system in the presence of inequality constraints. It is also obvious that the positive invariance property of $\mathscr{D}(G, g_1, g_2)$ is preserved.*

2.2.4 Reduction of the Algebraic Equation

In this section, we propose to use a reduction of the algebraic equation by means of the projection of system (2.63) in the subspace associated to the unstable open-loop eigenvalues. Consider system (2.57) and let the matrix A admit m controllable unstable or null eigenvalues, closed under complex conjugacy.

That means $l = n - m$ eigenvalues are stable, nonnull and closed by complex conjugacy. For this, define a subspace $\mathbb{E}_l$ associated to these eigenvalues and $\mathbb{E}_m$ its complementary subspace such that, $\mathbb{E}_l \oplus \mathbb{E}_m = \mathbb{R}^n$. It is well known that there always exists an orthonormal transformation $T \in \mathbb{R}^{nxn}$ which can be obtained from a *Schur* decomposition of A satisfying :

$$T = [T_l \ T_m], \text{ with } T^T T = TT^T = \mathbb{I}. \tag{2.73}$$

The open-loop system is then represented in the orthonormal basis formed by the columns of the matrix T by

$$\delta \begin{bmatrix} z_l(t) \\ z_m(t) \end{bmatrix} = \begin{bmatrix} A_l & A_2 \\ 0 & A_m \end{bmatrix} \begin{bmatrix} z_l(t) \\ z_m(t) \end{bmatrix} + \begin{bmatrix} B_l \\ B_m \end{bmatrix} u(t), \tag{2.74}$$

with

$$\begin{aligned} A_l &= T_l^T A T_l, \ (\in \mathbb{R}^{lxl}), \\ A_2 &= T_l^T A T_m, \ (\in \mathbb{R}^{lxm}), \\ A_m &= T_m^T A T_m, \ (\in \mathbb{R}^{mxm}), \\ 0 &= T_m^T A T_l, \ (\in \mathbb{R}^{mxl}), \\ B_l &= T_l^T B, \ (\in \mathbb{R}^{lxm}), \\ B_m &= T_m^T B, \ (\in \mathbb{R}^{mxm}), \end{aligned} \tag{2.75}$$

and z_l (resp. z_m) is the projection of x on $\mathbb{E}_l$ (resp. $\mathbb{E}_m$) along $\mathbb{E}_m$ (resp. $\mathbb{E}_l$).

In this projection, one can note that z_m is decoupled from z_l. This leads to have

$$\delta z_m(t) = A_m z_m(t) + B_m u(t), \tag{2.76}$$

with the following control law

$$u(t) = F\,[T_l \;\; T_m]\, z_m = [F_l \;\; F_m]\, z_m, \tag{2.77}$$

with $F_l \in \mathbb{R}^{mxl}$ and $F_m \in \mathbb{R}^{mxm}$.

The property of $\mathscr{K}er(F)$ given by (2.33) is preserved in the new basis by setting

$$\mathbb{E}_l = \mathscr{K}er(F),$$

that is, $F_l = 0$.

The desired m eigenvalues are to be assigned to the spectrum of the closed-loop matrix of the reduced order system

$$\delta z_m(t) = (A_m + B_m F_m) z_m(t). \tag{2.78}$$

It is interesting to examine the transformation of the *Algebraic Equation* and the positive invariance property of the domain $\mathscr{D}(F,\, q_1,\, q_2)$. The following result deals with this.

Theorem 2.6. *[60] The domain $\mathscr{D}(F,\, q_1,\, q_2)$ is positively invariant w.r.t. the system (2.63) if and only if the domain $\mathscr{D}(F_m,\, q_1,\, q_2)$ is positively invariant w.r.t. the reduced order system (2.78).*

Proof:

(Sufficiency): Let the domain $\mathscr{D}(F_m,\, q_1,\, q_2)$ be positively invariant w.r.t. the system (2.78). According to Theorem 1.2 in the case of discrete-time systems and Theorem 1.8 in the case of continuous-time systems, there exists a matrix $H \in \mathbb{R}^{mxm}$ such that

$$F_m(A_m + B_m F_m) = H F_m, \tag{2.79}$$

such that

$$\tilde{H}_{\delta q} \leq 0. \tag{2.80}$$

Post-multiplying (2.79) by T_l^T leads to

$$F_m(A_m T_l^T + B_m F_m T_l^T) = H F_m T_l^T,$$

and using (2.75) and (2.77) one can deduce that

$$F(A + BF) = HF, \tag{2.81}$$

with (2.80) unchanged. Consequently, the domain $\mathscr{D}(F,\, q_1,\, q_2)$ is positively invariant w.r.t. the system (2.63).

(Necessity:) Equation (2.81) can be rewritten by using (2.75) and (2.77) as follows:

$$H[\,0 \;\; F_m\,]\begin{bmatrix} T_l^T \\ T_m^T \end{bmatrix} = [\,0 \;\; F_m\,]\begin{bmatrix} T_l^T \\ T_m^T \end{bmatrix}\left(A + B[\,0 \;\; F_m\,]\begin{bmatrix} T_l^T \\ T_m^T \end{bmatrix}\right).$$

Post-multiplying the latter by T, one can obtain

$$H[\,0\ F_m\,] = [\,0\ F_m\,]\left(T^T A T + T^T B[\,0\ F_m\,]\right),$$

which implies that

$$[\,0\ HF_m\,] = [\,0\ F_m A_m\,] + [\,0\ F_m B_m F_m\,],$$

which is the same as equation (2.79) with condition (2.80) unchanged. □

This technique is given in [60] for a matrix A with r unstable eigenvalues where r is subjected to the condition $r \leq m$. However, the case $r < m$ leads to $rank(F) < m$ which is in contradiction with our initial assumption $rank(F) = m$.

Thus the order reduction cannot be less than m and in this case, condition (2.80) remains the same.

In this section, the solution to equation $XA + XBX = HX$ is presented. This solution is related to the pole assignment problem. It is shown that this problem, in the special case of $\Lambda = \Lambda_1 \cup \Lambda_2$, with $\Lambda = \sigma(H)$ and $\Lambda_2 \subset \sigma(A)$ is equivalent to the resolution of equation $XA + XBX = HX$.

2.3 Controller Design Using LMIs

In the previous section, a controller design method based on the resolution of equation $XA + XBX = HX$ was presented. This technique allows the whole set $\mathscr{D}(F,\ q_1,\ q_1)$ to be positively invariant w.r.t the system in closed loop while guaranteeing a linear behavior. Many attempts were done to write the algebraic equation $XA + XBX = HX$ under Linear Matrix Inequality (LMI) [24] or as Sylvester equation available in softwares as Matlab [21]. However, the simplest way to design the controller with LMIs is to use the initial idea of [97] which is based on quadratic Lyapunov functions. This is the purpose of this section.

The saturation function is assumed here to be symmetric and normalized,

$$sat(u_j) = \begin{cases} 1, \text{if}\, u_j \succeq 1 \\ u_j, \text{if} -1 \prec u_j \prec 1 \\ -1, \text{if}\, u_j \preceq -1 \end{cases},\ j = 1, \ldots, m.$$

For this, define the following subsets of $\mathbb{R}^n$:

$$\varepsilon(P,\rho) = \{x \in \mathbb{R}^n / x^T P x \preceq \rho, \rho \succ 0\}, \tag{2.82}$$

$$\mathscr{L}(F) = \{x \in \mathbb{R}^n / |F_j x| \preceq 1,\ j = 1, \ldots, m\}, \tag{2.83}$$

with $P = P^T$ a positive definite matrix, $F \in \mathbb{R}^{m \times n}$ and F_j the jth line of matrix F. Thus $\varepsilon(P,\rho)$ is an ellipsoid , while $\mathscr{L}(F)$ is a polyhedral consisting of states for which the saturation does not occur.

Theorem 2.7. *[97]*
Given an ellipsoid $\varepsilon(P,\rho)$, if there exists a matrix $F \in \mathbb{R}^{m\times n}$ such that:

- *for the discrete-time system (d.t.s.):*

$$[A+BF]^T P[A+BF]-P<0, \tag{2.84}$$

- *for the continuous-time system (c.t.s.):*

$$[A+BF]^T P+P[A+BF]<0, \tag{2.85}$$

and $\varepsilon(P,\rho) \subset \mathscr{L}(H)$, then $\varepsilon(P,\rho)$ is a contractively invariant set for the closed-loop system with saturation (1.7).

This result even ensures that only a restrictive symmetric region of linear behavior for the system can be used as LMIs for controller design as presented by the following result.

Corollary 2.5. *For a given positive scalar ρ, if there exist matrices $Y \in \mathbb{R}^{mxn}$ and a positive definite matrix $X = X^T \in \mathbb{R}^{nxn}$, solutions of the following LMIs:*

$$\begin{cases} \begin{bmatrix} X & [AX+BY]^T \\ * & X \end{bmatrix} > 0, \textit{ for } \textit{ d.t.s.} \\ [AX+BY]+[AX+BY]^T < 0, \textit{for } \textit{ c.t.s.}, \end{cases} \tag{2.86}$$

$$\begin{bmatrix} 1/\rho & Y_j \\ * & X \end{bmatrix} > 0, \tag{2.87}$$

$$j = 1,\ldots,m,$$

then, the system in closed loop (1.7) is asymptotically stable at the origin $\forall x_o \in \varepsilon(P,\rho)$ with

$$F = YX^{-1}, \tag{2.88}$$
$$P = X^{-1}. \tag{2.89}$$

Proof: Inequalities (2.86) are easily obtained by using $X = P^{-1}, Y = FP^{-1}$. For the discrete-time case, pre and post multiply inequality $[A+BF]^T P[A+BF]-P<0$ by P^{-1}. It follows $X-(AX+BY)^T X(AX+BY) > 0$, which is equivalent by Schur complement to LMI (2.86) for discrete-time case. Besides, the set of linear behavior is given by $\mathscr{L}(F)$, which satisfies $\mathscr{L}(F) \subset \varepsilon(P,\rho)$. Using a related result of [53], the inclusion $\mathscr{L}(F) \subset \varepsilon(P,\rho)$ is satisfied if and only if $1/\rho - F_i P^{-1} F_i^T > 0$. This inequality can be rewritten equivalently as $1/\rho - (FP^{-1})_i P (FP^{-1})_i^T > 0$. By letting $X = P^{-1}, Y = FP^{-1}$ and using Schur complement, LMI (2.87) follows. □

The controller designed with this technique will be called unsaturating controller since the saturation of the control is not allowed with this approach.

Comment 2.2. *In order to design unsaturating controllers by using LMIs, Corollary 2.5 can be used. In this case, the obtained set of linear systems is only a symmetric bounded region. Nevertheless, the design method developed in the previous sections, based on the resolution of equation* $XA + XBX = HX$ *leads to unbounded asymmetric regions of positive invariance.*

Remark 2.1. *The saturation method presented in this section can be extended to non-normalized saturation where the control is saturated as follows:*

$$-q_i \le u_i \le q_i, q_i > 0, i = 1,\dots,m, \tag{2.90}$$

by defining an auxiliary control as

$$-1 \le w_i \le 1, w_i = \frac{u_i}{q_i}, i = 1,\dots,m. \tag{2.91}$$

In this case, the saturated system can be rewritten as follows:

$$\dot{x}(t) = Ax(t) + Bsat(u(t)) = Ax(t) + \tilde{B}sat(w(t)), \tag{2.92}$$

where

$$\tilde{B} = B \begin{bmatrix} q_1 & 0 & 0 & \dots \\ 0 & q_2 & 0 & \dots \\ \dots & \dots & \dots & \dots \\ 0 & 0 & \dots & q_m \end{bmatrix}, \tag{2.93}$$

with $sat(w(t))$ *is a normalized saturation.*

Example 2.6. *Consider the same example as Example 1.2:*

$$A = \begin{bmatrix} 1 & 1 \\ 0 & 1 \end{bmatrix}, \; B = \begin{bmatrix} 0.5 \\ 1 \end{bmatrix},$$

with a set of admissible controls

$$\Omega = \{u \in R \,/ -10 \le u \le 10,\}$$

In this example, the constraint is not normalized as assumed in this section. For this, the system can be rewritten as: $x_{k+1} = Ax_k + Bu_k = Ax_k + \tilde{B}w_k$, *where* $\tilde{B} = 10B, w_k = u_k/10$. *The LMIs (2.86)-(2.87) are feasible for* $\rho = 100$ *leading to the following solutions:*

$$F = \begin{bmatrix} -0.0179 & -0.0461 \end{bmatrix}, A_c = \begin{bmatrix} 0.9104 & 0.7693 \\ -0.1792 & 0.5386 \end{bmatrix}, P = \begin{bmatrix} 0.2592 & 0.1836 \\ 0.1836 & 0.7663 \end{bmatrix}.$$

Note that in this case, the effective control is $u_k = 10Fx_k$.

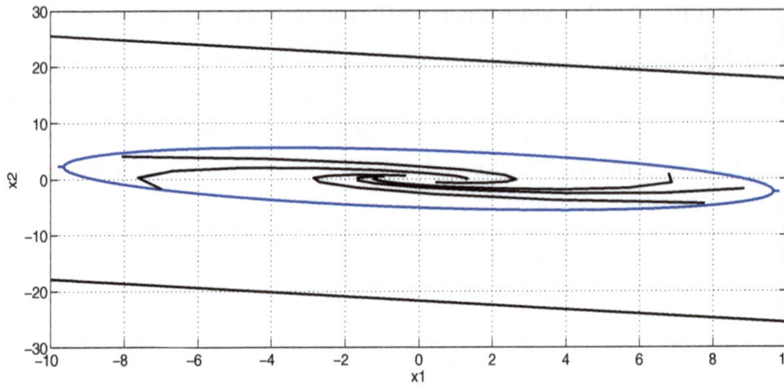

Fig. 2.1 Trajectories obtained with unsaturating controller with the polyhedral set $\mathscr{L}(F)$ in solid line

In this section, a technique of designing unsaturated controllers by using LMIs is presented. Even the obtained regions of linear behavior are symmetric and bounded, one can use new tools of computations available in all softwares as Matlab and Scilab. This technique is based on the first idea of [97] developed in 1985.

2.4 Stabilization of Linear Systems Subject to Actuator Saturation

The second approach recalled in this chapter concerns the design of saturating controllers which allow the saturations to take effect while guaranteeing asymptotic stability (see [103, 104] and the references therein). This approach, allowing the control to be saturated, leads to a bounded region of stability which, although can be obtained easily by the resolution of a set of LMIs, is ellipsoidal and symmetric. The main challenge in this approach is to obtain a large enough domain of initial states which ensures asymptotic stability for the system despite the presence of saturations [17, 19, 23, 26, 94, 103, 104].

This section is devoted to the study of linear systems described by equation (1.40) as presented by Figure 1.1:

$$\delta x(t) = Ax(t) + Bsat(u(t)), \tag{2.94}$$

where the operator denoted here δ is defined as follows:

$$\delta x(t) = \begin{cases} x(t+1), \\ \dot{x}(t), \end{cases}$$

for discrete-time case (d.t.c.) and continuous-time case (c.t.c.) respectively; $x \in \mathbb{R}^n$ is the state vector, and $u \in \mathbb{R}^m$ is the constrained control. The saturation function is assumed here to be symmetric and normalized.

Matrices A and B are constant, of appropriate size and satisfy the following assumptions:

- (AS1): The pair (A,B) is controllable.

In this subsection, we are interested in the synthesis of a stabilizing controller for this class of linear systems subject to saturations on the control, using a state feedback control law:

$$u(t) = Fx(t), \tag{2.95}$$

which writes the closed-loop system as,

$$\delta x(t) = Ax(t) + Bsat(Fx(t)). \tag{2.96}$$

The main objective is to design stabilizing controllers by saturating state feedback control.

The following result concerns an equivalent writing of the saturation function.

Lemma 2.1. *[103], [104]*
For all $u \in \mathbb{R}^m$ and $\vartheta \in \mathbb{R}^m$ such that $|\vartheta_j| \prec 1, j \in [1,m]$,

$$sat(u) \in co\{D_i u + D_i^- \vartheta, i \in [1,\eta]\}, \tag{2.97}$$

where co denotes the convex hull.

Here, D_i is an m by m diagonal matrix with elements either 1 or 0 and $D_i^- = \mathbb{I}_m - D_i$. There are $\eta = 2^m$ possible such matrices. These matrices D_i and D_i^- were introduced by [103] to model the saturation function as a linear one: Consequently, there exist scalars $\gamma_1 \succeq 0, \ldots, \gamma_\eta \succeq 0$ with $\sum_{i=1}^{\eta} \gamma_i = 1$ such that

$$sat(u) = \sum_{i=1}^{\eta} \gamma_i [D_i u + D_i^- \vartheta]. \tag{2.98}$$

This lemma leads to the following result.

Theorem 2.8. *[103], [104]*
Given an ellipsoid $\varepsilon(P,\rho)$, if there exists a matrix $H \in \mathbb{R}^{m \times n}$ such that:

- *for the discrete-time system:*

$$[A + B(D_i F + D_i^- H)]^T P [A + B(D_i F + D_i^- H)] - P < 0, \quad \forall i \in [1,\eta], \tag{2.99}$$

- *for the continuous-time system:*

$$[A + B(D_i F + D_i^- H)]^T P + P[A + B(D_i F + D_i^- H)] < 0, \quad \forall i \in [1,\eta], \tag{2.100}$$

and $\varepsilon(P,\rho) \subset \mathscr{L}(H)$, where the sets $\varepsilon(P,\rho), \mathscr{L}(H)$ are defined by (2.82)-(2.83), then $\varepsilon(P,\rho)$ is a contractively invariant set for the closed-loop system with saturation (2.96).

Proof: If there exist matrices $P = P^T$ and H satisfying $\varepsilon(P,\rho) \subset \mathscr{L}(H)$, then, by virtue of Lemma 2.1, there exists scalars $\gamma_i, i = 1,\ldots,\eta$ such that

$$sat(Fx(t)) = \sum_{i=1}^{\eta} \gamma_i(t)(D_i F + D_i^- H)x(t),$$
$$\gamma_i(t) \succeq 0; \sum_{i=1}^{\eta} \gamma_i(t) = 1, \forall t. \quad (2.101)$$

This allows to rewrite system (2.96) as follows:

$$\delta x(t) = \sum_{i=1}^{\eta} \gamma_i(t)[A + B(D_i F + D_i^- H)]x(t). \quad (2.102)$$

It is now easy to establish the Lyapunov stability condition for system (2.102) leading to inequality (2.99) for discrete-time systems and (2.100) for continuous-time systems. □

Conditions (2.99)-(2.100) can be written in an LMI form as follows:

Theorem 2.9. *For a given positive scalar ρ, if there exist matrices $Y \in \mathbb{R}^{mxn}$ and $Z \in \mathbb{R}^{mxn}$ and a positive definite matrix $X = X^T \in \mathbb{R}^{nxn}$, solutions of the following LMIs:*

$$\begin{cases} \begin{bmatrix} X & [AX + B(D_i Y + D_i^- Z)]^T \\ * & X \end{bmatrix} > 0, \; for \;\; d.t.c. \\ [AX + B(D_i Y + D_i^- Z)] + [AX + B(D_i Y + D_i^- Z)]^T < 0, for \;\; c.t.c., \end{cases} \quad (2.103)$$

$$\begin{bmatrix} 1/\rho & Z_j \\ * & X \end{bmatrix} > 0, \quad (2.104)$$

$$i = 1,\ldots,\eta; \;\; j = 1,\ldots,m,$$

then the saturated system in closed loop (2.96) is asymptotically stable at the origin $\forall x_o \in \varepsilon(P,\rho)$ with

$$F = YX^{-1}, \quad (2.105)$$
$$H = ZX^{-1}, \quad (2.106)$$
$$P = X^{-1}. \quad (2.107)$$

One can also consult the work of [26] for more details and other extensions to linear systems with both constraints on the control and the increment or rate of the control.

As a particular case, one can deduce unsaturating controllers developed in the previous sections by letting $D^+ = \mathbb{I}, D^- = 0$. Recall that these controllers work in a linear behavior and do not allow saturation to take effect.

Example 2.7. *Consider the same example as Example 1.2:*

$$A = \begin{bmatrix} 1 & 1 \\ 0 & 1 \end{bmatrix}, B = \begin{bmatrix} 0.5 \\ 1 \end{bmatrix},$$

with a set of admissible controls

$$\Omega = \{u \in R / -10 \leq u \leq 10, \}.$$

In this example, the constraint is not normalized as assumed in this section. The same transformation can be used as in the previous section.The LMIs (2.86)-(2.87) are feasible for $\rho = 100$ *leading to the following solutions:*

$$F = \begin{bmatrix} -0.0527 & -0.1265 \end{bmatrix}, H = \begin{bmatrix} -0.0472 & -0.0941 \end{bmatrix},$$
$$A_c = \begin{bmatrix} 0.7364 & 0.3677 \\ -0.5271 & -0.2647 \end{bmatrix}, P = \begin{bmatrix} 1.2348 & 0.4825 \\ 0.4825 & 1.2923 \end{bmatrix}.$$

Note that in this case, the effective control is $u_k = 10Fx_k$. *It is worth noting that for* $\rho = 100$, *the set of saturation* $\mathscr{L}(H)$ *is larger than the set of linear behavior* $\mathscr{L}(F)$ *obtained together with the saturation approach presented in this section. However, the set* $\mathscr{L}(F)$ *obtained in this section is less larger than the set* $\mathscr{L}(F)$ *obtained in the previous section.*

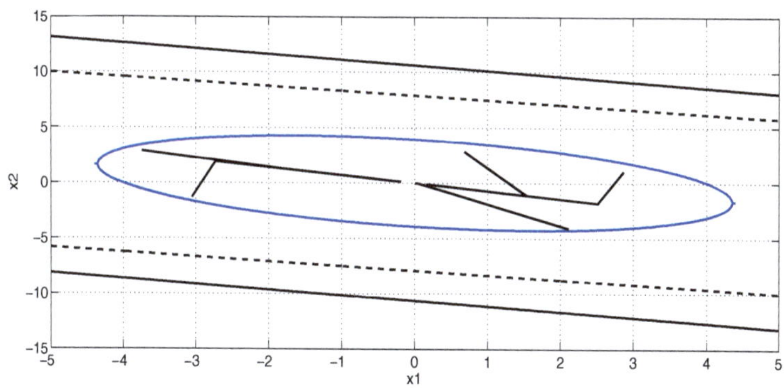

Fig. 2.2 Trajectories obtained with saturating controller with the polyhedral set $\mathscr{L}(F)$ in dotted line and $\mathscr{L}(H)$ in solid line

In this section, a technique of designing saturated controllers by using LMIs is presented. These controllers allow the control to be saturated contrary to unsaturating controllers developed in the previous section. However, to realize this advantage, one has to solve a set of $2^m + m$ LMIs. A comparison between these two techniques will be addressed in the subsequent chapters.

2.5 Saturated Singular Systems

This section studies the pole assignment problem by state feedback for singular linear systems by using the solution of the algebraic equation $XA + XBXE = HXE$. This technique is then applied to design constrained controller for such class of systems by using the concept of positive invariance. These results can be found in [42].

2.5.1 Solution of Equation XA+XBXE=HXE

The first objective of this subsection is to solve the nonlinear algebraic equation

$$XA + XBXE = HXE \tag{2.108}$$

where matrices A, B, E are given by the singular system (1.74) and matrix H is chosen with some desired spectrum to be assigned to the closed-loop singular system. This equation is involved in the constrained control problem for singular linear systems as detailed in the previous section. The solution of this equation is also related to the pole assignment problem as formulated before and more particularly to the result of Theorem 1.10. It will be shown that this approach leads to the desired spectrum with only simple computations and less conditions.

For this, consider a matrix $H \in \mathbb{R}^{m \times m}$ and note its corresponding eigenvalues and eigenvectors by, $H\theta_i = \lambda_i \theta_i, i = 1, \ldots, m$. Let H be given according to the following assumptions

$$\{\theta_1, \ldots, \theta_m\} \text{ are linearly independent} \tag{2.109}$$

$$\sigma(H) \bigcap \sigma((A, E)) = \emptyset \tag{2.110}$$

where $\sigma(H)$ denotes the spectrum of H and $\sigma((A, E))$ denotes the generalized spectrum of the pencil (A, E). Note that assumption (2.109) means that matrix H is chosen diagonalizable, that is the eigenvalue λ_i is not necessarily simple (of multiplicity 1). By noting matrix Θ_r as follows, $\Theta_r = [\theta_1, \ldots, \theta_r]$, one can write

$$H\Theta_r = \Theta_r \Lambda. \tag{2.111}$$

The problem of pole placement described above leads obviously to the following relations linking the desired generalized eigenvectors of the pencil $(A + BF, E)$ and the associated eigenvalues in $\sigma(H)$,

$$(A + BF)\xi_i = \lambda_i E \xi_i, i = 1, \ldots, r. \tag{2.112}$$

These eigenvectors are not yet known since matrix F is still unknown. Further, by virtue of assumption H1), the pencil (A,E) is regular and impulse free, which means that it contains only r finite generalized eigenvalues. The aim of this subsection is to place arbitrarily these r eigenvalues. However, the spectrum of matrix H which contains the eigenvalues to be placed has m elements. Without loss of generality, first assume that $m \geq r$. The case $m < r$ will be treated later. Equations (2.112) mean that only r eigenvalues from the spectrum of matrix H are chosen to be placed in closed loop.

Choose matrix F such that $F = XE$ and post-multiply each side of the equation (2.108) by vector ξ_i; the following developments follow by using (2.112).

$$\begin{aligned} X(A+BF)\xi_i &= HXE\xi_i, i = 1, \ldots, r. \\ &= \lambda_i XE\xi_i \end{aligned} \tag{2.113}$$

This can be arranged as

$$H(XE\xi_i) = \lambda_i(XE\xi_i), i = 1, \ldots, r. \tag{2.114}$$

Since λ_i is an eigenvalue of matrix H, one can deduce

$$XE\xi_i = \theta_i, i = 1, \ldots, r. \tag{2.115}$$

Define $V_r = [\xi_1, \ldots, \xi_r]$. Thus, equation (2.115) can be written equivalently in a condensed form as

$$XEV_r = \Theta_r. \tag{2.116}$$

At this step, Equation (2.112) can also be rewritten in a condensed form as follows

$$AV_r + B\Theta_r = EV_r\Lambda. \tag{2.117}$$

Equation (2.115) has to be completed to lead to a solvable linear equation system. The idea to complete with S_∞ is taken from [108].

$$XES_\infty = W. \tag{2.118}$$

Recall that $S_\infty \in \mathbb{R}^{n\times(n-r)}$ and $W \in \mathbb{R}^{m\times(n-r)}$. According to (1.76) where matrix $[S_0 \ S_\infty]$ is orthogonal, that is, $E[S_0 \ S_\infty] = [R \ 0]$ one should have $ES_\infty = 0$. It follows that $W = 0$. That is,

$$XES_\infty = 0. \tag{2.119}$$

To see the impact of equation (2.119) on the algebraic equation (2.108), post-multiplying both the sides of this equation by matrix S_∞ leads to

$$XAS_\infty = 0. \tag{2.120}$$

Assembling Equations (2.116) and (2.120), one obtains the following equation

$$X[EV_r \ AS_\infty] = [\Theta_r \ 0] \tag{2.121}$$

where $V_r = [\xi_1, \ldots, \xi_r]$ and $\Theta_r = [\theta_1, \ldots, \theta_r]$. At this level, note that the generalized eigenvectors of the pencil $(A+BF, E)$ are not yet known. Their expressions can be computed, by virtue of relations (2.115), as,

$$\begin{aligned}(A+BF)\xi_i &= \lambda_i E\xi_i, i = 1, \ldots, r \\ &= A\xi_i + B\theta_i\end{aligned}$$

or in a matrix form, $(A+BF)\Lambda = EV_r\Lambda = AV_r + B\Theta_r$.

Thus, the eigenvectors ξ_i can be obtained before computing matrix F by

$$\xi_i = (\lambda_i E - A)^{-1} B\theta_i, i = 1, \ldots, r \tag{2.122}$$

In our context, matrix V_r is computed according to the given matrix H by means of some of its eigenvalues and eigenvectors. For this, assume that matrix H is also chosen such that

$$[V_r \ S_\infty] \text{ is nonsingular.} \tag{2.123}$$

Assumption (2.123) implies that any vector ξ_i cannot be null, which is realized if $B\theta_i \neq 0, i = 1, \ldots, m$. In consequence, the assumption of full rankness of matrix B is not needed in this work. It is worth noting that these vectors exist and are nonnull by virtue of assumptions (2.109)-(2.110) and (2.123). With these developments, we can now give the main result of this work by the following theorem.

Theorem 2.10. *For given matrices A, B, E, H according to assumptions H1-H2, (2.109)-(2.110) and (2.123), there exists a unique solution of Equation (2.108) if and only if*

$$\Sigma = [EV_r \ AS_\infty] \ \textit{in nonsingular.} \tag{2.124}$$

In this case, the solution is given by

$$X = [\Theta_r \ 0]\Sigma^{-1}. \tag{2.125}$$

Proof: Necessity: Assume that Equation (2.108) has a solution X. Following the same developments from (2.113) to (2.120), Equation (2.121) is obtained. A unique solution of this equation exists if and only if condition (2.124), with $\Sigma \in \mathbb{R}^{n \times n}$, is satisfied and the solution is given by (2.125).

Sufficiency: Let condition (2.124) be satisfied. Hence, the unique solution of Equation (2.121) is given by (2.125). We have to show that this matrix is a solution of Equation (2.108). Bearing in mind relations (2.116), (2.117) and (2.120), post-multiplying matrix $\Delta = XA + XBXE - HEX$ by $[V_r \ S_\infty]$, the following developments are obtained:

$$\begin{aligned}\Delta\,[V_r\ S_\infty] &= XA\,[V_r\ S_\infty] + XBXE\,[V_r\ S_\infty] - HXE\,[V_r\ S_\infty]\\ &= XA\,[V_r\ S_\infty] + XB\,[\Theta_r\ 0] - H\,[\Theta_r\ 0]\\ &= X\,[AV_r + B\Theta_r\ AS_\infty] - [H\Theta_r\ 0]\\ &= X\,[EV_r\Lambda\ AS_\infty] - [\Theta_r\Lambda\ 0]\\ &= [\Theta_r\Lambda\ XAS_\infty] - [\Theta_r\Lambda\ 0] = 0.\end{aligned} \tag{2.126}$$

Since matrix H is also chosen according to (2.123), it is easy to conclude that Δ is identically null. Let matrix X^* also be a solution of (2.108). Following the same developments from (2.113) to (2.120), Equation (2.121) is obtained with X^*. Condition (2.124) enables one to conclude that $X = X^*$. □

Comment 2.3.

- *Note that if $r \le m$, $rank\,[\Theta_r\ 0] = r$, since matrix Θ_r is made up by r linearly independent vectors θ_i.*
- *In order to give the analogy with the results of Theorem 1.10, note that the first condition (1.83) is always satisfied, that is, $U_1^T(AV_r - EV_r\Lambda) = U_1^T B\Theta_r = 0$ since $U_1^T B = 0$ by construction of B. While condition (1.84), which concerns only matrices A, B, E, is obtained by assumption H1 in our context.*
- *Note that condition (1.84) is necessary and sufficient to have $rank[B, E + AS_\infty S_\infty^T] = n$, which is the necessary condition to ensure a regular closed-loop system by feedback control and complete controllability of the open-loop infinite poles [108]. This condition has been the subject of a great discussion in the literature of singular systems. It is obvious that assumption H1 ensures the full rankness of matrix $[B, E + AS_\infty S_\infty^T]$ independently of matrix B.*

2.5.2 Controller Design for Singular Systems

In this subsection, the obtained solution of equation $XA + XBXE = HEX$ is used to design the stabilizing controller for saturated singular systems. The controller design is given by the next result.

Theorem 2.11. *For given matrices A, B, E, H according to assumptions H1-H2, (2.109)-(2.110) and (2.123), matrix F satisfying (1.82) and such that the pencil $(A + BF, E)$ is regular is given by*

$$F = [\Theta_r\ 0][V_r\ S_\infty]^{-1}. \tag{2.127}$$

Proof: The proof follows directly by considering Equations (2.115) and (2.119) and taking account of assumption (2.123). Besides, to ensure that the closed-loop singular system remains regular and impulse free, that is, $det(\lambda E - (A + BF)) \neq 0$ and $deg(det(\lambda E - (A + BF))) = r$, the necessary and sufficient condition $rank[E + (A + BF)S_\infty S_\infty^T] = n$ must hold. Or equivalently, taking account of (2.118), $rank[E + AS_\infty S_\infty^T + U_0 Z_B F S_\infty S_\infty^T] = rank[E + AS_\infty S_\infty^T + U_0 Z_B W S_\infty^T] = rank[E + AS_\infty S_\infty^T] = n$, by virtue of assumption H1, matrix W being chosen null. □

Corollary 2.6. *The solution X of Equation (2.108) given by (2.125) and F given by (2.127), satisfy $F = XE$.*

Proof: Let X be solution of Equation (2.108) given by (2.125) while matrix F is given by (2.127). Equations (2.125) and (2.127) give: $XEV_r = FV_r = \Theta_r$, since matrix V_r is of full column rank, one can obtain $F = XE$. □

Remark 2.2.

- *According to Equation (2.127) where matrix Θ_r is composed of r linearly independent eigenvectors of H, we have $rank(F) = rank(\Theta_r)$. It follows that matrix F is of full rank row if and only if $m \leq r$.*
- *In Theorem 1.10, with our design method, the expression*

$$Z_B^{-1}U_0^T(EV_r\Lambda - AV_r) = Z_B^{-1}U_0^T B\Theta_r = \Theta_r$$

since $Z_B^{-1}U_0^T B = \mathbb{I}_m$ by construction of B. Thus, matrix $F = [\Theta_r\ Z_B^{-1}W][V_r\ S_\infty]^{-1}$, which is the same as in our case (2.127), matrix W being chosen null. Hence, while in our construction method, matrix F is unique, Theorem 1.10 proposes a set of solutions F even if the spectrum and its directions are fixed, by means of matrix W. This difference is imposed by the structure of Equation (2.108).

Comment 2.4. *Besides the solution of Equation (2.108) presented by Theorem 2.10, Theorem 2.11 proposes a solution to the pole assignment problem discussed in this paper. This result realizes the same objective as Theorem 1.10 but in a different and simple way. In fact, all the conditions (1.78)-(1.83) and (1.84) are satisfied by construction in our case as mentioned in Comment 2.3. Note also that in our case, the restriction on matrix B to be of full rank is relaxed.*

In case of $m < r$, one can use a different way to compute the required solution of Equation (2.108) given by the following approach. Rewrite the system (1.74) under the equivalent form:

$$E\delta x(t) = Ax(t) + B_a w(t), \tag{2.128}$$

with matrix B_a given by

$$B_a = \begin{bmatrix} B\ 0 \end{bmatrix}.$$

This augmentation technique leads to the introduction of $r - m$ fictitious entries. In this case, the control law is also modified and becomes

$$w(t) = \begin{bmatrix} u(t) \\ v(t) \end{bmatrix} = Kx(t) = \begin{bmatrix} F \\ F_f \end{bmatrix} x(t), \tag{2.129}$$

with the fictitious feedback control $v(t) = F_f x(t)$.

Note that the closed-loop system given by the augmented control $w(t)$ remains the same as (1.79). The same developments are then followed as before to obtain the solution X or K of Equation (2.108). One has to precise that the effective solution has to be extracted from the obtained one according to (2.129). It is also worth to note that the assumption of full rankness of matrix B_a is not satisfied here as required by [108].

The corresponding algorithm to this technique is given as follows:

Algorithm 2.1.

- *Step 1: If $m < r$, augment the system entries by adding $r - m$ null columns in B.*
- *Step 2: Verify assumptions H1-H2 while decomposing matrix E to have S_∞.*
- *Step 3: Choose a matrix $H \in \mathbb{R}^{m\times m}$ according to the conditions (2.109), (2.110) and (2.123).*
- *Step 3: If condition (2.124) is satisfied, continue, else go to Step2 to change matrix H.*
- *Step 4: Compute the solutions X and F of the algebraic equation (2.108) by using (2.125) and (2.127).*
- *Step 5: If the augmentation technique is used, extract the compatible solutions.*

2.5.2.1 Example

In order to apply the results of Theorem 2.10, we consider the following continuous-time linear singular system with

$$A = \begin{bmatrix} 1 & -1 & 0 \\ 0 & 1 & -1 \\ 1 & -2 & 0 \end{bmatrix}; E = \begin{bmatrix} 1 & 0 & 0 \\ 0 & 1 & 1 \\ 0 & 0 & 0 \end{bmatrix} \text{ with } S_\infty = \begin{bmatrix} 0 \\ -0.7071 \\ 0.7071 \end{bmatrix}.$$

The pencil (A,E) has two finite generalized eigenvalues: $\sigma((A,E)) = \{0.5; -1\}$, and matrix E is singular with $rank(E) = 2$.

Case m = r:

$$B = \begin{bmatrix} 0 & 0 \\ 1 & 0 \\ 0 & 1 \end{bmatrix}.$$

Choose matrix H according to assumptions (2.109)-(2.110) and (2.123) as follows:

$$H = \begin{bmatrix} -4 & -1 \\ 0 & -5 \end{bmatrix}.$$

Assumptions H1-H2 are all satisfied. Matrix V_r and the solutions of Equation (2.108) are given by

$$V_r = \begin{bmatrix} 0 & 0.0643 \\ 0 & 0.3857 \\ -0.3333 & -0.4018 \end{bmatrix}; X = \begin{bmatrix} -11.75 & -3 & 2.875 \\ 11 & 0 & -5.5 \end{bmatrix}; F = \begin{bmatrix} -11.75 & -3 & -3 \\ 11 & 0 & 0 \end{bmatrix}.$$

The test error gives $norm(XA + XBXE - HXE) = 1.4429e - 13$. The generalized eigenvalues in closed loop are $\sigma((A+BF,E)) = \{-4; -5\}$. Note that for this case,

matrix Σ is nonsingular which leads to a unique solution X of Equation (2.108).
Case $m > r$:

$$B = \begin{bmatrix} 0 & 0 & 0 \\ 1 & 1 & 0 \\ 0 & 1 & 1 \end{bmatrix}.$$

Choose matrix H according to assumptions (2.109)-(2.110) and (2.123) as follows:

$$H = \begin{bmatrix} -4 & 0 & 1 \\ 0 & -5 & 1 \\ 0 & 0 & -6 \end{bmatrix}.$$

Assumptions H1-H2 are all satisfied. In this case, only two eigenvalues from the spectrum of H are to be placed. Matrix V_r computed with only the two first eigenvalues of matrix H, the solution of Equation (2.108) and the corresponding matrix F are given by

$$V_r = \begin{bmatrix} -0.0440 & -0.0355 \\ -0.2198 & -0.2130 \\ -0.0989 & -0.2781 \end{bmatrix}; X = \begin{bmatrix} -20 & -3 & 7 \\ 11 & 0 & -5.5 \\ 0 & 0 & 0 \end{bmatrix}; F = \begin{bmatrix} -20 & -3 & -3 \\ 11 & 0 & 0 \\ 0 & 0 & 0 \end{bmatrix}.$$

The test error gives $norm(XA + XBXE - HXE) = 8.6749e - 14$. The generalized eigenvalues in closed loop are $\sigma((A+BF,E)) = \{-4; -5\}$. It is worth noting that also in this case, matrix Σ is non singular which leads to a unique solution X of Equation (2.108). Nevertheless, matrix F is not of full rank $rank(F) = 2$ as mentioned in Remark 2.2.

Case $m < r$: $B = [0 \ 1 \ 0]^T$. Augmenting the entries of the system with one fictitious entry, matrix B becomes

$$B_a = \begin{bmatrix} 0 & 0 \\ 1 & 0 \\ 0 & 0 \end{bmatrix}.$$

Note that $rank(B_a) = 1$. Choose the same matrix H as in the first case. The corresponding matrix V_r and the solutions of Equation (2.108) are given by

$$V_r = \begin{bmatrix} -0.0440 & -0.0296 \\ -0.2198 & -0.1775 \\ -0.0989 & -0.0651 \end{bmatrix}; X = \begin{bmatrix} 29 & 6 & -8.5 \\ -110 & -33 & 22 \end{bmatrix}; K = \begin{bmatrix} 29 & 6 & 6 \\ -110 & -33 & -33 \end{bmatrix}.$$

The test error gives $norm(XA + XBXE - HXE) = 2.2428e - 12$. The generalized eigenvalues in closed loop are $\sigma((A+BF,E)) = \{-4; -5\}$. Again in this case, matrix Σ in not singular, hence the solution X is unique. Matrix F has to be extracted from matrix K: $F = [29 \ 6 \ 6]$ with $rank(F) = m = 1$.

In order to apply the results of Theorem 1.11, we assume that it is constrained as follows:

$$u \in \Omega = \{u \in \mathbb{R}^m | -u_{min} \leq u \leq u_{max}\}; g = [30\ 25\ 40\ 30]^T.$$

The same corresponding matrix H is also chosen. Note that

$$\tilde{H}_c g = [-95\ \ -125\ \ -130\ \ -150]^T < 0.$$

Assumptions H1-H2 are all satisfied. The solution of Equation (2.108) and the corresponding matrix in closed loop are given by

$$F = \begin{bmatrix} -11.75 & -3 & -3 \\ 11 & 0 & 0 \end{bmatrix}; A_c = \begin{bmatrix} 1 & -1 & 0 \\ -11.75 & -2 & -4 \\ 12 & -2 & 0 \end{bmatrix}.$$

The generalized eigenvalues in closed loop are $\sigma(A_c, E) = \{-4; -5\}$. It is worth noting that this example is given for $m = r$ to illustrate the approach; however, as it is given in the previous section, one can take differently $m < r$ or $m > r$, in the third case, conditions (1.88)-(1.89) will be only sufficient. In order to plot the trajectories of the obtained singular system in closed-loop (1.79), the following equations are obtained:

$$\begin{aligned} \dot{x}_1(t) &= -5x_1(t); \\ \dot{x}_2(t) &= -30x_1(t); \\ \dot{x}_3(t) &= -4x_3(t) + 6.25x_1(t). \end{aligned}$$

The integration of these equations leads to

$$\begin{aligned} x_1(t) &= e^{-5t}x_1(0); \\ x_2(t) &= 6e^{-5t}x_1(0) \\ x_3(t) &= e^{-4t}x_3(0) + 6.25e^{-4t}x_1(0) - 6.25e^{-5t}x_1(0). \end{aligned}$$

These trajectories depend only on the two independent modes linked to the two assigned poles of matrix H. The evolution of the states and controls in the time domain are presented by Figure 2.3 and Figure 2.4 respectively. We can note that the singular closed-loop system obtained with a controller designed by this approach is regular and impulse free while the asymmetric constraints on the control are respected.

In this section, the pole assignment problem for singular linear systems is presented by solving the nonlinear algebraic equation (2.108). This equation is the direct extension of the one used for regular systems and studied in Section 2.2. The link with this equation and the pole assignment method given by [108] is shown. The necessary and sufficient condition of existence of a solution is given. It is also indicated how to place any r finite number of eigenvalues in closed loop given together with their directions with matrix H, even of size m. The case of $r > m$ can

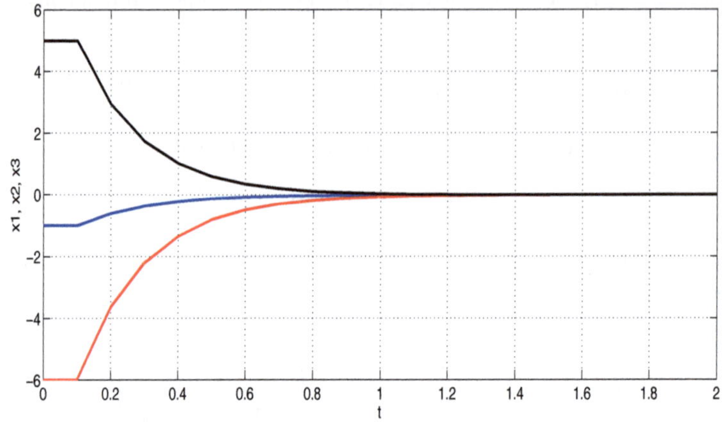

Fig. 2.3 The evolution of the three states of the singular system in closed loop emanating from $x_o = [-1 \ -6 \ 5]^t$

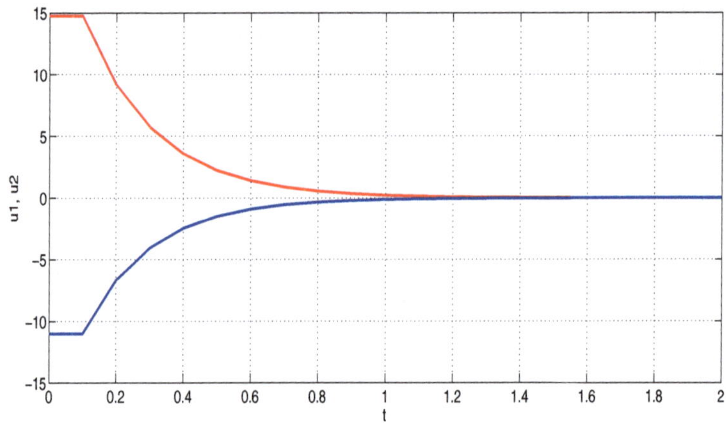

Fig. 2.4 The evolution of the two controls of the singular system in closed loop obtained with the initial state $x_o = [-1 \ -6 \ 5]^t$

also be taken into account by using the well-known technique of augmentation of the system entries. An example is used for each case $m = r$, $m > r$ and $m < r$. Finally, the practical use of this equation in constrained control of continuous-time singular linear systems is also presented. Conditions of positive invariance for this class of systems are given. The synthesis of the constrained controller is based on the solution of Equation (2.108). An example is used to show the interest of the proposed method. The discrete-time case can be obtained straightforwardly. This technique can be easily used to build an observer for singular systems even with saturated control as presented in [70] based on the work of [130].

2.6 Conclusion

In this chapter, the regulator problem of linear discrete-time and continuous-time systems with asymmetric constrained control is studied. Besides, the solution of equation $XA+XBX=HX$ is also presented. This solution is related to a pole assignment problem. It is shown that this problem, in the special case of $\Lambda = \Lambda_1 \cup \Lambda_2$, with $\Lambda = \sigma(H)$ and $\Lambda_2 \subset \sigma(A)$ is equivalent to the resolution of equation $XA+XBX=HX$. The resolution of equation $XA+XBX=HX$ is used to design unsaturating controllers working inside a large non-symmetrical region of linear behavior. This technique can be easily extended to the problem of saturated control with constraints on the increment or rate of the control as developed in [129]. Finally, in order to design controllers by using LMIs, two techniques are developed, the so-called positive invariance approach based on the idea of working inside linear behavior regions without tolerating the control to be saturated and the technique of saturation tolerating the control to be saturated while guaranteeing the asymptotic stability of the system. The problem of designing stabilizing controllers for saturated singular linear systems is also solved by using the resolution of equation $XA+XBXE=HXE$.

Chapter 3
Introduction to Switched Systems

3.1 Introduction

Switched systems are a class of hybrid systems encountered in many practical situations which involve switching between several subsystems depending on various factors. Generally, a switching system consists of a family of continuous-time subsystems and a rule that supervises the switching between them. The literature on this subject is abundant. Some references can be consulted to have an overview as [75] and [119].

3.2 Physical Examples

The highly simplified longitudinal dynamics of an aircraft can take the following form:

$$\begin{bmatrix} \dot{q}(t) \\ \dot{\theta}(t) \end{bmatrix} = \begin{bmatrix} -1 & -10 \\ 1 & -1 \end{bmatrix} \begin{bmatrix} q(t) \\ \theta(t) \end{bmatrix} + \begin{bmatrix} -1 \\ 0.1 \end{bmatrix} u_{\alpha}(t).$$

The output equation is

$$\begin{bmatrix} \theta(t) \\ n_z(t) \end{bmatrix} = \begin{bmatrix} 0 & 1 \\ 0 & -300 \end{bmatrix} \begin{bmatrix} q(t) \\ \theta(t) \end{bmatrix} + \begin{bmatrix} 0 \\ 30 \end{bmatrix} u_{\alpha}(t).$$

where $\theta \leq \theta_{max}$ is the constrained angle of attack and q is the pitch rate. $\alpha(t) \in \{1;2\}$, n_z is the normal acceleration, and the control variable $u_{\alpha}(t)$ is the angle of the elevator measured down from the horizontal with the aircraft. The control objective is twofold: track the pilot's reference normal acceleration while maintaining the safety constraint that the angle of attack must be less than θ_{max} .

The simplified dynamics of a car (mass) with an automatic transmission having velocity on a road inclined at an angle is

A. Benzaouia: Saturated Switching Systems, LNCIS 426, pp. 77–94.
springerlink.com

$$\dot{v}(t) = -\frac{K}{m}v^2(t)sign(v(t)) - gsin(\theta(t)) + \frac{G_{\alpha(t)}}{m}T$$
$$\omega(t) = G_{\alpha(t)}v(t)$$

where the discrete state $G_{\alpha(t)} \in \{G_1, G_2, G_3, G_4\}, G_1 > G_2 > G_3 > G_4$ are the transmission gear ratios normalized by the wheel radius R, K is an appropriate constant, ω is the angular velocity of the motor and T is the torque generated by the engine, an input to the model.

3.3 Continuous-Time Systems

3.3.1 Problem Formulation

We consider the class of switched systems described by

$$\dot{x}(t) = A_\alpha x(t) \tag{3.1}$$

where $x \in \mathbb{R}^n$ is the state, α is a switching rule which takes its values in the finite set $\mathscr{I} = \{1, \ldots, N\}$.

For the switching rule α, two cases can be encountered in practice: The switching rule α is not known *a priori* but its value is available at each sampling period, that is each t. As reported by the cited references, this case corresponds to practical implementations where the switched system is supervised by a discrete-event system or operator allowing for the value of α to be known at only each sampling period in real time. The switching rule α can be designed to realize the control objective as stability or stabilizability.

3.3.2 Unstability of the Switched System: Example 1

Consider the following switching system composed of two modes:

$$A_1 = \begin{bmatrix} -1 & 10 \\ -100 & -1 \end{bmatrix}; A_2 = \begin{bmatrix} -1 & 100 \\ -10 & -1 \end{bmatrix}.$$

Note that $\sigma(A_1) = \{-1.0000 + 31.6228i;\ -1.0000 - 31.6228i\}$ and $\sigma(A_2) = \{-1.0000 + 31.6228i;\ -1.0000 - 31.6228i\}$
That is, each subsystem is stable.

Consider the following switching strategy:

If $x_1x_2 \le 0$ and $\alpha = 2$, choose mode 1;
If $x_1x_2 > 0$ and $\alpha = 1$, choose mode 2.

The evolution of the switching system with this strategy is given in Fig. 3.1, where one can notice that the system is unstable even if each subsystem is stable. This instability is then due to the chosen switching strategy.

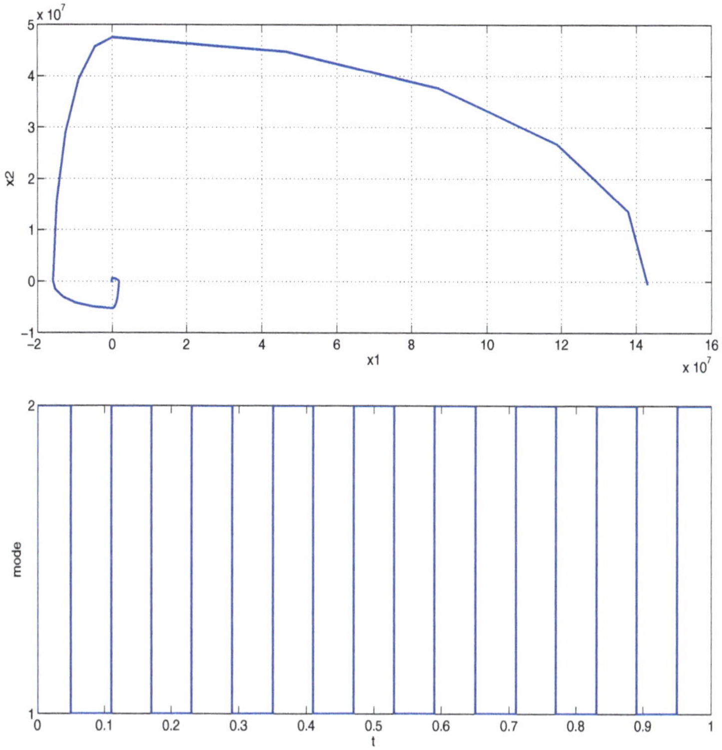

Fig. 3.1 The evolution of a trajectory of the switching system starting at $x_0 = [1; 1]^T$ together with the switching rule α

3.3.3 Stability of the Switched System: Example 2

Consider the following switching system composed of two modes:

$$A_1 = \begin{bmatrix} 0 & 10 \\ 0 & 0 \end{bmatrix}; A_2 = \begin{bmatrix} 1.5 & 2 \\ -2 & -0.5 \end{bmatrix}.$$

Note that $\sigma(A_1) = \{0;\ 0\}$ and $\sigma(A_2) = \{0.5000 + 1.7321i;\ 0.5000 - 1.7321i\}$ That is, each subsystem is unstable.

Consider the following switching strategy:

If $x_1 x_2 \leq 0$ and $\alpha = 2$, choose mode 1;
If $x_1 x_2 > 0$ and $\alpha = 1$, choose mode 2.

The evolution of the switching system with this strategy is given in Fig. 3.2, where one can notice that the system is stable even if each subsystem is unstable. This stability is then due to the chosen switching strategy.

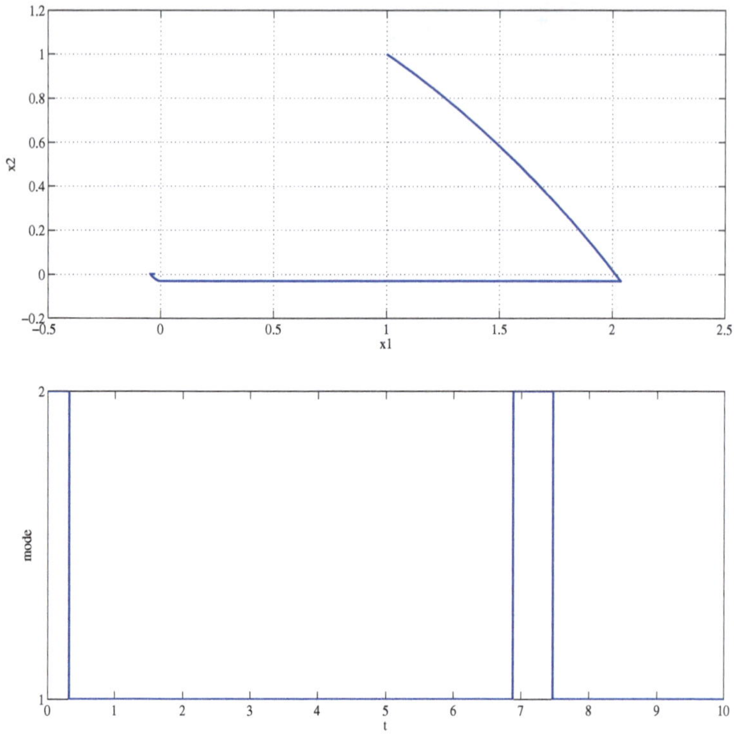

Fig. 3.2 The evolution of a trajectory of the switching system starting at $x_0 = [1;1]^T$ and the switching rule α

Consequence:
In consequence of these examples, two main problems arise in the literature of switched systems:

- What is the class of switching systems which are stable in the presence of any arbitrary switching rule?
- What are the switching rules which lead to stable switched systems even if the subsystems are unstable?

3.4 Common Lyapunov Quadratic Function

It was shown that with a judicious choice of a switching strategy, a switching system can be stable even if each subsystem is unstable. The problem is more difficult when the switching sequence is arbitrary. How can we study the stability of the switching/switched system?

A strong tool to study the stability of dynamical systems is Lyapunov functions. It is well known in control theory that to guarantee the asymptotic stability of a variable system, it suffices to associate it to a common Lyapunov function . This can be applied to the case of switching systems. Recall their definitions:

Definition 3.1. *We say that V is a candidate Lyapunov function if V is a continuous positive definite function with continuous partial derivatives and $V(0) = 0$ satisfying:*

$$\dot{V}(x(t)) = \left[\frac{\partial}{\partial x}V(x(t))\right]A_j x(t) < 0, \forall j \in \mathscr{I}. \tag{3.2}$$

Note

$$\nabla V(x(t)) = \left[\frac{\partial}{\partial x}V(x(t))\right].$$

However, it is very difficult to find such common Lyapunov functions.

3.4.1 Equivalence with Polytopic Uncertain Structure

Assume that a common Lyapunov function V exists such that:

$$\dot{V}(x(t)) = \nabla V(x(t))A_j x(t) < -\beta V(x(t)), \forall j \in \mathscr{I}, \beta > 0. \tag{3.3}$$

One can write equivalently for a set of scalars ε_j satisfying:$\sum_{j=1}^{N} \varepsilon_j = 1; \varepsilon_j \geq 0$

$$\begin{aligned}\dot{V}(x(t)) &= \left[\sum_{j=1}^{N} \varepsilon_j \nabla V(x(t))A_j x(t)\right] \\ &= \nabla V(x(t))\left[\sum_{j=1}^{N} \varepsilon_j A_j x(t)\right] \\ &< -\beta V(x(t)), \forall j \in \mathscr{I}, \beta > 0.\end{aligned} \tag{3.4}$$

Consequence:
The stability of the switching system with arbitrary sequences of switching is equivalent to the stability of an uncertain linear system with polytopic uncertainties:

$$\dot{x}(t) = \sum_{j=1}^{N} \varepsilon_j A_j x(t) \tag{3.5}$$

This is a polytopic uncertain structure which is a standard representation in robust control of standard linear systems for which efficient tools are available as LMIs.

3.5 Stability Analysis of Switched Systems

We assume here that we can act on the switching sequence to stabilize the switched system. How can we choose the switching strategy? Assume that each matrix A_i is unstable.

Proposition 3.1. *[155] Assume that there exists a system*

$$\dot{x}(t) = \hat{A}x(t); \hat{A} \in co\{A_1, \ldots, A_N\}$$

which is stable with a Lyapunov function V satisfying

$$\dot{V}(x(t)) = \nabla V(x(t))\hat{A}x(t) < -\beta V(x(t)), \beta > 0,$$

then there exists a stabilizing switching strategy of the form

$$\alpha(x) = arg \min_i \nabla V(x(t))A_i x(t).$$

Recall that $\hat{A} \in conv\{A_1, \ldots, A_N\}$ if and only if there exist scalars ε_j satisfying: $\sum_{j=1}^{N} \varepsilon_j = 1; \varepsilon_j \geq 0$ such that $\hat{A} = \sum_{j=1}^{N} \varepsilon_j A_j$.

Remark 3.1. *It is easy to see that:*

$$\min_i \nabla V(x(t))A_i x(t) \leq \nabla V(x(t))\hat{A}x(t) < -\beta V(x(t)).$$

Theorem 3.1. *[155] Choose $V(x) = x^T Px$ and assume that there exists $\hat{A}$ such that*

$$\hat{A}^T P + P\hat{A} \leq -Q, Q > 0 \tag{3.6}$$

then there exists a stabilizing switching strategy of the form

$$\alpha(x) = arg \min_i \nabla V(x(t))A_i x(t) \tag{3.7}$$

$$= arg \min_i \left[x^T (A_i^T P + PA_i)x\right] \tag{3.8}$$

which assures the condition $\dot{V}(x(t))) \leq -x^T Qx$.

Remark 3.2. *If $N = 2$, the condition of the Proposition 3.1 is also necessary (see [155]).*

In [146], the necessary and sufficient condition of existence of a common Lyapunov quadratic function for a switching system composed of only two subsystems of second order are given:

$$\gamma(A_1, A_2), \gamma(A_1, A_2^{-1}) \text{ are stable,} \tag{3.9}$$

where,

$$\gamma(A_1, A_2) = \{A/A = \varepsilon A_1 + (1-\varepsilon)A_2, \varepsilon \in [0,\ 1]\}$$

While other sufficient conditions are given in the literature as in [155]: For matrices A_1, A_2 both unstable, there exists a switching sequence $\alpha(t)$ such that the switching system is quadratically stable ($i.e.; A_i^T P + PA_i = -Q_i; Q_i \geq 0$) if and only if there exists $\varepsilon \in]0,1[$ such that $\gamma(A_1, A_2)$ is a stability matrix.

Consequence:

- Assume $\gamma(A_1, A_2) = \varepsilon A_1 + (1-\varepsilon)A_2$ is stable $\forall \varepsilon \in [0,\ 1]$
- Let $\varepsilon = 1$, then $\gamma(A_1, A_2) = A_1$
- Let $\varepsilon = 0$, then $\gamma(A_1, A_2) = A_2$
- Consequence: When the switching rule is arbitrary, a necessary condition of stability of the switching system is that each subsystem is stable.

3.5.1 Unstability of Switched Systems: Example 3

Consider the same switching system of Example 1 composed of two stables modes. It is easy to show with a simple program that condition (3.9) is not satisfied in this case:

$$\gamma(A_1, A_2^{-1}) = \begin{bmatrix} -0.0020 & -0.0898 \\ -0.0900 & -0.0020 \end{bmatrix}; \varepsilon = 0.001;$$
$$\sigma(\gamma(A_1, A_2^{-1})) = \{0.0879;\ -0.0919\}.$$

However, a simple way to know whether a common Lyapunov quadratic function exists is to solve the following LMI problem: Find $X = P^{-1}$ s.t.

$$XA_1^T + A_1 X < 0$$
$$XA_2^T + A_2 X < 0.$$

In this case, the LMI is not feasible.

3.5.2 Stability of Switched Systems: Example 4

Consider the following switching system composed of two modes:

$$A_1 = \begin{bmatrix} -1 & -1 \\ 1 & -1 \end{bmatrix}; A_2 = \begin{bmatrix} -1 & -10 \\ 0.1 & -1 \end{bmatrix}$$

Note that $\sigma(A_1) = \sigma(A_2) = \{-1.0000 + 1.0000i;\ -1.0000 - 1.0000i\}$. That is, each subsystem is stable.

Consider the following switching strategy:
If $x_1 \leq 0$ and $\alpha = 2$, choose mode 1;
If $x_1 > 0$ and $\alpha = 1$, choose mode 2.

According to condition (3.9), a common Lyapunov quadratic function does not exist since:

$$\gamma(A_1, A_2^{-1}) = \begin{bmatrix} -0.5585 & 4.2980 \\ 0.0728 & -0.5585 \end{bmatrix}; \varepsilon = 0.1170;$$
$$\sigma(\gamma(A_1, A_2^{-1})) = \{0.0011; -1.1181\}.$$

One can notice that the system is stable even if a common Lyapunov quadratic function does not exist as shown in Fig. 3.3.

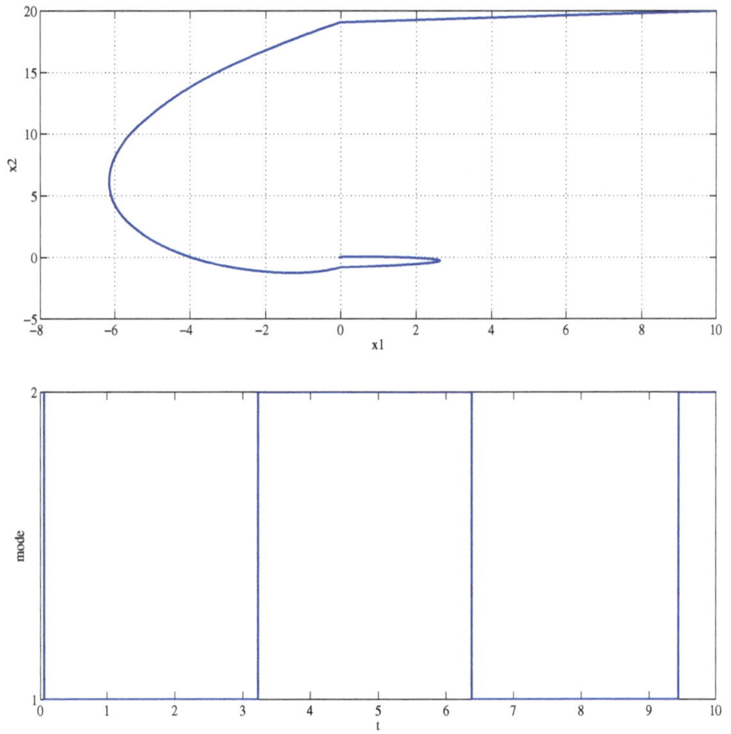

Fig. 3.3 Trajectory of the switching system and the corresponding $\alpha(t)$

3.5.3 Stability Analysis of Switched Systems

A switching or switched system can be stable without the existence of a common Lyapunov quadratic function. What is the adequate tool to study the stability of switching/switched systems? This tool is called Multiple Lyapunov functions.

3.6 Multiple Lyapunov Functions

3.6.1 Time Basis

Definition 3.2. *[126] A hybrid time basis τ is a finite or infinite sequence of sets $I_s = \{t \in \mathbb{R}_+ : t_s \leq t \leq \bar{t}_s\}$, with $\bar{t}_s = t_{s+1}$ for $s \in \mathscr{L} = \{0, \ldots, L\}$, $\bigcup_{s \in \mathscr{L}} I_s \subseteq \mathbb{R}_+$, where the system (3.1) evolves with $\alpha = s$. If the sequence is finite, one may take $card(\tau) = L + 1 < \infty$ and $\bar{t}_L = \infty$, then, I_L may be of the form $I_L = \{t \in \mathbb{R}_+ : t_L \leq t < \infty\}$.*

If $\bar{t}_s - t_s \geq \delta$, $\forall s \in \mathscr{L}$, then δ is called dwell-time.
$\delta = max(\delta_s), s \in \mathscr{L}$ is called maximum dwell-time.

We consider the class of switching systems described by (3.1):

$$\dot{x}(t) = A_\alpha x(t), \alpha \in \mathscr{I}.$$

Definition 3.3. *[54] Given a strictly increasing sequence of times $\mathscr{T}$ in $\mathbb{R}$, we say that $V_j : \mathbb{R} \times \mathbb{R}^n \longrightarrow \mathbb{R}_+$ are Lyapunov-like functions or multiple Lyapunov functions with $V_j(x(t)), t \in I_s; j \in \mathscr{I}$, for the switching system (3.1) if the following hold:*

$$\dot{V}_j(x(t)) = \left[\frac{\partial}{\partial x} V_j(x(t))\right] A_j x(t) < 0, \forall t \in I_j; \forall j \in \mathscr{I} \tag{3.10}$$

$$V_{j+1}(x(t_{j+1})) < V_j(x(t_j)), \forall j \in \mathscr{I}, \forall t_j \in I_j. \tag{3.11}$$

3.6.2 Example 5

The second condition, which is very strong, can be replaced by the following relaxed one [75]: For $i < j$, let $t_i < t_j$ be switching times for which $\alpha(t_i) = \alpha(t_j)$:

$$V_{\alpha(t_j)}(x(t_{j+1})) \leq V_{\alpha(t_i)}(x(t_{i+1})). \tag{3.12}$$

Consider the following switched system composed of two modes:

$$A_1 = \begin{bmatrix} 0 & 10 \\ 0 & 0 \end{bmatrix}; A_2 = \begin{bmatrix} 1.5 & 2 \\ -2 & -0.5 \end{bmatrix}.$$

Consider the following Lyapunov function candidate for this continuous-time switched system:
$V_j(t) = x^T(t) P_j x(t); j = 1, 2$

$$P_1 = \begin{bmatrix} 0.46875 & -1.875 \\ -1.875 & 15 \end{bmatrix}; P_2 = \begin{bmatrix} 1 & 1.2 \\ 1.2 & 1.6 \end{bmatrix}.$$

3.6.3 Example 6

Define the following subsets:

$$\Omega_i = \{x | x^T (A_i^T P_i + P_i A_i) x \leq 0\}.$$

Consider the following switching sequence:
If $x_1 \in \Omega_1$ and $\alpha = 2$, choose mode 1;
If $x_2 \in \Omega_2$ and $\alpha = 1$, choose mode 2.

The chosen switching sequence satisfies Definition 3.3, that is the sequence $\{V_i(x(tj))\}$ is nonincreasing inside (t_i, t_{i+1}).
Further, condition (3.12) is also satisfied as shown in Figure 3.4.

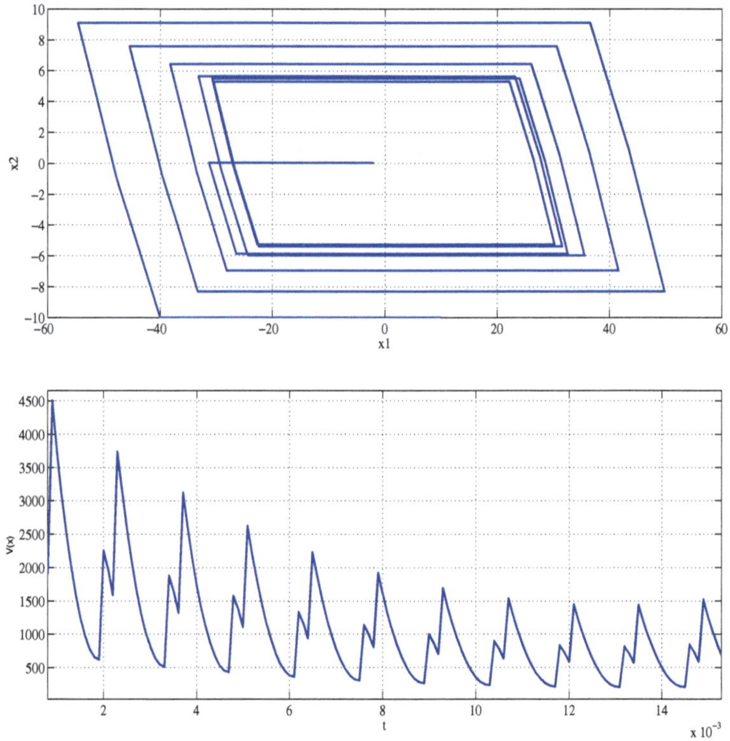

Fig. 3.4 The evolution of the switched system and the multiple Lyapunov-like function

3.6.4 Design of the Switching Scheme

The objective is to design a switching index function $\alpha(x)$ such that the resulting switched system is asymptotically stable. For this, we use a multiple Lyapunov function candidate $V_i(x) = x^T P_i x$. This technique is based on the construction of the following sets for chosen matrices Q_i.

Define the following subsets:

$$\Omega_i = \left\{ x | x^T Q_i x \geq 0 \right\}, i \in \mathscr{I}$$
$$\Omega_{i,j} = \left\{ x | x^T (Q_j - Q_i) x = 0 \right\}, i \in \mathscr{I}, j \in \mathscr{I}.$$

Satisfying:
Covering property: $\Omega_1 \bigcup \Omega_2 \bigcup \ldots \Omega_N = \mathbb{R}^n$
Switching property: $\Omega_{i,j} \subseteq \bar{\Omega}_i \bigcap \bar{\Omega}_j$,
where $\bar{\Omega}_j$ is the closure of the set Ω_j.

The switching strategy is then defined by

$$\alpha(x) = arg \left\{ \max_{i \in \mathscr{I}} x^T Q_i x \right\}.$$

The first property says that there are no regions of the state space where none of the subsystems is activated.

The second property means that a switch from subsystem i to j occurs only for states where the regions Ω_i, Ω_j are adjacent. Consequently, switching occurs on the switching surface $x^T Q_i x = x^T Q_j x$.

Lemma 3.1. *(Covering condition) [138]*
If there exist scalars $\theta_i > 0, i \in \mathscr{I}$, such that for every $x \in \mathbb{R}^n$,

$$\Sigma_{i=1}^N \theta_i x^T Q_i x \geq 0, \forall x \in \mathbb{R}^n$$

then,

$$\Omega_1 \bigcup \Omega_2 \bigcup \ldots \Omega_N = \mathbb{R}^n.$$

Theorem 3.2. *[138] Suppose there exist symmetric matrices $P_i > 0$, Q_i, scalars $\delta_i \geq 0$, $\mu_{i,j}$ and $\theta_i > 0$ such that*

$$A_i^T P_i + P_i A_i + \delta_i Q_i < 0,$$
$$P_i = P_j + \mu_{i,j}(Q_j - Q_i),$$
$$\Sigma_{i=1}^N \theta_i x^T Q_i x \geq 0$$
$$\forall i, j \in \mathscr{I}^2.$$

Then, the switched system resulting from the switching strategy $\alpha(x) = arg \left\{ \max_{i \in \mathscr{I}} x^T Q_i x \right\}$ is asymptotically stable if no sliding motion occurs.

It is worth noting that to the best of our knowledge, for continuous-time switching systems with arbitrary switching sequences $\alpha(x)$ no stability result is available in the literature with multiple Lyapunov functions.

3.7 Discrete-Time Switched Systems

3.7.1 Problem Formulation

We consider the class of switched systems described by

$$x(t+1) = A_{\alpha}x(t) \tag{3.13}$$

where $x \in \mathbb{R}^n$ is the state, $t \in \mathbb{N}$ means tT where T is the sampling period, α is a switching rule which takes its values in the finite set $\mathscr{I} = \{1,\dots,N\}$. Let $I_s = \{t \in \mathbb{Z}_+ : t_s \leq t \leq \bar{t}_s\}$.

Remark 3.3. *All the previous notions established on continuous-time switched systems remain valid for discrete-time switched systems. Only the Lyapunov tools will be adapted.*

3.7.2 Common Lyapunov Function

Definition 3.4. *We say that V is a candidate Lyapunov function if V is a continuous positive definite function with continuous partial derivatives and $V(0) = 0$ satisfying:*

$$\triangle V(x(t)) = V(x(t+1)) - V(x(t)) < 0, \forall i \in \mathscr{I}. \tag{3.14}$$

Recall that it is very difficult to find such common Lyapunov functions . Let $V(x) = x^T Px$, the condition of stability of the switched system is:

$$\triangle V(x(t)) = x^T(t)(A_i^T PA_i - P)x(t) < 0, \forall i \in \mathscr{I} \tag{3.15}$$

which is satisfied if,

$$A_i^T PA_i - P < 0, \forall i \in \mathscr{I}. \tag{3.16}$$

3.7.3 Multiple Lyapunov Functions

We consider the class of switching systems described by

$$x(t+1) = A_{\alpha}x(t), \alpha \in \mathscr{I}. \tag{3.17}$$

Definition 3.5. *Given a strictly increasing sequence of times $\mathscr{T}$ in $\mathbb{R}$, we say that $V_j : \mathbb{N} \times \mathbb{R}^n \longrightarrow \mathbb{R}_+$ are Lyapunov-like functions or multiple Lyapunov functions with $V_j(x(t)), t \in I_s; j \in \mathscr{I}$, for the switching system (3.17) if the following hold:*

$$\triangle V_j(x(t)) < 0, \forall t \in I_j; \forall j \in \mathscr{I} \tag{3.18}$$

$$V_{j+1}(x(t_{j+1})) < V_j(x(t_j)), \forall j \in \mathscr{I}, \forall t_j \in I_j \tag{3.19}$$

The second condition, which is very strong, can be replaced by the following relaxed one [65]: For $i < j$, let $t_i < t_j$ be switching times for which $\alpha(t_i) = \alpha(t_j)$:

$$V_{\alpha(t_j)}(x(t_{j+1})) \leq V_{\alpha(t_i)}(x(t_{i+1})). \tag{3.20}$$

Recall two useful lemmas where the first is called the Schur complement Lemma:

Lemma 3.2. *Define the following matrix:*

$$M = \begin{bmatrix} X & Y \\ Y^T & Z \end{bmatrix}.$$

Matrix $M > 0$ *if and only if one of the following set of inequalities holds:*

$$Z > 0; X - YZ^{-1}Y^T > 0$$
$$X > 0; Z - Y^T X^{-1} Y > 0.$$

Lemma 3.3. *[57]*

$$(\sum_{i=1}^{N} \xi_i A_i)^T P (\sum_{i=1}^{N} \xi_i A_i) \leq \sum_{i=1}^{N} \xi_i \xi_j \, (A_i^T P A_i)$$
$$\sum_{i=1}^{N} \xi_i = 1; \xi_i \geq 0; P \geq 0.$$

A smart idea is used in the literature to introduce a specific multiple Lyapunov function only adequate for discrete-time switching systems with arbitrary sequences of switching. Let ξ_i be an indicator function satisfying: $\xi_i = 1$, if the system is in mode i and 0 elsewhere. Note that $\sum_{i=1}^{N} \xi_i = 1$. With this function, system (3.17) can be represented equivalently by

$$x(t+1) = \sum_{i=1}^{N} \xi_i A_i x(t) \tag{3.21}$$

Let

$$V(x) = x^T (\sum_{i=1}^{N} \xi_i P_i) x \, . \tag{3.22}$$

With this choice, the Lyapunov function candidate is considered as a common Lyapunov one while in reality it is a multiple one. The following result, first used for linear time-varying systems [64], presents a sufficient condition of asymptotic stability for switching discrete-time systems.

Theorem 3.3. *[65, 85] The switching system (3.21) is asymptotically stable if there exist N symmetric definite positive matrices* $P_1, \ldots, P_N$ *satisfying,*

$$A_i^T P_j A_i - P_i < 0, \forall (i, j) \in \mathscr{I}^2 \tag{3.23}$$

Proof: *Consider the following developments:*

$$\begin{aligned}\triangle V(x(t)) &= V(x(t+1)) - V(x(t))\\ &= x^T(t+1)(\sum_{j=1}^{N}\xi_j P_j)x(t+1) - x^T(t)(\sum_{i=1}^{N}\xi_i P_i)x(t)\\ &= x^T(t)(\sum_{i=1}^{N}\xi_i A_i)^T(\sum_{j=1}^{N}\xi_j P_j)(\sum_{i=1}^{N}\xi_i A_i)x(t) - x^T(t)(\sum_{i=1}^{N}\xi_i P_i)x(t)\ .\end{aligned}$$

Using Lemma 3.3, one can obtain

$$V(x(t)) \leq \sum_{j=1}^{N}\sum_{i=1}^{N}\xi_i\xi_j x^T(t)\left(A_i^T P_j A_i - P_i\right)x(t)\ .$$

Condition (3.23) follows obviously. □

3.7.3.1 Multiple Lyapunov Function: LMI

Pose $X_i = P_i^{-1}$, post and pre multiply by X_i. Inequality (3.23) can be equivalently rewritten as:

$$\begin{aligned}X_i^{-1} - A_i^T X_j^{-1} A_i &> 0,\\ X_i - X_i A_i^T X_j^{-1} A_i X_i &> 0\\ X_i - (A_i X_i)^T X_j^{-1}(A_i X_i) &> 0\ .\end{aligned} \tag{3.24}$$

Applying directly Lemma 3.2 to inequality (3.24), it follow that

$$\begin{bmatrix} X_i & (A_i X_i)^T \\ * & X_j \end{bmatrix} > 0, \forall (i,j) \in \mathscr{I}^2\ . \tag{3.25}$$

3.7.4 Stabilization by Multiple Lyapunov Functions

We consider the class of switched/switching systems described by:

$$x(t+1) = A_\alpha x(t) + B_\alpha u(t) \quad \text{for discrete-time case} \tag{3.26}$$

$$\dot{x}(t) = A_\alpha x(t) + B_\alpha u(t) \quad \text{for continuous-time case} \tag{3.27}$$

where $u \in \mathbb{R}^m$ is the control, the same notations are used as before.

The objective is to built a stabilizing control law by state feedback given by.

$$u(t) = F_\alpha x(t) \tag{3.28}$$

The system in closed loop is then given by:

$$x(t+1) = Ac_{\alpha}x(t) \quad \text{for discrete-time case} \tag{3.29}$$

$$\dot{x}(t) = Ac_{\alpha}x(t) \quad \text{for continuous-time case} \tag{3.30}$$

$$Ac_{\alpha} = A_{\alpha} + B_{\alpha}F_{\alpha} \ . \tag{3.31}$$

Remark 3.4. *The problem can be solved if one uses the results of Theorem 3.2 for switched systems and Theorem 3.3 for switching systems, while replacing matrix A_i by Ac_i.*

Consider the discrete-time switching system case with arbitrary switching .

The system in closed loop (3.29) can also be represented equivalently by:

$$x(t+1) = \sum_{i=1}^{N} \xi_i \,(A_i + B_iF_i)\,x(t) \ . \tag{3.32}$$

The controller design can be obtained by using the following result presented under LMI form.

Theorem 3.4. *[65] The switching system (3.32) is asymptotically stable for any sequence of switching α, if there exist N symmetric definite positive matrices $X_1,\ldots,X_N$, matrices $Y_1,\ldots,Y_N$ satisfying,*

$$\begin{bmatrix} X_i & (A_iX_i + B_iY_i)^T \\ * & X_j \end{bmatrix} > 0, \forall (i,j) \in \mathscr{I}^2 \tag{3.33}$$

with $F_i = Y_iX_i^{-1}$ and $P_i = X_i^{-1}$.

Proof: Condition (3.23) becomes:

$$(A_i + B_iF_i)^T P_j (A_i + B_iF_i) - P_i < 0, \forall (i,j) \in \mathscr{I}^2 \ . \tag{3.34}$$

Using the Schur complement, one obtains equivalently:

$$\begin{bmatrix} X_i & [(A_i + B_iF_i)X_i]^T \\ * & X_j \end{bmatrix} > 0, \forall (i,j) \in \mathscr{I}^2 \ .$$

The following developments can be used:

$$\begin{bmatrix} X_i & (A_iX_i + B_iF_iX_i)^T \\ * & X_j \end{bmatrix} > 0, \forall (i,j) \in \mathscr{I}^2 \ .$$

Pose $Y_i = F_iX_i$, it follow that

$$\begin{bmatrix} X_i & (A_iX_i + B_iY_i)^T \\ * & X_j \end{bmatrix} > 0, \forall (i,j) \in \mathscr{I}^2 \ . \tag{3.35}$$

□

This result can be relaxed by introducing some slack variables as presented by the following result.

Corollary 3.1. *[65] The switching system (3.32) is asymptotically stable for any sequence of switching α, if there exist N symmetric definite positive matrices $X_1,\ldots,X_N$, matrices $Y_1,\ldots,Y_N$ and matrices $G_1,\ldots,G_N$ satisfying,*

$$\begin{bmatrix} G_i+G_i^T-X_i & (A_iG_i+B_iY_i)^T \\ * & X_j \end{bmatrix} > 0, \forall (i,j) \in \mathscr{I} \times \mathscr{I}, \tag{3.36}$$

with $F_i = Y_iG_i^{-1}$ and $P_i = X_i^{-1}$.

Proof:
By noting that inequality (3.34) is equivalent to:

$$G_i^T Ac_i^T P_j Ac_i G_i - G_i^T P_i G_i < 0, \forall (i,j) \in \mathscr{I} \times \mathscr{I}, \tag{3.37}$$

where $Ac_i = A_i + B_iF_i$ for any nonsingular matrix G_i. Inequality (3.37) is equivalent by Schur complement to

$$\begin{bmatrix} G_i^T X_i^{-1} G_i & G_i^T Ac_i^T \\ * & X_j \end{bmatrix} > 0, \forall (i,j) \in \mathscr{I} \times \mathscr{I} . \tag{3.38}$$

By using the fact that $(G_i - X_i)^T X_i^{-1}(G_i - X_i) \geq 0$, implies $G_i^T X_i^{-1} G_i \geq G_i + G_i^T - X_i$. Inequality (3.38) is also in turn equivalent to the following inequality, generally used to relax the previous one (3.33)[65, 66]:

$$\begin{bmatrix} G_i+G_i^T-X_i & G_i^T Ac_i^T \\ * & X_j \end{bmatrix} > 0, \forall (i,j) \in \mathscr{I} \times \mathscr{I} \tag{3.39}$$

where matrices G_i, called slack variables, are nonsingular matrices and X_i the positive definite matrices. By letting $Y_i = F_iG_i$, LMI (3.36) follows. □

3.7.5 Remarks

- The solution of these types of LMIs is not unique since it is only based on sufficient conditions.
- With this result, the assumption of stability of each subsystem (needed in open loop for switching systems with arbitrary switching) is not needed in the stabilization problem. In fact, each subsystem in closed loop will be stable.
- The case of common quadratic Lyapunov function can be seen as a particular case of this problem by setting $X_i = X_j$ in the LMI.

Example 3.1. *Consider a numerical switching discrete-time system with two modes given by the following matrices:*

$$A_1 = \begin{bmatrix} -0.7 & 1 \\ -0.5 & -1.5 \end{bmatrix}; B_1 = \begin{bmatrix} 1 \\ 0 \end{bmatrix};$$
$$A_2 = \begin{bmatrix} 0.9 & -1 \\ 1.7 & -1.5 \end{bmatrix}; B_2 = \begin{bmatrix} 0 \\ -1 \end{bmatrix}.$$

Note that matrix A_1 is unstable while A_2 is stable,

$$\sigma(A_1) = \{-1.1000 + 0.5831i; -1.1000 - 0.5831i\}$$

$$\sigma(A_2) = \{-0.3000 + 0.5099i; -0.3000 - 0.5099i\}$$

$$P_1 = \begin{bmatrix} 0.4797 & 1.2112 \\ 1.2112 & 3.9758 \end{bmatrix}; F_1 = \begin{bmatrix} 1.3731 & 1.0359 \end{bmatrix};$$
$$Ac_1 = \begin{bmatrix} 0.6731 & 2.0359 \\ -0.5000 & -1.5000 \end{bmatrix};$$

$$P_2 = \begin{bmatrix} 0.2764 & -0.2522 \\ -0.2522 & 0.5106 \end{bmatrix}; F_2 = \begin{bmatrix} 1.7663 & -1.5740 \end{bmatrix};$$
$$Ac_2 = \begin{bmatrix} 0.9000 & -1.0000 \\ -0.0663 & 0.0740 \end{bmatrix}.$$

The same system is now studied with LMIs (3.36), the following results are obtained:

$$P_1 = \begin{bmatrix} 0.0029 & 0.0074 \\ 0.0074 & 0.0231 \end{bmatrix}; F_1 = \begin{bmatrix} 1.3284 & 0.8852 \end{bmatrix};$$
$$Ac_1 = \begin{bmatrix} 0.6284 & 1.8852 \\ -0.5000 & -1.5000 \end{bmatrix};$$

$$P_2 = \begin{bmatrix} 0.0016 & -0.0015 \\ -0.0015 & 0.0031 \end{bmatrix}; F_2 = \begin{bmatrix} 1.7773 & -1.5858 \end{bmatrix};$$
$$Ac_2 = \begin{bmatrix} 0.9000 & -1.0000 \\ -0.0773 & 0.0858 \end{bmatrix}.$$

Figure 3.5 presents the ellipsoid sets obtained as stability domains for the switching discrete-time system. These results are derived by using LMIs (3.33), while Fig. 3.6 presents the ellipsoid sets of stability obtained with LMIs (3.36). One can notice that for the same $\rho = 20$, the sets of Fig. 3.6 are larger than those of Fig. 3.5.

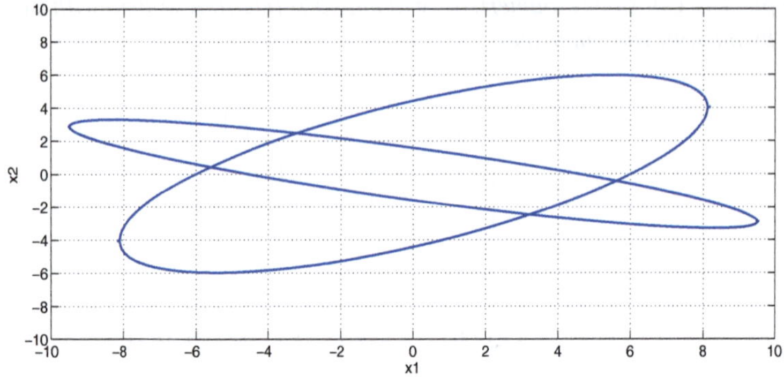

Fig. 3.5 The ellipsoid sets for the switching discrete-time linear system computed with LMIs (3.33) and $\rho = 20$

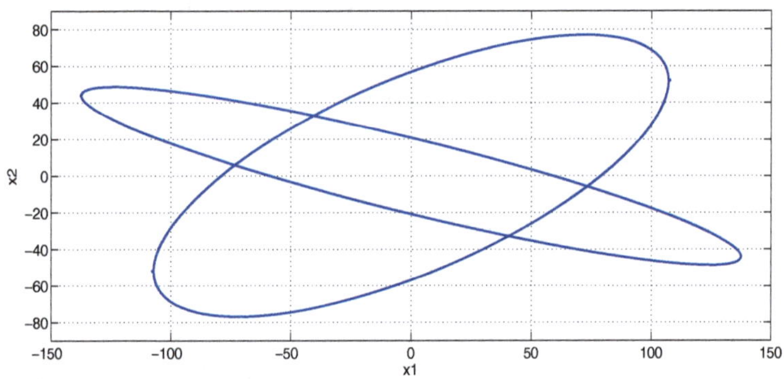

Fig. 3.6 The ellipsoid sets for the switching discrete-time linear system computed with LMIs (3.36) and $\rho = 20$

Obviously this fact is not very important for unsaturated switching systems where the stability is global, but it will play an important criterion of conservatism for saturated switching systems as will be seen in Chapter VI.

3.8 Conclusion

In this chapter, an introduction to switched and switching systems is presented. The different particular cases of stability and instability are raised. The tool of Lyapunov function is presented to be the more adequate one to study stability of switched/switching systems. Besides, the limitation of common Lyapunov function is shown. The introduction of a new tool called multiple Lyapunov function is developed. The last results on this subject based on LMIs are presented. Different examples are treated to show the complexity of the problem and the applicability of the proposed results in the literature.

Chapter 4
Saturated Control Problem of Switching Systems

4.1 Introduction

In this chapter, we study a switching system with linear discrete-time dynamics under unknown arbitrary switches and control saturation. The studied system is taken under complete modeling taking into account reset functions at each switch and a different system's dimension. A feedback controller, using a linear systems technique, is derived. By this technique, we are able to provide tractable conditions ensuring asymptotic stability inside the union of all the positive invariant sets associated to each subsystem. The extension to saturated switching discrete-time singular systems is also presented. The main results of this chapter can be found in [27] with a different proofs.

4.2 Saturated Switching Linear Systems

In this section, necessary and sufficient conditions of positive invariance are presented for discrete-time switching systems with saturated control. The controller design is based on the solution of equation $XA+XBX=HX$ which is developed in Chapter *II*.

4.2.1 Problem Statement

In this section, we present the definitions and the properties of the switching systems as introduced in [76].

Definition 4.1. *A discrete-time linear switching system is a tuple*

$$\mathscr{H}=(\mathscr{E},Q,U_d,U_i,S_d,S,E,R) \tag{4.1}$$

where,

- $\mathscr{E}=\bigcup_{i\in\mathscr{I}}\{q_i\}\times\mathbb{R}^{n_i}$ *is the hybrid space.*
- $Q=\{q_i,i\in\mathscr{I}\}$ *is the set of discrete states,* $\mathscr{I}=\{1,\ldots,N\}$*;*
- U_d *is the set of external uncontrollable events;*

A. Benzaouia: Saturated Switching Systems, LNCIS 426, pp. 95–116.
springerlink.com

- *$U_i \subset \mathbb{R}^m$ is the input space.*
- *S_d is a subclass of discrete-time linear dynamical systems such that, $S_i \in S_d$ is defined by the state representation as follows:*

$$\begin{array}{l} x(t+1) = A_i x(t) + B_i u(t), \quad i \in \mathscr{I} \\ x(t) \in \mathbb{R}^{n_i}, \\ u(t) \in U_i \subset \mathbb{R}^m \end{array} \tag{4.2}$$

Matrices $A_i \in \mathbb{R}^{n_i \times n_i}$ and $B_i, \in \mathbb{R}^{n_i \times m}$ are constant with $n_i \geq m$.
- *$S : Q \longrightarrow S_d$ is an application associating to each discrete state a discrete-time dynamical system;*
- *$E = Q \times U_d \times Q$ is the set of the discrete transitions;*
- *$R : E \times \mathscr{E} \longrightarrow \mathscr{E}$ is the reset function. We suppose that R is linear, i.e., for $e = (q_i, \sigma, q_s) \in E$,$x \in \mathbb{R}^{n_i}$,$\xi(t_j, j) = (q_i, x) \in \mathscr{E}, R(e, \xi) = (q_s, L_{is} x), L_{is} \in \mathbb{R}^{n_s \times n_i}$.*

As introduced in [76], we define the hybrid time basis τ as follows:

Definition 4.2. *An hybrid time basis τ is an infinite or finite sequences of sets $I_j = \{t \in \mathbb{R} : t_j \leq t \leq \bar{t}_j\}$, with $\bar{t}_j = t_{j+1}$ for $j \in \mathscr{L} = \{0, \ldots, L\}$, and if $card(\tau) = L+1 < \infty$ then $\bar{t}_L$ can be finite or infinite.*

We denote by $\mathscr{T}$ the set of all hybrid time basis . In order to describe the state evolution of the switching system, we define the notion of the switching system execution.

Definition 4.3. *An execution of the switching system $\mathscr{H}$ is a collection $\mathscr{X} = (\xi_0, \tau, \sigma, q, u, \xi)$ with,*

- *$\xi_0 = (\hat{q}, x_0) \in \mathscr{E}$ is the initial state,*
- *$\tau \in \mathscr{T}$ is the hybrid time basis,*
- *$\sigma : \tau \longrightarrow U_d$ is a map associating to each sequence an event,*
- *$q : \tau \longrightarrow Q$*
- *$\xi : \mathscr{T} \times \mathbb{N} \longrightarrow \mathbb{R}^{n_i}$ is defined as,*

$$\begin{array}{l} \xi(t_0, 0) = \xi_0 = (\hat{q}, x_0) \\ \xi(t_{j+1}, j+1) = R(e_j, \xi(\bar{t}_j, j)) \quad , \\ e_j = (q(I_j), \sigma(I_{j+1}), q(I_{j+1})) \\ x(t, j) = x_i(t, x(t_j, j), u), \end{array}$$

where $x_i(t, x(t, j), u)$ is the unique solution at time t of the dynamical system $S_i = S(q(I_j))$ with initial time t_j and initial state $x(t_j, j)$ and input $u|_{[t_j, t)}$.

Throughout the chapter, we make the following assumptions:

Assumption 4.1. *Given the switching system $\mathscr{H}$, we suppose that:*

- *$t_{j+1} \geq t_j + 1, \forall j \in \mathscr{L}$,*
- *The switching system is stabilizable,*
- *Each matrix $A_i, i \in \mathscr{I}$ admits $n_i - m$ non-null stable eigenvalues.*

It will be shown later that this assumption is not restrictive since one can always avoid it by using a technique of augmentation of the system entries.

Our goal in this chapter is the study of the stabilization problem of the switching system $\mathscr{H}$ under input constraints. For this, we apply the positive invariance concept.

We suppose that the set U_i representing the input constraints is given by

$$U_i = \{u \in \mathbb{R}^m |\ |u| \leq \rho_i, \rho_i > 0, i \in \mathscr{I}\} \tag{4.3}$$

and the feedback control law is written as follows:

$$u(t) = F(q_i)x(t,j), \quad t \in I_j; j \in \mathscr{L}, rank(F_i) = m. \tag{4.4}$$

Our objective is then to design stabilizing linear state feedback controllers for the switching system $\mathscr{H}$ among those of full rank gains and stabilizing subsystems (A_i, B_i) while respecting the imposed constraints on the system.

4.2.2 Linear Systems Analysis

This section is devoted to the study of the problem of positive invariance and asymptotic stability of the autonomous switching system. Necessary and sufficient conditions of positive invariance of a given domain with respect to (w.r.t.) the trajectories of this switching system are established. A link between the conditions of contractivity and a Lyapunov-like function is also presented.

We consider the discrete-time linear autonomous switching system,

$$\mathscr{H}_1 = (\mathscr{E}, Q, U_d, S_{d1}, S, E, R)$$

as described in the previous section. First, we recall the definition of the asymptotic stability for switching systems.

Definition 4.4. *Let,* $\mathscr{B} = \bigcup_{i \in \mathscr{I}} \{q_i\} \times \mathscr{B}_0$, $\mathscr{B}_0 = \{x \in \mathbb{R}^n : \|x\| \leq 1\}$.
The switching autonomous system $\mathscr{H}_1$ *is asymptotically stable if,*
$\forall \varepsilon > 0, \forall \xi_0 \in \alpha\mathscr{B}$ *with arbitrarily large* $\alpha, \exists \mu \in \mathbb{R}$, *s.t* $\xi_1(t,j) \in \varepsilon\mathscr{B}_0, \forall t \in I_j \cap [\mu, \infty), \forall j \geq j_1$, *with* $j_1 = inf\{j,\ j \in I_j\}$, *for all execution of the switching system with initial state* ξ_0.

Definition 4.5. *A hybrid region* $\Pi = \bigcup_{i \in \mathscr{I}} \{q_i\} \times \mathscr{D}_i$ *is said to be positive invariant with respect to all execution* $\mathscr{X}$ *of the switching autonomous system* $\mathscr{H}_1$ *if for each initial state* $\xi_0 \in \Pi$, *one has* $\xi(t,j) \in \Pi$, *for all* $t \in I_j, I_j \in \tau$ *and* $j \in \mathscr{L}$.

Proposition 4.1. *A set* $\Pi \subset \mathscr{E}$ *is positively invariant for the switching autonomous system* $\mathscr{H}_1$ *if and only if:*

i) $A_s\mathscr{D}_s \subset \mathscr{D}_s, \forall s \in \mathscr{I}$,
ii) $L_{si}A_s\mathscr{D}_s \subset \mathscr{D}_i, \forall s, i \in \mathscr{I}, s \neq j$.

Proof: Sufficiency:
Given any $\xi(t_j, j)$ in $\{q_i\} \times \mathscr{D}_i$, since *i*) holds, $\xi(t, j) \in \{q_i\} \times \mathscr{D}_i$, $\forall t \in I_j$, and $\xi(t_{j+1}, j+1) \in \{q_j\} \times \mathscr{D}_j$, because *ii*) holds and because *i*) implies that $A_i^k \mathscr{D}_i \subset \mathscr{D}_i$, for any $k > 1$. Therefore, by induction, Π is positively invariant w.r.t. the system.

Necessity: If *i*) is false, then Π cannot be positively invariant, because there exists an execution with initial state in $\{q_i\} \times \mathscr{D}_i$, and with $\tau = \{I_0\}$, such that the continuous state escapes from $\mathscr{D}_i$. If *ii*) is false, then Π cannot be positively invariant, because there exists an execution with initial state in $\{q_i\} \times \mathscr{D}_i$, and with $\bar{t}_0 - t_0 = 1$, such that the continuous state after the transition (q_i, q_s) does not belong to $\mathscr{D}_s$. □

Several works were interested to find the necessary and sufficient conditions of asymptotic stability for switching systems.

Hereafter, we present a necessary and sufficient condition of asymptotic stability of the switching autonomous system $\mathscr{H}_1$ based on a contractivity reasonning.

Proposition 4.2. *A switching autonomous system $\mathscr{H}_1$ is asymptotically stable if and only if there exists a set $\Pi = \bigcup_{i \in \mathscr{I}} \{q_i\} \times \mathscr{D}_i \subset \mathscr{E}$ with $0 \in int(\mathscr{D}_i)$ and $\mathscr{D}_i$ is bounded $\forall i \in \mathscr{I}$ and a real $\lambda \in (0,1)$ such that:*

i) $A_s \mathscr{D}_s \subset \lambda \mathscr{D}_s, \forall s \in \mathscr{I}$,
ii) $L_{si} A_s \mathscr{D}_s \subset \lambda \mathscr{D}_i, \forall s, i \in \mathscr{I}$.

Proof: The sufficiency is obvious. Note that the stability of the system is obtained for $\lambda = 1$. The above conditions are equivalent to the positive invaraince of the set Π. Since $0 \in int(\mathscr{D}_i), \forall i \in \mathscr{I}$, this property is equivalent to the stability of the system. For the necessity, if the system $\mathscr{H}_1$ is asymptotically stable, then there exists a real $\lambda \in (0,1)$ such that the set $\mathscr{H}_1'$ obtained from $\mathscr{H}$ by substituting each matrix A_s by $\lambda^{-1} A_s$, is stable. Therefore, there exists a set $\Pi = \bigcup_{i \in \mathscr{I}} \{q_i\} \times \mathscr{D}_i \subset \mathscr{E}$, with $0 \in int(\mathscr{D}_i), \forall i \in \mathscr{I}$, positively invariant for $\mathscr{H}'$. From Proposition 4.1, $\lambda^{-1} A_s \mathscr{D}_i \subset \mathscr{D}_i$ and $\lambda^{-1} L_{si} A_s \mathscr{D}_s \subset \mathscr{D}_i, \forall s, i \in \mathscr{I}$. Therefore the conditions of the theorem follow. □

As a consequence of the above Propositions, we can state the following results for a given box defined by

$$D_s = \{x \in \mathbb{R}^{n_i} : |x| \leq \omega_s, s \in \mathscr{I}\} . \tag{4.5}$$

The corresponding hybrid region is

$$\mathscr{D}_s = \bigcup_{i \in \mathscr{I}} \{q_i\} \times D_s . \tag{4.6}$$

Theorem 4.1. *: The set $\mathscr{D}_s \subset \mathscr{E}$ is positively invariant for the switching autonomous system $\mathscr{H}_1$ if and only if*

- $$|A_s| \omega_s \leq \omega_s, \forall s \in \mathscr{I} \tag{4.7}$$

- $$|L_{si} A_s| \omega_s \leq \omega_i, \forall s, i \in \mathscr{I} . \tag{4.8}$$

Proof: The condition (4.7) is equivalent to positive invariance of the box D_s and the condition (4.8) implies that $L_{si}A_sD_s \subset D_i, \forall s,i \in \mathscr{I}, s \neq j$. □

Theorem 4.2. *If there exist positive vectors $\omega_1, \dots, \omega_N$ such that conditions (4.7) and (4.8) hold then, the switching autonomous system $\mathscr{H}_1$ is stable. If the inequalities strictly hold, then $\mathscr{H}_1$ is asymptotically stable.*

Proof: The proof follows readily from Proposition 4.2. □

Remark 4.1. *In the case of $L_{si} = \mathbb{I}, i,s \in \mathscr{I}$, the condition of the asymptotic stability for the switching autonomous system $\mathscr{H}_1$ is of the form*

$$|A_s|\omega_s < \omega_s \ , \ \forall s \in \mathscr{I} \ \textit{for some} \ \omega_s > 0$$

As it was seen in Chapter *III*, recall that the use of the Lyapunov functions revealed two ways:

- The existence of a common Lyapunov function to the various subsystems guaranteed the asymptotic stability of the switching system. In general, the search for such function is not always obvious [146].
- The Lyapunov-like functions, and the multiple Lyapunov functions were introduced by Branicky [54]. They are considered as a strong tool in the analysis of the stability of the hybrid systems and in particular the switching systems.

Definition 3.5 can now be adapted to the switching system studied in this section by the following definition.

Definition 4.6. *[54] If there exists a Lyapunov-like function $V_{t_j}(\xi,t), t \in I_j$, with $V_{t_j} : \mathbb{R}^{n_i} \longrightarrow \mathbb{R}_+$, such that the following hold:*

- $\Delta V_{t_j}(\xi(t,j)) < 0$ *for any time* $t \in I_j$
- $V_{t_k}(\xi(t_k,k)) < V_{t_j}(\xi(t_j,j)), \forall k > j \in \mathscr{L}$.

Then, the switching autonomous system $\mathscr{H}_1$ is asymptotically stable.

The results of Theorem 4.2 can be easily linked to the following Lyapunov-like function.

$$V_{t_j}(\xi(t,j)) = max_i \frac{|\xi(t,j)|_i}{\omega_s(i)}, \qquad t \in I_j, j \in \mathscr{L}, s \in J, q(I_j) = q_s. \tag{4.9}$$

where $|\xi(t,j)| = |x(t,j)|$. In this case, any box can be equivalently written with function $V_{t_j}(\xi(t,j))$ given by (4.9) as follows:

$$\begin{aligned} D_s &= \{x \in \mathbb{R}^{n_i} : |x| \leq \omega_s, s \in \mathscr{I}\} \\ &= \left\{x \in \mathbb{R}^{n_i} : V_{t_j}(\xi(t,j)) \leq 1\right\} . \end{aligned} \tag{4.10}$$

It follows from above that necessary and sufficient conditions for function $V_{t_j}(\xi(t,j))$ to be a Lyapunov-like function for the switching system $\mathscr{H}_1$ in the sense of Definition 4.6, are exactly conditions (4.7) and (4.8).

In the control of switching systems, face to the constraint problem, several works were interested to construct the maximal invariant controlled set of the switching system. By definition, the maximal safe set is the set of all initial states guaranteeing the existence of a feedback control law respecting the constraints. In [76], a computation algorithm to build the maximal safe set for the switching system, is presented. However, the main difficulty of determination of the maximal controlled invariant set, even for discrete-time linear system, is that the convergence of the algorithm is not guaranteed in a finite number of steps. In this work, based on the previous results, we propose a synthesis methodology for linear state feedback controllers ensuring the asymptotic stability of the switching system $\mathscr{H}$ while respecting the constraints on the control.

4.2.3 Stabilization Problem with Constrained Control

The results obtained in the previous section are applied to the stabilization problem of the switching system $\mathscr{H}$ with constrained control.

4.2.3.1 Conditions of Stabilizability

Consider the switching system $\mathscr{H}$. Given the control law (4.4) and the set (4.3), we can write that,

$$u \in U \Longrightarrow x(t,j) \in \Omega_s,\ t \in I_j$$

where,

$$\Omega_s = \{x \in \mathbb{R}^{n_s} : |F_s x| \leq \rho_s, \rho_s > 0\} \tag{4.11}$$

with $F_s = F(q_s) \in \mathbb{R}^{m \times n_s}$. The set Ω_s is an unbounded polyhedral region of linear behavior for $m < n_s$, where the saturation does not occur.

The closed-loop execution is presented under the following form:

$$\mathscr{H}_c = (\mathscr{E}, Q, U_d, \Omega_s, S_{dc}, S, E, R) \tag{4.12}$$

where the subclass of discrete-time linear systems S_{dc} is given by $S_s \in S_{dc}$ and defined by,

$$\begin{aligned} &x(t+1) = (A_s + B_s F_s)x(t), \quad s \in \mathscr{I} \\ &x(t) \in \Omega_s \end{aligned}$$

Define the following set

$$\Omega = \bigcup_{s \in \mathscr{I}} \Omega_s\,. \tag{4.13}$$

Remark 4.2.

- *If for each $s \in \mathscr{I}$, there exists a state feedback F_s able to make the set Ω_s positively invariant w.r.t. all closed-loop execution of the switching system $\mathscr{H}$, the domain,*

$$\mathscr{O}_s = \bigcup_{i \in \mathscr{I}' \subset J} \{q_i\} \times \Omega_s \tag{4.14}$$

  is a controlled invariant for the switching system.
- *Considering the feedback control law (4.4), we have*

$$u(t) \in U_s \Longrightarrow x(t,j) \in \Omega_s = \{x \in \mathbb{R}^{n_s} \mid |F_s x| \leq \rho_s\} \ , \ F_s = F(q(I_j)) \ , \ t \in I_j \ .$$

  If we make the following change of variable $\xi_1(q_s, z(t,j))$ computed as follows:

$$\xi(q_s, x(t,j)) \Longrightarrow \xi_1(q_s, z(t,j)) = (q_s, F_s x(t,j)) \ , \ t \in I_j \tag{4.15}$$

  Each domain Ω_s in the state space can be transformed in to the following form:

$$Z_s = \left\{ z \in \mathbb{R}^m \mid |z| \leq \rho_s \ , \ z = F_s x \ \ x \in \mathbb{R}^{n_s} \ \ F_s \in \mathbb{R}^{m \times n_s} \right\} \tag{4.16}$$

  The corresponding hybrid region is given by

$$\mathscr{Z} = \bigcup_{s \in \mathscr{I}' \subset J} \{q_s\} \times Z; Z = \bigcup_{s \in \mathscr{I}' \subset J} \{q_s\} Z_s \ . \tag{4.17}$$

  Consider the new hybrid state given by

$$\xi_1(t,j) = (q_s, F_s x(t,j)), \quad t \in I_j \ , j \in \mathscr{L}, q_s = q(I_j) \ . \tag{4.18}$$

  This change of variable allows to obtain a free switching system enabling one to use the results of the previous section.

Define the following hybrid region

$$\mathscr{O} = \bigcup_{s \in \mathscr{I}' \subset J} \{q_s\} \times \Omega \ . \tag{4.19}$$

Theorem 4.3. *For given full rank matrices F_i stabilizing the subsystem $(A_i, B_i), i \in \mathscr{I}$, the set $\mathscr{O}$ is positively invariant w.r.t. any execution of the switching system $\mathscr{H}_c$, if and only if there exists, for all $s, i \in J$, non-singular matrices $H_s \in \mathbb{R}^{m \times m}$ and matrices $M_{si} \in \mathbb{R}^{m \times m}$ such that the following conditions hold:*

$$F_i A_i + F_i B_i F_i = H_i F_i, \forall i \in \mathscr{I} \tag{4.20}$$

$$F_i L_{si} = M_{si} F_s, \forall s, i \in \mathscr{I} \tag{4.21}$$

$$|H_s| \rho_s \leq \rho_s, \forall s \in \mathscr{I} \tag{4.22}$$

$$|M_{si} H_s| \rho_s \leq \rho_i, \forall s, i \in \mathscr{I} \tag{4.23}$$

Proof: (Sufficiency):
Let conditions (4.20)–(4.23) be satisfied and $\xi(t,j) \in \mathscr{O}_s, t \in I_j, q_s = q(I_j)$. Compute $\xi(t+1,j)$ for $t+1 \in I_j$.

$$\xi(t+1,j) = (q_s, F_s x(t+1,j)) = (q_s, F_s(A_s + B_s F_s)x(t,j)), t \in I_j\ , j \in \mathscr{L}; q_s = q(I_j)$$

It follows, according to condition (4.20) that

$$F_s x(t+1,j) = F_s(A_s + B_s F_s)x(t,j) = H_s F_s x(t,j), \quad t \in I_j\ , j \in \mathscr{L}; q_s = q(I_j)\ .$$

Since $\xi(t,j) \in \mathscr{O}_s$, $|F_s x(t,j)| \leq \rho_s$ and by virtue of (4.22),

$$|F_s x(t+1,j)| \leq |H_s||F_s x(t,j)| \leq |H_s|\rho_s \leq \rho_s, \quad t \in I_j\ , j \in \mathscr{L}; q_s = q(I_j)\ .$$

Then, $\xi(t+1,j) \in \mathscr{O}_s$.

Now let $x(\bar{t}_{k-1}, k-1) \in \Omega_s$; $q_s = q(I_{k-1}), q_i = q(I_k)$. What happens at the switching time $\bar{t}_{k-1}$?

For this compute, $x(t_k,k)$. We have the following developments:

$$\begin{aligned}\xi(t_k,k) &= (q_i, x(t_k,k)) \\ &= R(e_{k-1}, \xi(\bar{t}_{k-1}, k-1))\ ,\ ; e_{k-1} = (q(I_{k-1}), \sigma(I_k), q(I_k)) \\ &= (q_i, L_{si} x(\bar{t}_{k-1}, k-1))\ .\end{aligned} \tag{4.24}$$

Hence, $x(t_k,k) = L_{si} x(\bar{t}_{k-1}, k-1)$. It follows that $F_i x(t_k,k) = F_i L_{si} x(\bar{t}_{k-1}, k-1)$. The switch to the next region Ω_i will be seen a unit time after

$$F_i x(t_k+1,k) = F_i L_{si} x(\bar{t}_{k-1}+1, k-1)$$

Thus we have

$$F_i x(t_k+1,k) = F_i L_{si}(A_s + B_s F_s)x(\bar{t}_{k-1}, k-1)\ . \tag{4.25}$$

Taking account of condition (4.20) and (4.21), one obtains,

$$\begin{aligned}F_i x(t_k+1,k) &= M_{si} F_s (A_S + B_s F_s) x(\bar{t}_{k-1}, k-1) \\ &= M_{si} H_s F_s x(\bar{t}_{k-1}, k-1)\end{aligned} \tag{4.26}$$

It follows that

$$|F_i x(t_k+1,k)| \leq |M_{si} H_s||F_s x(\bar{t}_{k-1}, k-1)|.$$

Since $x(\bar{t}_{k-1}, k-1) \in \Omega_s$, one has, $|F_s x(\bar{t}_{k-1}, k-1)| \leq \rho_s$.

Taking account of (4.23), one should obtain,

$$|F_i x(t_k+1,k)| \leq |M_{si} H_s| \rho_s \leq \rho_i.$$

That is, $x(t_k+1,k) \in \Omega_i$.

Consequently, the set $\mathscr{O}$ is positively invariant w.r.t. the switching system $\mathscr{H}_c$ if conditions (4.20)–(4.23) hold.

(Necessity):

Assume that the set $\mathscr{O}$ is positively invariant w.r.t. the switching system $\mathscr{H}_c$. Two cases may be presented:

i) Each set $\mathscr{O}_s, s \in \mathscr{I}$ is positively invariant w.r.t. the system. In this case, the results obtained for linear discrete-time systems of [47], [13] can be used and conditions (4.20) and (4.22) are also necessary.

ii) At a switching time t_k, we have the following implications:

$$x(\bar{t}_{k-1}, k-1) \in \Omega_s \Longrightarrow x(t_k, k) \in \Omega_i, q_s = q(I_{k-1}), q_i = q(I_k), \forall k \in \mathscr{L} .$$

Using the same change of variable (4.15), it follows that

$$z(\bar{t}_{k-1}, k-1) \in Z_s \Longrightarrow z(t_k, k) \in Z_i, q_s = q(I_{k-1}), q_i = q(I_k), \forall k \in \mathscr{L}$$

Recall from (4.24), $\xi(t_k, k) = (q_i, x(t_k, k)) = (q_i, L_{si}x(\bar{t}_{k-1}, k-1))$. Then, one can obtain,

$$\xi_1(t_k, k) = (q_i, F_i x(t_k, k)) = (q_i, F_i L_{si} x(\bar{t}_{k-1}, k-1)) . \tag{4.27}$$

Since $\xi_1(t_k, k) \in \mathscr{Z}_s$, defined by (4.17), then there exists a reset function between the states $\xi_1(\bar{t}_{k-1}, k-1)$ and $\xi_1(t_k, k)$. That is, there exists matrices $M_{si} \in \mathbb{R}^{m \times m}$ such that

$$\xi_1(t_k, k) = (q_i, M_{si} z(\bar{t}_{k-1}, k-1)) . \tag{4.28}$$

Comparing (4.27) and (4.28) while using $z(\bar{t}_{k-1}, k-1)) = F_s x(\bar{t}_{k-1}, k-1)$, it follows that

$$\begin{aligned} z(t_k, k) &= F_i L_{si} x(\bar{t}_{k-1}, k-1) = M_{si} z(\bar{t}_{k-1}, k-1) \\ &= M_{si} F_s x(\bar{t}_{k-1}, k-1)), \forall k \in \mathscr{L} . \end{aligned} \tag{4.29}$$

Two cases can arise:

- $x(\bar{t}_{k-1}, k-1) \notin \mathscr{K}er(F_s)$: In this case, we have $F_i L_{si} = M_{si} F_s, \forall k \in \mathscr{L}$.

- $x(\bar{t}_{k-1}, k-1) \in \mathscr{K}er(F_s)$: In this case, one should have

$$L_{si} x(\bar{t}_{k-1}, k-1) \in \mathscr{K}er(F_i) . \tag{4.30}$$

Recall an important property for linear systems: for each subsystem (A_s, B_s), the set Ω_s is positively invariant w.r.t. the system if and only if the set $\mathscr{K}er(F_s)$ is positively invariant w.r.t. the system [13].

Note that if $x(\bar{t}_{k-1}, k-1) \in \mathscr{K}er(F_s)$, then, $x(t_{k-1}, k-1) \in \mathscr{K}er(F_s)$ according to (4.20). In fact,

$$F_s x(\bar{t}_{k-1}, k-1) = F_s (A_s + B_s F_s)^{\alpha_{k-1}} x(t_{k-1}, k-1) . \tag{4.31}$$

By virtue of condition (4.20), one can obtain $H_s^{\alpha_{k-1}} F_s = F_s(A_s + B_s F_s)^{\alpha_{k-1}}$. Hence,

$$F_s x(\bar{t}_{k-1}, k-1) = H_s^{\alpha_{k-1}} F_s x(t_{k-1}, k-1) \tag{4.32}$$

Relation (4.30) implies then that the set $\bigcup_{s=1}^{N} \mathscr{K}er(F_s)$ is positively invariant w.r.t. the switching system $\mathscr{H}_c$.

Develop equation (4.29) by using the solution of the subsystem (A_s, B_s). It follows that

$$F_i L_{si} x(\bar{t}_{k-1}, k-1) = F_i L_{si} (A_s + B_s F_s)^{\alpha_{k-1}} x(t_{k-1}, k-1) = 0 \tag{4.33}$$

The set $\mathscr{K}er(F_s)$ will be positively invariant w.r.t. the corresponding subsystem only if there exist matrices N_{si} such that

$$F_i L_{si} (A_s + B_s F_s)^{\alpha_{k-1}} = N_{si} F_s \ . \tag{4.34}$$

In this case, by substituting (4.34) into (4.33), it follows that

$$F_i L_{si} x(\bar{t}_{k-1}, k-1) = N_{si} F_s x(t_{k-1}, k-1) = 0 \tag{4.35}$$

Since matrices H_s are nonsingular, one can always compute matrices M_{si} such that $N_{si} = M_{si} H_s^{\alpha_{k-1}}$. Equation (4.34) becomes, by using (4.20) again,

$$\begin{aligned} F_i L_{si} (A_s + B_s F_s)^{\alpha_{k-1}} &= M_{si} H_s^{\alpha_{k-1}} F_s \\ &= M_{si} F_s (A_s + B_s F_s)^{\alpha_{k-1}} \end{aligned} \tag{4.36}$$

Recall that the spectrum of each matrix in the closed-loop $A_s + B_s F_s$, which is formed by the $n_s - m$ nonnull and stable eigenvalues of A_s (the third assumption) and the spectrum of the non-singular matrix H_s, does not contain a null eigenvalue. That is, matrices $A_s + B_s F_s$ are non-singular. Consequently, there always exist matrices M_{si} such that $F_i L_{si} = M_{si} F_s, \forall s, i \in \mathscr{I}$. Now with conditions (4.20)-(4.22) established, the same reasoning as before leads to obtain the switching autonomous system $\mathscr{H}$ of the previous section. Applying Theorem 4.1, the set $\mathscr{L}$ is positively invariant w.r.t. the system $\mathscr{H}$ only if condition (4.23) holds. □

Comment 4.1. *For stability purposes, we can only ensure the contractivity property of the set $\mathscr{O}$ w.r.t. the switching system $\mathscr{H}$ when conditions (4.22)- (4.23) hold strictly. However, the asymptotic stability is not obvious since the set $\mathscr{O}$ is unbounded. This is due to the possibility that the state goes to infinity throughout the sets $\mathscr{K}(F_s)$. A Lyapunov-like function for this system is then needed.*

Remark 4.3. *A result slightly different from the one of Theorem 4.3, while using a different proof, is presented in [27].*

In order to establish this result, we proceed to the following technique of augmentation. Without loss of generality, it was assumed that each matrix $A_i, i \in \mathscr{I}$

possesses $(n_i - m)$ non-null stable eigenvalues. Even this assumption can be compatible with stabilizability assumption, it is always possible to avoid it. The idea of [10] consists in augmenting the representation of the switching system $\mathscr{H}$ as : let $v_i \in \mathbb{R}^{n_i - m}, i \in \mathscr{I}$ be a vector of fictitious inputs such that

$$-\delta_i \leq v_i \leq \delta_i \; \delta_i > 0, i \in \mathscr{I}$$

where δ_i are fictitious constraints which are chosen with a great degree of freedom. In this case, the augmented system is then given by

$$\begin{aligned} x(t+1) &= A_i x(t) + \begin{bmatrix} B_i & 0 \end{bmatrix} \begin{bmatrix} u(t) \\ v_i(t) \end{bmatrix} \\ &= A_i x(t) + B_{a_i} w_i(t), t \in I_j, j \in \mathscr{L}, i \in \mathscr{I} . \end{aligned} \tag{4.37}$$

It is true that this augmentation limits the domain of linear behavior of the closed-loop system, but it is always possible to soften the fictitious limitations to enlarge the domain. It is easy to see that for the obtained square system the assumption of $(n_i - m) = 0$ stable eigenvalues is fulfilled.

The state feedback control becomes

$$w_i(t) = K(q_i)x(t,j), K_i \in \mathbb{R}^{n_i \times n_i}, t \in I_j, j \in \mathscr{L}, i \in \mathscr{I} \tag{4.38}$$

where matrices K_i are full rank. The set of linear behavior Ω_i which was unbounded becomes bounded as follows:

$$\Omega_{a_i} = \{x \in \mathbb{R}^{n_i} : |K_i x| \leq \gamma_i, \gamma_i > 0\} \tag{4.39}$$

with γ_i given by

$$\gamma_i = \begin{bmatrix} \rho_i \\ \delta_i \end{bmatrix}, \delta_i \in \mathbb{R}^{n_i - m}, i \in \mathscr{I} . \tag{4.40}$$

With the augmented system, we can present the following result.

Corollary 4.1. *For given full rank matrices $K_i, i \in \mathscr{I}$, if there exist, for all $i \in \mathscr{I}$, matrices $H_i \in \mathbb{R}^{n_i \times n_i}$ such that:*

$$K_i A_i + K_i B_{a_i} K_i = H_i K_i, \forall i \in \mathscr{I} \tag{4.41}$$

$$|H_s| \gamma_s < \gamma_s, \forall s \in \mathscr{I} \tag{4.42}$$

$$|K_i L_{si} K_s^{-1} H_s| \gamma_s < \gamma_i, \forall s, i \in \mathscr{I} \tag{4.43}$$

then the switching system $\mathscr{H}$ with the state feedback control (4.38) is asymptotically stable at the origin $\forall \xi_0 \in \mathscr{O}_a$.

Proof: Consider the following function:

$$V_{t_j}(\xi(t,j)) = \max_l \frac{|K_i\xi(t,j)|_l}{\gamma_i(l)} \quad , t \in I_j, j \in \mathscr{L}, i \in \mathscr{I}, q(I_j) = q_i. \qquad (4.44)$$

where $|\xi(t,j)| = |x(t,j)|$. We have to prove that it is a Lyapunov-like function of the switching system $\mathscr{H}$. For this, compute the hybrid state $\xi(t+1,j)$ for $t+1 \in I_j$,

$$\xi(t+1,j) = (q_s, x(t+1,j)), \quad t \in I_j\ , j \in \mathscr{L}; q_s = q(I_j)\ . \qquad (4.45)$$

That is,

$$\xi(t+1,j) = (q_s, (A_s + B_{a_s}K_s)x(t,j)), \quad t \in I_j, t+1 \in I_j\ , j \in \mathscr{L}\ . \qquad (4.46)$$

If $x(t,j) \in \Omega_{a_s}$, then we have, $|K_s x(t,j)| \leq \gamma_s$, which means that, $V_{t_j}(\xi(t,j)) < 1$ this implies, condition (4.41) kept in mind,

$$\begin{aligned} |K_s x(t+1,j)| &= |K_s(A_s + B_{a_s}K_s)x(t,j))| \\ &\leq |H_s K_s x(t,j))|, \\ &\leq |H_s||K_s x(t,j))|, \\ &\leq |H_s|\gamma_s \end{aligned}$$
$$t \in I_j, t+1 \in I_j\ , j \in \mathscr{L}$$

In terms of Lyapunov-like equation,

$$\begin{aligned} \Delta V_{t_j}(\xi(t,j)) &= V_{t_j}(\xi(t+1,j)) - V_{t_j}(\xi(t,j)) \\ &\leq \max_l \frac{[|H_s|\gamma_s]_l}{\gamma_s(l)} - 1 \end{aligned}$$
$$t \in I_j, t+1 \in I_j\ , j \in \mathscr{L}\ .$$

Taking account of condition (4.42), $\Delta V_{t_j}(\xi(t,j)) < 0$.

Consider now a switching time t_k, the reasoning which leads to (4.25) is still used. We can write

$$K_i x(t_k+1,k) = K_i L_{si}(A_s + B_{a_s}K_s)x(\bar{t}_{k-1}, k-1)\ . \qquad (4.47)$$

Let $x(\bar{t}_{k-1}, k-1) \in \Omega_{a_s}$, that is, $|K_s x(\bar{t}_{k-1}, k-1)| \leq \gamma_s$, it follows that

$$|K_i x(t_k+1,k)| \leq |K_i L_{si}(A_s + B_{a_s}K_s)K_s^{-1}||K_s x(\bar{t}_{k-1}, k-1)|$$

According to condition (4.41), we can write

$$K_i L_{si}(A_s + B_{a_s}K_s)K_s^{-1} = K_i L_{si}K_s^{-1}H_s = P.$$

Or equivalently, one can develop as follows:

$$\begin{aligned}
|K_i x(t_k+1,k)|_r &\leq \sum_{l=1}^{n_i} |P(r,l)|\gamma_s(l) \frac{|K_s x(\bar{t}_{k-1},k-1)|_l}{\gamma_s(l)} \\
&\leq \left[\sum_{l=1}^{n_i} |P(r,l)|\gamma_s(l)\right] \max_l \frac{|K_s x(\bar{t}_{k-1},k-1)|_l}{\gamma_s(l)} \\
&\leq [|P|\gamma_s]_r \max_l \frac{|K_s x(\bar{t}_{k-1},k-1)|_l}{\gamma_s(l)} .
\end{aligned}$$

Using the Lyapunov-like function candidate,

$$V_{t_k}(\xi(t_k+1,k)) \leq \max_l \frac{[|K_i L_{si} K_s^{-1} H_s|\gamma_s]_l}{\gamma_i(l)} V_{t_{k-1}}(\xi(\bar{t}_{k-1},k-1)) .$$

Then, condition (4.43) leads to,

$$V_{t_k}(\xi(t_k+1,k)) < V_{t_{k-1}}(\xi(\bar{t}_{k-1},k-1)), \forall t_k \in I_j .$$

Consequently, according to Definition 4.6, the switching system $\mathscr{H}$ is asymptotically stable inside the region $\mathscr{O}_a$ while the constraints on the control are respected with the feedback control law given by (4.38). □

4.2.3.2 Design of the Controller

Our approach of designing the controller for the switching system is based on the technique of resolution of equation $XA+XBX=HX$ given in Chapter *II*. This link makes possible to solve the problem stated above. Remember that we assume, without loss of generality, that each matrix $A_i, i \in \mathscr{I}$ admits $(n_i - m)$ stable eigenvalues. The resolution of equation

$$X_i A_i + X_i B_i X_i = H_i X_i, i \in \mathscr{I} \tag{4.48}$$

gives a state feedback assigning spectrum of matrix H_i $(\sigma(H_i))$ together with the stable part of spectrum of matrix A_i in closed loop. For this equation, to have a valid solution, each matrix H_i must satisfy :

$$\begin{cases} \sigma(H_i) \cap \sigma(A_i) = \emptyset \\ B_i \theta_l^i \neq 0,\ l = 1,\dots,m; i \in \mathscr{I} \\ \theta_l^i, l = 1,\dots,m \text{ are linearly independent} \end{cases} \tag{4.49}$$

for θ_l^i eigenvectors of matrix H_i. There exists a unique solution to equation (4.48) if and only if

$$\{\chi_1^i \cdots \chi_{n_i}^i\} \text{ are linearly independent}$$

where $\chi_l^i, i = m+1, ..., n_i; i \in \mathscr{I}$ are eigenvectors associated to stable eigenvalues of matrix A_i, and $\chi_l^i, l = 1, ..., m; i \in \mathscr{I}$ are computed by

$$\chi_l^i = (\lambda_l^i \mathbb{I}_{n_i} - A_i)^{-1} B_i \theta_l^i, \ l = 1, ..., m; i \in \mathscr{I} .$$

Hence, the unique full rank solution of equation (4.48) is given by

$$F_i = [\,\theta_1^i \, ... \, \theta_m^i \, 0 \, ... \, 0\,][\chi_1^i \cdots \chi_m^i \, \chi_{m+1}^i \cdots \chi_{n_i}^i\,]^{-1} . \tag{4.50}$$

Besides, the obtained solutions of equations (4.48) will be nonsingular in the case of the augmentation technique followed before. Furthermore, the computation of the solutions F_i of equations (4.48) can be given more easily by using simply a Sylvester equation obtained as follows [21]: Let $Y_i = F_i^{-1}$, then equation (4.48) can be transformed easily to the following Sylvester equation:

$$A_i Y_i - Y_i H_i = -B_{a_i}, i \in \mathscr{I} \tag{4.51}$$

where B_{a_i} is the augmented matrix.

Notice that the augmentation technique of the system can be used even if each matrix A_i admits $n_i - m$ stable eigenvalues to relax the problem computations.

The different steps of the design technique of stabilizing controller are summarized by the following algorithm.

Algorithm 4.1.

Step 1: Augment each matrix B_i by adding $n_i - m$ null columns.
Step 2: Give N matrices $H_i \in \mathbb{R}^{n_i \times n_i}$ satisfying the required conditions (4.49) and (4.42) by choosing the fictitious constraints δ_i.
Step 3: Compute the solutions of equations (4.48) by using (4.50) or the Sylvester equation (4.51).
Step 4: If conditions (4.43) are satisfied, stop if not go to Step 2 to change matrices H_i.
Step 5: Extract the effective gains F_i.

It is worth noting that this algorithm is based on a trial and error technique. However, the augmentation technique leads to a great relaxation of this procedure by accepting the reduction of the sets of positive invariance. This drawback is compensated by the fact that the union of all the positive invariant sets constitutes our region of asymptotic stability.

Example 4.1. *Consider the following example:*

$$A_1 = \begin{bmatrix} 0 & 1 \\ 1 & 1.5 \end{bmatrix}, \ B_1 = \begin{bmatrix} 1 \\ 0.5 \end{bmatrix}; L_{12} = \begin{bmatrix} 0 & 1.1 \\ 0 & -0.5 \end{bmatrix}; L_{11} = \mathbb{I}_2; \rho_1 = 5.$$

$$A_2 = \begin{bmatrix} 1 & 3 \\ 0 & 2.3 \end{bmatrix}, \quad B_2 = \begin{bmatrix} 0 \\ 1.5 \end{bmatrix}; L_{21} = \begin{bmatrix} 0.2 & -0.1 \\ 0 & 2 \end{bmatrix}; L_{22} = \mathbb{I}_2; \rho_2 = 6.$$

The augmentation technique is used by introducing fictitious entries and fictitious constraints $\delta_1 = 8, \delta_2 = 6$. *In this case,*

$$\gamma_1 = [5\ 8]^T; \gamma_2 = [6\ 6]^T$$

.

Give two matrices H_1, H_2 *satisfying conditions (4.42) and (4.49).*

$$H_1 = \begin{bmatrix} 0.5 & 0.1 \\ 0 & 0.6 \end{bmatrix}, \quad H_2 = \begin{bmatrix} -0.4 & -0.1 \\ 0 & 0.58 \end{bmatrix}.$$

Note that $|H_1|\gamma_1 = [3.3\ 4.8]^T < \gamma_1$ *and* $|H_2|\gamma_2 = [3\ 3.48]^T < \gamma_2$.

The resolution of equations (4.48) leads to the following nonsingular solutions:

$$K_1 = \begin{bmatrix} 0.1667 & -1.1333 \\ -12.8333 & -5.1333 \end{bmatrix}, \quad K_2 = \begin{bmatrix} -0.1307 & -2.0800 \\ -2.2475 & -4.8160 \end{bmatrix}. \tag{4.52}$$

These two matrices are stabilizing the corresponding subsystems respectively. For this, compute the matrices in closed loop for each subsystem $A_{c_i} = A_i + B_{a_i} K_i$, $i = 1, 2$.

$$A_{c_1} = \begin{bmatrix} 0.1667 & -0.1333 \\ 1.0833 & 0.9333 \end{bmatrix}, \sigma(A_{c_1}) = \{0.5;\ 0.6\} = \sigma(H_1) \tag{4.53}$$

$$A_{c_2} = \begin{bmatrix} 1.0000 & 3.0000 \\ -0.1960 & -0.8200 \end{bmatrix}, \sigma(A_{c_2}) = \{-0.4;\ 0.58\} = \sigma(H_2)\ . \tag{4.54}$$

Now, we check conditions (4.43) for only $i \neq s$ *since* $L_{11} = L_{22} = \mathbb{I}$,

$$|K_1 L_{21} K_2^{-1} H_2|\gamma_2 = \begin{bmatrix} 4.2409 \\ 7.0660 \end{bmatrix} < \gamma_1 \tag{4.55}$$

$$|K_2 L_{12} K_1^{-1} H_1|\gamma_1 = \begin{bmatrix} 2.5113 \\ 0.1799 \end{bmatrix} < \gamma_2\ . \tag{4.56}$$

For the simulation work, we consider the following time basis and switching strategy: $[t_0 = 0, t_1 = 5); [t_1 = 5, t_2 = 15); [t_2 = 15, t_3 = 25); [t_3 = 25, t_4 = 40); [t_4 = 40, t_5 = 50)$. *The state is initialized with subsystem* (A_2, B_2) *and at each switching time, the system switches to a different subsystem.*

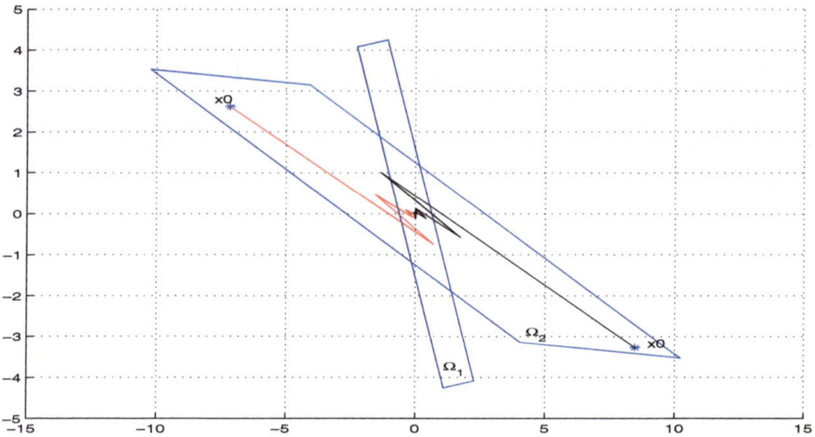

Fig. 4.1 The region $\mathscr{O}_a$ as a set of asymptotic stability for the switching system $\mathscr{H}$ in the state space (x_1, x_2)

4.3 Saturated Switching Singular Linear Systems

In this section, new necessary and sufficient conditions of positive invariance are presented for discrete-time switching singular system with saturated control. These conditions are obtained directly without use of any decomposition of the initial system. These results allow to design a stabilizing feedback controller by state feedback ensuring the admissibility of the switching singular system in closed loop (regularity and impulsion free) while respecting constraints on the control. This technique is based on the solution of equation $XA + XBXE = HXE$ which is developed in Chapter *II*.

4.3.1 Problem Formulation

In this section, we give the problem formulation related to switching singular linear systems studied in this section.

Consider the following switching singular linear system described by

$$\begin{aligned} E_{\alpha}(t)x(t+1) &= A_{\alpha}(t)x(t) + B_{\alpha}(t)u(t) \\ x(0) &= x_0 \end{aligned} \tag{4.57}$$

where t is a non-negative integer, α a switching rule which takes its values in the finite set $\mathscr{I} := \{1, \ldots, N\}$, $x \in \mathbb{R}^n$ is the state, $u \in \mathbb{R}^m$ is the control. Matrices A_i, B_i, E_i are real of appropriate size with E_i a square matrix such that $Rank(E_i) = r_i \leq n$. The control is assumed here to be constrained as follows:

$$u \in \Omega = \{u \in \mathbb{R}^m | -\rho_i \leq u \leq \rho_i; \rho_i \geq 0\}. \tag{4.58}$$

We assume for system (4.57) that:

H1) The switching rule α is not known *a priori* but its value is available at each sampling period, that is at each t.
H2) Each pencil $[E_i, A_i]$ is regular and impulse free, *i.e.*, $rank[E_i + A_i S_i S_i^T] = n$.
H3) The switching singular system (4.57) is controllable.

Consider a stabilizing state feedback control given by

$$u(t) = F_i x(t), rank(F_i) = m, \tag{4.59}$$

the corresponding closed-loop switching singular linear system is then obtained as

$$E_i x(t+1) = (A_i + B_i F_i)x(t) = Ac_i x(t). \tag{4.60}$$

Each pencil $[A_i + B_i F_i, E_i]$ is assumed to be regular and impulse free: $rank(E_i + Ac_i S_i S_i^T) = n$. The induced set of saturation is defined by

$$\mathscr{D}_i = \{x \in \mathbb{R}^n | -\rho_i \leq F_i x \leq \rho_i\} \tag{4.61}$$

The objective of this section is to present:

- Necessary and sufficient condition for the set $\mathscr{D} = \bigcup_{i \in \mathscr{I}} \mathscr{D}_i$ to be positively invariant with respect to (w.r.t.) the switching singular linear system in closed-loop (4.60).
- The synthesis of the required regulator by solving equations $X_i A_i + X_i B_i X_i E_i = H_i X_i E_i$ for given matrices H_i.

4.3.2 Analysis and Controller Design

4.3.2.1 Analysis

In this section, necessary and sufficient conditions of positive invariance of the set $\mathscr{D}$ w.r.t. the switching singular linear system in closed-loop (4.60) are presented. Assume that there exists matrix $\Gamma_i \in \mathbb{R}^{m \times m}$ of full row rank such that $F_i = \Gamma_i E_i$.

Theorem 4.4. *The set $\mathscr{D} = \bigcup_{i=1}^{N} \mathscr{D}_i$ is positively invariant w.r.t. the discrete-time switching singular system in closed-loop (4.60) if and only if there exist matrices $H_i \in \mathbb{R}^{m \times m}$ such that,*

$$\Gamma_i A_i + \Gamma_i B_i \Gamma_i E_i = H_i \Gamma_i E_i \tag{4.62}$$

$$|H_i| \rho_i \leq \rho_j, \forall (i,j) \in \mathscr{I}^2 \,. \tag{4.63}$$

Proof: (Sufficiency): Let conditions (4.62)–(4.63) be satisfied. Consider the following change of coordinates $z(t) = F_i x(t)$ leading to:

$$\begin{aligned} z(t+1) &= F_i x(t+1) = \Gamma_i E_i x(t+1) \\ &= \Gamma_i (A_i + B_i F_i) x(t) \\ &= H_i z(t) \, . \end{aligned} \tag{4.64}$$

With this transformation, the switching singular system (4.60) is transformed into an ordinary switching system (4.64) while the set $\mathscr{D}_i$ is also transformed into

$$\mathscr{S}_i = \{ z \in \mathbb{R}^m | -\rho_i \leq z \leq \rho_i \} \, . \tag{4.65}$$

According to Theorem 4.3, condition (4.63) is the necessary and sufficient condition of the set $\mathscr{S} = \bigcup_{i=1}^N \mathscr{S}_i$ to be positively invariant w.r.t. the switching system (4.64). That means that for any $z(t) \in \mathscr{S}$, $z(t+1) \in \mathscr{S}$ w.r.t. the system (4.64). This implies that, for any $x(t) \in \mathscr{D}$, $x(t+1) \in \mathscr{D}$. To complete the proof, it must be shown that the positive invariance of the set $\mathscr{S}$ w.r.t. the system (4.64) implies the positive invariance of the set $\mathscr{D}$ w.r.t. the system (4.60). Rewriting system (4.64) equivalently by using (4.62), the following developments can be addressed:

$$\begin{aligned} z(t+1) &= H_i z(t) \\ F_i x(t+1) &= H_i F_i x(t) = H_i \Gamma_i E_i x(t) \\ \Gamma_i E_i x(t+1) &= \Gamma_i (A_i + B_i F_i) x(t) \, . \end{aligned} \tag{4.66}$$

Since matrix Γ_i is assumed to be of full row rank m, one can deduce from equality (4.66) that the positive invariance of the set $\mathscr{D}$ is realized w.r.t. the switching singular system (4.60). Finally, let $x(t) \in \mathscr{D}_i$ for $t \in [t_i, \bar{t}_i]$ with $\bar{t}_i = t_{i+1}$ and asssume that a switch occurs at $\bar{t}_i$ to mode j. Then, by using the same transformation as above and equation (4.62), one has $|z(t)| \leq \rho_i$. The switch will appear one sample time later, for this compute $z(t+1) = H_i z(t)$ one can ask in which domain $x(t+1)$ will belong. One may write $|z(t+1)| \leq |H_i| \rho_i$. By virtue of (4.63), one has $|z(t+1)| \leq \rho_j$ which means that $z(t+1) \in \mathscr{S}_j$ or equivalently $x(t+1) \in \mathscr{D}_j$. Consequently, the switching singular system can be initialized in any set $\mathscr{D}_i$, that is $x_0 \in \mathscr{D} = \bigcup_{i \in \mathscr{I}} \mathscr{D}_i$.

(Necessity:) Assume that the set $\mathscr{D}$ is positively invariant w.r.t. the switching singular system in closed-loop (4.60) and $rank(F_i) = m$. Consider a vector $\omega \in Ker(F_i)$, that is $F_i \omega(t) = 0$. Decompose without loss of generality ω, F_i and $\Gamma_i A_{ci}$ as follows:

$$\omega = \begin{bmatrix} \omega_1 \\ \omega_2 \end{bmatrix} ; F_i = \begin{bmatrix} F_1^i & F_2^i \end{bmatrix} ; \Gamma_i A_{ci} = \begin{bmatrix} M_1^i & M_2^i \end{bmatrix} \tag{4.67}$$

with $\omega_1 \in \mathbb{R}^m$, $F_1^i \in \mathbb{R}^{m \times m}$ and $rank(F_1^i) = m$. Then, $F_i \omega(t) = F_1^i \omega_1(t) + F_2^i \omega_2(t) = 0$, and $\omega_1(t) = -(F_1^i)^{-1} F_2^i \omega_2(t)$. That means that with such $\omega_1(t)$, for any $\omega_2(t) \in \mathbb{R}^{n-m}$, $\omega(t) \in Ker(F_i)$. According to (4.57), one can write, $F_i \omega(t+1) = \Gamma_i A c_i \omega(t) = (M_2^i - M_1^i (F_1^i)^{-1} F_2^i) \omega_2(t) = G_i \omega_2(t)$. Assume that matrix G_i is nonnull, that is, it contains at least one component $g_{sj} \neq 0$. Since ω_2 is arbitrarly chosen, take a vector with only one non-null component $\omega_2^j = \delta \rho_i^s / g_{sj}, \delta > 1$. It follows obviously that $G_i \omega_2 = \delta \rho_i^s > \rho_i^s$, consequently, $[F_i \omega(t+1)]_s > \rho_i^s$ which means that the set $\mathscr{D}_i$ is not positively invariant w.r.t. the system. This contradicts the assumption. Besides, $G_i = 0$ implies by noting $H_i = M_1^i (F_1^i)^{-1}$, $M_2^i = H_i F_2^i$ and $M_1^i = H_i F_1^i$ or in a condensed form, $\Gamma_i A c_i = [M_1^i \ M_2^i] = H_i [F_1^i \ F_2^i] = H_i F_i$, which is the same as (4.62). The second part of the proof is then easily derived: the existence of matrix H_i solution of equation (4.62) enables one to use the same transformation as in the sufficiency to obtain the reduced order dynamic system (4.64) which is a standard system. Applying the result of Theorem 4.3, condition (4.63) follows. □

The following result of admissibility of the singular switching system can be addressed.

Corollary 4.2. *If there exist matrices $H_i \in \mathbb{R}^{m \times m}$ satisfying (4.62) such that*

$$|H_i|\rho_i < \rho_j, \forall (i,j) \in \mathscr{I}^2, \tag{4.68}$$

then the switching singular system (4.60) is admissible (stable and impulse free) $\forall x(t) \in \mathscr{D}$ for any sequence of switching $\alpha(t) \in \mathscr{I}$.

Proof: The admissibility (in the sense of stability and impulsion free) of the switching singular system is realized since there exist matrices H_i satisfying (4.62), enabling one to transform the initial switching singular system (4.57) to the reduced order switching system (4.64). Besides, condition (4.68) guarantees that the set (4.65) is positively invariant and contractive w.r.t. the system (4.64). That is, the set (4.61) is also positively invariant w.r.t. the switching singular system in closed-loop (4.57) which always has a linear comportment inside this region. According to the assumptions on matrix F_i, this switching singular linear system is admissible $\forall x(t) \in \mathscr{D} = \bigcup_{i=1}^N \mathscr{D}_i$ for any sequence of switching $\alpha(t) \in \mathscr{I}$. □

4.3.2.2 Synthesis of the Controller

In this section, the synthesis of the controller ensuring that the switching singular system in closed-loop (4.57) is admissible is presented for a particular case of $r_i = m$ by using the solution of equation (4.62) given in Chapter *II*. For this assume that:
H4) $r_i = m, i \in \mathscr{I}$.

Define the following two matrices:
$\Theta_i = [\theta_1^i, \dots \theta_m^i]$, θ_s^i presents the eigenvector of matrix H_i associated to the eigenvalue $\lambda_s, 1 \le s \le m$ of matrix H_i, for $i \in \mathscr{I}$.
$V_i = [\xi_1^i, \dots \xi_m^i]$ where $\xi_s^i = (\lambda_s E_i - A_i)^{-1} B_i \theta_s^i$ presents the generalized eigenvector of matrix $A_i + B_i F_i$ in closed loop associated to the eigenvalue $\lambda_s, 1 \le s \le m$ of matrix H_i, for $i \in \mathscr{I}$.

Let matrices H_i be chosen such that:

$$\{\theta_1^i, \dots, \theta_m^i\} \quad \text{are} \quad \text{linearly} \quad \text{independent} \tag{4.69}$$

$$\sigma(H_i) \bigcap \sigma([A_i, E_i]) = \emptyset, i \in \mathscr{I} \tag{4.70}$$

$$[V_i \ S_i] \quad \text{is} \quad \text{nonsingular}, i \in \mathscr{I} \tag{4.71}$$

$$[E_i V_i \ A_i S_i] \quad \text{in nonsingular}, i \in \mathscr{I}, \tag{4.72}$$

where matrix S_i is computed following the decomposition (1.76) for each matrix E_i.

Corollary 4.3. *Let matrices H_i be given according to conditions (4.68)-(4.72), the switching singular closed-loop system (4.57) with the gain controllers given by*

$$\Gamma_i = [\Theta_i \ 0][E_iV_i \ A_iS_i]^{-1}, i \in \mathscr{I} \tag{4.73}$$

$$F_i = \Gamma_i E_i = [\Theta_i \ 0][V_i \ S_i]^{-1}, i \in \mathscr{I} \tag{4.74}$$

is admissible $\forall x(t) \in \mathscr{D}$ for any sequence of switching $\alpha(t) \in \mathscr{I}$.

Proof: According to Theorems 2.10 and 2.11, for each subsystem, conditions (4.69)–(4.72) guarantee the existence of the solutions Γ_i and F_i of equation (4.62) by placing the eigenvalues of matrix H_i as finite eigenvalues in closed loop for the corresponding singular system. These gains ensure in addition that the obtained singular system in closed loop is admissible (stable and impulse free). Since the solutions of equations (4.62) lead to obtain matrices Γ_i and F_i of rank r_i while Theorem 4.4 requires these matrices to be of full row rank m, assumption H4) is needed. Recall that equations (4.62) allow to transform each singular subsystem (4.57) into an ordinary subsystem (4.64). The presence of constraints on the control has to be also taken into account, then condition (4.68) is required to ensure to the control to be always admissible (respecting the constraints) in the sense of a switching system as shown in Theorem 4.3, for any sequence of switching $\alpha(t) \in \mathscr{I}$. □

4.3.3 Example

In order to apply the results of Theorem 4.4, consider the following example:

$$A_1 = \begin{bmatrix} 1 & -1 & 0 \\ 0 & 1 & -1 \\ 1 & -2 & 0 \end{bmatrix}, B_1 = \begin{bmatrix} 0 & 0 \\ 1 & 0 \\ 0 & 1 \end{bmatrix}, E_1 = \begin{bmatrix} 1 & 0 & 0 \\ 0 & 1 & 1 \\ 0 & 0 & 0 \end{bmatrix}$$

$$A_2 = \begin{bmatrix} 1 & 0 & 0 \\ 0 & 1 & -1 \\ 1 & 0 & 0 \end{bmatrix}; B_2 = \begin{bmatrix} 1 & 1 \\ 0 & 0 \\ 0 & 0 \end{bmatrix}, E_2 = \begin{bmatrix} 0 & 0 & 0 \\ 0 & 0 & 1 \\ 0 & 1 & 0 \end{bmatrix}.$$

The control is assumed here to be constrained as follows:

$$\rho_1 = \begin{bmatrix} 5 & 7 \end{bmatrix}^T; \rho_2 = \begin{bmatrix} 6 & 4 \end{bmatrix}^T.$$

Decomposing matrices E_1 and E_2 leads to

$$S_1 = \begin{bmatrix} 0 \\ -0.7071 \\ 0.7071 \end{bmatrix}, S_2 = \begin{bmatrix} 1 \\ 0 \\ 0 \end{bmatrix}.$$

The pencil $[A_1, E_1]$ has two finite generalized eigenvalues: $\sigma([A_1, E_1]) = \{0.5; -1\}$, and matrix E_1 is singular with $rank(E_1) = 2$. The pencil $[A_2, E_2]$ has two finite generalized eigenvalues: $\sigma([A_2, E_2]) = \{0; -1\}$, and matrix E_2 is singular with $rank(E_2) = 2$. Choose matrices H_1 and H_2 according to assumptions (4.69)-(4.72) as follows:

$$H_1 = \begin{bmatrix} -0.4 & -0.1 \\ 0 & -0.5 \end{bmatrix}, H_2 = \begin{bmatrix} 0.4 & 0.1 \\ 0 & 0.5 \end{bmatrix}.$$

Note that $|H_1|\rho_1 = [2.7\ 3.5]^T$ that is, $|H_1|\rho_1 < \rho_1$ and $|H_1|\rho_1 < \rho_2$. $|H_2|\rho_2 = [2.8\ 2]^T$ that is, $|H_2|\rho_2 < \rho_2$ and $|H_2|\rho_2 < \rho_1$. Assumptions H2-H4 are all satisfied. The solution of equations (4.62) and the corresponding matrices in closed loop are given by

$$F_1 = \begin{bmatrix} -1 & 0.6 & 0.6 \\ -1 & 0 & 0 \end{bmatrix}; \Gamma_1 = \begin{bmatrix} -1 & 0.6 & 1.1 \\ -1 & 0 & 0.5 \end{bmatrix};$$
$$F_2 = \begin{bmatrix} 0 & 1.85 & -3.15 \\ 0 & -3.75 & 5.25 \end{bmatrix}; \Gamma_2 = \begin{bmatrix} -1.85 & -3.15 & 1.85 \\ 3.75 & 5.25 & -3.75 \end{bmatrix};$$
$$Ac_1 = \begin{bmatrix} 0 & -1 & 0 \\ -1 & 1.6 & -0.4 \\ 1 & -2 & 0 \end{bmatrix}; Ac_2 = \begin{bmatrix} 1 & -1.9 & 2.1 \\ 0 & 1 & -1 \\ 1 & 0 & 0 \end{bmatrix}.$$

The generalized eigenvalues in closed loop are, $\sigma(Ac_1, E_1) = \{-0.4; -0.5\}$ and $\sigma(Ac_2, E_2) = \{0.4; 0.5\}$. It is worth noting that this example is given for $m = r_1 = r_2 = 2$ and with $rank(B_2) = 1$. Thus, this technique does not assume that matrix B_i is of full row rank. The necessary and sufficient condition for each singular system in closed loop to be regular and impulse free (admissible) are also satisfied: $rank(E_1 + Ac_1 S_1 S_1^T) = 3$ and $rank(E_2 + Ac_2 S_2 S_2^T) = 3$. Note also that the solution of equations (4.62) are obtained with a great precision: $norm(\Gamma_1 Ac_1 - H_1 F_1) = 2.5579E - 16$ and $norm(\Gamma_2 Ac_2 - H_2 F_2) = 1.3613E - 14$.

The obtained singular switching system in closed-loop (4.60) is given by the following equations :

For mode 1:

$$\begin{aligned} x_1(t+1) &= -x_2(t); \\ x_2(t+1) &= -0.5x_2(t) + 0.5x_1(t); \\ x_3(t+1) &= -0.4x_3(t) - x_1(t) + 2.1x_2(t) \end{aligned}$$

For mode 2:

$$\begin{aligned} x_1(t+1) &= 1.9x_1(t) - 2.1x_2(t) + 2.1x_3(t); \\ x_2(t+1) &= 1.9x_2(t) - 2.1x_3(t); \\ x_3(t+1) &= x_2(t) - x_3(t)\ . \end{aligned}$$

4.4 Conclusion

In this chapter, necessary and sufficient conditions of positive invariance of the union of all the positive invariant sets w.r.t. the subsystems, is presented. However, these conditions which allow the contractivity property when they hold strictly, do not ensure the asymptotic stability since the set of contractivity is unbounded. A technique of augmentation of the continuous representation of the switching system leads to bounded contractive regions where asymptotic stability is guaranteed by the use of Lyapunov-like function. An example illustrating the proposed technique is also studied. Besides, the problem of stabilizability of switching singular linear systems with constrained control is studied. Necessary and sufficient conditions of positive invariance are established. Among these conditions, the nonlinear algebraic equation (4.62) arises. In order to synthesize the required controllers, the solution of this equation is used in the particular case of $rank(E_i) = m$. This solution was already presented in Chapter *II* for singular linear systems with constrained control. An example is used to show the interest of the proposed method.

Chapter 5
Saturated Markovian Switching Systems

5.1 Introduction

Linear systems with Markovian jumping parameters, which are a class of switching systems with a switching sequence of Markovian nature, offer the advantage to model large varieties of physical phenomena. This class of systems has been used successfully to model manufacturing systems, power systems, economic systems, etc. We refer the reader, for example to [52] and references therein. It is well known that all these physical systems admit inputs limitation which are modeled by constraints of inequality type. The problem of stochastic stability of discrete-time linear systems with both Markovian jumping parameters and constrained control have been first investigated by [51] for systems without delay while systems with delay were studied in [74].

The aim of this chapter is to study the regulator problem for both discrete-time and continuous-time systems with Markovian jumping parameters and constrained control by using the positive invariance approach.

5.2 Discrete-Time Systems

In this section, the discrete-time system class is first studied. We begin our study with only symmetrical constraints. A necessary and sufficient condition allowing the control law to be always admissible despite the stochastic character of the system is presented. A sufficient condition for stochastic stability of the system is also obtained.

5.2.1 Problem Statement

Let us consider a system described by the following difference equations :

$$\begin{aligned} x_{k+1} &= A(r(k))x_k + B(r(k))u_k \\ x(0) &= x_0 \end{aligned} \tag{5.1}$$

A. Benzaouia: Saturated Switching Systems, LNCIS 426, pp. 117–144.
springerlink.com

where $x_k = x(kT) \in \mathbb{R}^n$ is the system state vector and $u_k = u(kT) \in \mathscr{U} \in \mathbb{R}^m$ is the system constrained input vector at every sample period T, $m \leq n$. $A(r(k))$ and $B(r(k))$ are given matrices with appropriate dimensions for each $r(k)$. $r(k)$ represents a homogeneous discrete-time Markov chain, referred as the mode of the system, taking values in a finite set $\mathscr{I} = \{1,2,\ldots,N\}$ with transition probability matrix $\Pi = [p_{\alpha\beta}], \alpha, \beta \in \mathscr{I}$ given as:

$$p_{\alpha\beta} = Pr\{r(k+1) = \beta | r(k) = \alpha\} \tag{5.2}$$

with $0 \leq p_{\alpha\beta} \leq 1$ and $\sum_{\beta=1}^{N} p_{\alpha\beta} = 1$.

The constraint control set, $\mathscr{U}(\alpha)$ for each mode is given as:

$$\mathscr{U}(\alpha) = \{u \in \mathbb{R}^m | -\bar{u}(\alpha) \leq u_k \leq \bar{u}(\alpha), \bar{u}(\alpha) > 0\}, \forall \alpha \in \mathscr{I} . \tag{5.3}$$

This domain is a symmetric polyhedral set.

Equations (5.1)-(5.3) describe the dynamics of the class of systems with Markovian jumping parameters and constrained control which is assumed to be stochastically stabilizable.

In this chapter, we will restrict ourselves to the following class of stabilizing control law:

$$u_k = F(r(k))x_k, rankF(r(k)) = m, \forall r(k) \in \mathscr{I} . \tag{5.4}$$

Using this control law, the closed-loop system becomes

$$x_{k+1} = A_c(r(k))x_k \tag{5.5}$$

with $A_c(r(k)) = A(r(k)) + B(r(k))F(r(k))$.

Our objective is to determine the gain matrices $F(r(k)), \forall r(k) \in \mathscr{I}$ that stabilizes the closed-loop system, i.e.:

$$\rho(A_c(r(k)) < 1, \forall r(k) \in \mathscr{I} \tag{5.6}$$

where $\rho(.)$ denotes the spectrum radius.

The aim of this work is to develop an approach that helps us to compute the feasible gain matrices $F(r(k)), r(k) \in \mathscr{I}$ ensuring the stochastic stability of the discrete-time system with Markovian jumping parameters despite the presence of constraint on control.

5.2.2 Preliminary Results

We begin this section by stating our stability results of the free discrete-time system with Markovian jumping parameters. In the second subsection, application of the concept of positive invariance to the previous system is proposed.

5.2.2.1 Stability

We will consider the stability problem of system (5.1) with $u_k = 0$.

Definition 5.1. *System (5.1) with $u_k = 0$ is said to be stochastically stable if, for all finite $x_0 \in \mathbb{R}^n$ and the initial mode $\alpha \in \mathscr{I}$, there exists a positive scalar $0 < \tilde{c} < \infty$ satisfying*

$$E\left\{\sum_{k=0}^{\infty} \|x(k,x_0,\alpha)\|\right\} \leq \tilde{c}\|x_0\| \tag{5.7}$$

where $\|(.)\|$ is any vectorial norm.

We now give the first result of this chapter which consists in a sufficient condition of stochastic stability for system (5.1) with $u_k = 0$ [51].

Theorem 5.1. *System (5.1) with $u_k = 0$ is stochastically stable if there exist positive vectors $\mu(\alpha) \in \mathbb{R}^n, \alpha \in \mathscr{I}$ such that*

$$\sum_{\beta=1}^{N} p_{\alpha\beta} \max_{1\leq i\leq n}\left\{\frac{[|A(\alpha)|\mu(\alpha)]_i}{\mu_i(\beta)}\right\} < 1, \forall \alpha \in \mathscr{I} \tag{5.8}$$

Proof: Let the stochastic Lyapunov function candidate be

$$V(x_k, r(k)) = \max_{1\leq i\leq n}\left\{\frac{|x_k|_i}{\mu_i(r(k))}\right\}, \tag{5.9}$$

which is positive definite, $\mu(r(k))$ is a given positive real vector. Compute now its rate of increase for $r(k) = \alpha$, i.e., $\Delta V(x,\alpha)$. Recall that at time $k+1$ the system may jump to any mode $r(k+1) = \beta$.

$$\Delta V(x,\alpha) = E\left[V(x_{k+1}, r(k+1))|x_k, r(k) = \alpha\right] - V(x_k,\alpha) \tag{5.10}$$

By applying the definition of the expectation of a function, one should obtain

$$\begin{aligned} E\left[V(x_{k+1}, r(k+1))|x_k, r(k) = \alpha\right] &= E\left[V(A(\alpha)x_k, r(k+1))|x_k, r(k) = \alpha\right] \\ &= \sum_{\beta=1}^{N} p_{\alpha\beta} \max_{1\leq i\leq n}\left\{\frac{[|A(\alpha)x_k|]_i}{\mu_i(\beta)}\right\} \end{aligned} \tag{5.11}$$

On the other hand, one can develop $\left\{\frac{[|A(\alpha)x_k|]_i}{\mu_i(\beta)}\right\}$ as follows:

$$\left\{\frac{[|A(\alpha)x_k|]_i}{\mu_i(\beta)}\right\} \leq \sum_{j=1}^{n} \frac{|a_{ij}(\alpha)||x_j(k)|}{\mu_i(\beta)} = \sum_{j=1}^{n} \frac{|a_{ij}(\alpha)|\mu_j(\alpha)}{\mu_i(\beta)} \frac{|x_j(k)|}{\mu_j(\alpha)}$$
$$\leq \sum_{j=1}^{n} \frac{|a_{ij}(\alpha)|\mu_j(\alpha)}{\mu_i(\beta)} \max_{1\leq j\leq n}\left\{\frac{|x(k)|_j}{\mu_j(\alpha)}\right\}$$
$$\leq \sum_{j=1}^{n} \frac{|a_{ij}(\alpha)|\mu_j(\alpha)}{\mu_i(\beta)} V(x_k,\alpha) \tag{5.12}$$

Using (5.11) and (5.12), Eq. (5.10) gives

$$\Delta V(x,\alpha) \leq \left\{\sum_{\beta=1}^{N} p_{\alpha\beta} \max_{1\leq i\leq n}\left\{\sum_{j=1}^{n} \frac{|a_{ij}(\alpha)|\mu_j(\alpha)}{\mu_i(\beta)}\right\} - 1\right\} V(x_k,\alpha),$$
$$\leq \left\{\sum_{\beta=1}^{N} p_{\alpha\beta} \max_{1\leq i\leq n}\left\{\frac{[|A(\alpha)|\mu(\alpha)]_i}{\mu_i(\beta)}\right\} - 1\right\} V(x_k,\alpha),$$
$$\forall \alpha \in \mathscr{I}. \tag{5.13}$$

$\forall x_k \neq 0$, (5.13) leads to

$$\frac{\Delta V(x,\alpha)}{V(x_k,\alpha)} \leq \sum_{\beta=1}^{N} p_{\alpha\beta} \max_{1\leq i\leq n}\left\{\frac{[|A(\alpha)|\mu(\alpha)]_i}{\mu_i(\beta)}\right\} - 1. \tag{5.14}$$

Let the scalar ν be defined by the following:

$$\nu = 1 - \max_{1\leq\alpha\leq N}\left[\sum_{\beta=1}^{N} p_{\alpha\beta} \max_{1\leq i\leq n}\left\{\frac{[|A(\alpha)|\mu(\alpha)]_i}{\mu_i(\beta)}\right\}\right] \tag{5.15}$$

According to condition (5.8), $\nu > 0$. Thus, inequality (5.14) can equivalently be rewritten as follows:

$$\Delta V(x,\alpha) \leq -\nu V(x_k,\alpha) . \tag{5.16}$$

By applying the definition of the rate of increase of the function $V(x_{k+1},\alpha)$, one should obtain

$$E\left[V(x_{k+1},r(k+1))|x_k,r(k)\right] \leq (1-\nu)V(x_k,r(k)) . \tag{5.17}$$

This latter leads recursively to the following inequality:

$$E\left[V(x_k,r(k))\right] \leq (1-\nu)^k V(x_0,\alpha) . \tag{5.18}$$

Note that $0 \leq 1-\nu < 1, \forall\alpha \in \mathscr{I}$, otherwise, if $\nu > 1$, we get $E\left[V(x_k,r(k))\right] \leq 0$ which contradicts the fact that the function (5.9) is definite positive.

Summing both sides from 0 to ∞ and using the fact that $\nu \leq 1$, we get

$$E\left\{\sum_{k=0}^{\infty} \max_{1\leq i\leq n} \frac{|x_i(k)|}{\mu_i(\alpha)}\right\} \leq \left\{\sum_{k=0}^{\infty}(1-\nu(\alpha))^k\right\} \max_{1\leq i\leq n} \frac{|x_i(0)|}{\mu_i(\alpha)}$$
$$\leq \frac{1}{\nu} \max_{1\leq i\leq n}\left[\frac{|x_i(0)|}{\mu_i(\alpha)}\right] . \tag{5.19}$$

Note that condition (5.8) ensures that $\nu \neq 0$. Recall that for two positive vectors a and b, we can write

$$\frac{max_i(a_i)}{max_i(b_i)} \leq \max_{1\leq i\leq n}\left\{\frac{a_i}{b_i}\right\} \leq \frac{max_i(a_i)}{min_i(b_i)} .$$

This leads to obtain the following expectation :

$$E\left\{\sum_{k=0}^{\infty} \max_{1\leq i\leq n} |x_i(k,x_0,\alpha)|\right\} \leq \tilde{c} \max_{1\leq i\leq n} |x_i(0)| \tag{5.20}$$

with

$$\tilde{c} = \frac{1}{\nu}\frac{1}{\min_{1\leq i\leq n, 1\leq \alpha\leq N} \mu_i(\alpha)} . \tag{5.21}$$

This inequality can equivalently be rewritten as (5.7) with $\|.\|_\infty$. Then condition (5.8) ensures to the system with Markovian jumping parameters to be stochastically stable. □

An important consequence of Theorem 5.1 is to obtain the union of all the level sets, associated to each mode, as a set of stochastic stability. For this define the following level set:

$$\mathscr{R}_{r(k)} = \left\{x \in \mathbb{R}^{n\times n} | V(x_k, r(k)) = \max_{1\leq i\leq n}\left\{\frac{|x_k|_i}{\mu_i(r(k))}\right\} < 1\right\} . \tag{5.22}$$

Theorem 5.2. *If condition (5.8) holds, then the set $\bigcup_{\alpha=1}^{N}\mathscr{R}_\alpha$ is a set of stochastic stability of system (5.1).*

Proof: Assume that at time k, $x(k) \in \mathscr{R}_\alpha$. Recall that one can only compute $E\left[V(x(k+1), r(k+1))\right]$, and at time $k+1$ the system may jump to any mode $r(k+1) = \beta$. It follows from (5.17) that

$$E\left[V(x_{k+1}, r(k+1)) | x_k, r(k)\right] \leq (1-\nu)V(x_k, r(k)), \tag{5.23}$$

with ν given by (5.15). If condition (5.8) holds, $1-\nu < 1$ which means that $E\left[V(x_{k+1}, r(k+1)) | x_k, r(k)\right] \leq V(x_k, r(k))$. Since $x(k) \in \mathscr{R}_\alpha$, $V(x_k, r(k)) < 1$. It follows that $E\left[V(x_{k+1},r(k+1))|x_k,r(k)\right] < 1$. Consequently, $x(k+1) \in \mathscr{R}_\beta$. Finally, the set $\bigcup_{\alpha=1}^{N} \mathscr{R}_\alpha$ is a set of stochastic stability of system (5.1). □

5.2.2.2 Positive Invariance

Now we state and prove our second result concerning the concept of positive invariance. Let us consider the following discrete-time system with Markovian jumping parameters

$$E\left[z_{k+1}|r(k),z_k\right] = H(r(k))z_k, z_k \in \mathbb{R}^m, r(k) \in \mathscr{I}$$
$$z(0) = z_0 \,. \tag{5.24}$$

Definition 5.2. *A subset $\mathscr{D}$ of $\mathbb{R}^m$ is said to be stochastically positively invariant with respect to (w.r.t) system (5.24) if, for all $z_0 \in \mathscr{D}$ and the initial mode $\alpha \in \mathscr{I}$, $E(z_k) \in \mathscr{D}$ for every $k \in \mathbb{N}$.*

Consider the following domain $\mathscr{D}(\alpha)$ given as

$$\mathscr{D}(\alpha) = \{z_k \in \mathbb{R}^m | -w(\alpha) \le z_k \le w(\alpha), w(\alpha) > 0,\}, \forall \alpha \in \mathscr{I} \,. \tag{5.25}$$

Assume that, for every mode, the set $\mathscr{D}(\alpha)$ is positively invariant w.r.t the corresponding system (5.24), i.e.,

$$|H(\alpha)|w(\alpha) \le w(\alpha), \forall \alpha \in \mathscr{I} \,. \tag{5.26}$$

Let the domain $\mathscr{D}$ be defined as follows:

$$\begin{aligned} \mathscr{D} &= \bigcap\nolimits_{\alpha=1}^{N} \mathscr{D}(\alpha) \\ &= \{z_k \in \mathbb{R}^m | -\varphi \le z_k \le \varphi\} \end{aligned} \tag{5.27}$$
$$\text{with, } \varphi_i = \min_{1 \le \alpha \le N} \{w_i(\alpha)\} \,.$$

Corollary 5.1. *The set $\mathscr{D}$ is stochastically positively invariant w.r.t system (5.24) if and only if:*

$$|H(\alpha)|w(\alpha) \le w(\beta), \forall \alpha, \forall \beta \in \mathscr{I} \tag{5.28}$$

Proof: (Sufficiency): Let $z_k \in \mathscr{D} \subseteq \mathscr{D}(\alpha)$, i.e.,

$$-w(\alpha) \le -\varphi \le z_k \le \varphi \le w(\alpha) \,. \tag{5.29}$$

Decompose matrix H as follows: $H = H^+ - H^-$, with $H^+ = sup(H,0)$ and $H^- = sup(-H,O)$ and note that $\mid H \mid = H^+ + H^-$. By multiplying (5.29) successively by matrices H^+ and $-H^-$ and adding the two obtained inequalities, one can obtain:

$$-\mid H(\alpha)\mid w(\alpha) \leq -\mid H(\alpha)\mid \varphi \leq E[z_{k+1}|r_k=\alpha, z_k] = H(\alpha)z_k \\ \leq\mid H(\alpha)\mid \varphi \leq\mid H(\alpha)\mid w(\alpha)\ . \quad (5.30)$$

Note that at the time $k+1$, the system may jump to any mode $r(k+1)=\beta$, $\forall\beta \in \mathscr{I}$. On the other hand

$$E[z_{k+1}] = \sum_{\alpha\in\mathscr{I}} P[r_k=\alpha]E[z_{k+1}|r_k=\alpha]\ . \quad (5.31)$$

If condition (5.28) holds, we have

$$-w(\beta) = -\sum_{\alpha\in\mathscr{I}} p_k(\alpha)w(\beta) \leq E[z_{k+1}] \leq \sum_{\alpha\in\mathscr{I}} p_k(\alpha)w(\beta) = w(\beta) \quad (5.32)$$

where $p_k(\alpha) = P[r(k)=\alpha]$. Therefore, the state $E[z_{k+1}]$ will belong to the set $\mathscr{D}$ given by (5.27).

(Necessity:) Let domain $\mathscr{D}$ be stochastically positively invariant w.r.t system (5.24) and condition (5.28) does not hold. That is, there exist $i\in[1,m]$, α and $\beta\in\mathscr{I}$ such that

$$[|H(\alpha)|w(\alpha)]_i > w_i(\beta)\ . \quad (5.33)$$

Let $w_i(\alpha) = \varphi_i$ and consider the following vector:

$$\zeta_k = \begin{cases} \varphi_i, & \text{if, } h_{ii}(\alpha) > 0 \\ -\varphi_i, & \text{if, } h_{ii}(\alpha) < 0 \\ 0, & \text{if, } h_{ii}(\alpha) = 0 \\ 0, & \text{elsewhere} \end{cases}\ . \quad (5.34)$$

Note that the vector $E(\zeta_k) = \zeta_k \in \mathscr{D}$ and consider the corresponding vector at time $k+1$.

$$E[(\zeta_{k+1})_i|r_k=\alpha] = [|H(\alpha)|\varphi_i]_i\ .$$

According to hypothesis (5.33), $E[(\zeta_{k+1})_i|r_k=\alpha] > w_i(\beta)$. This latter shows that if the parameters of the system jump to the mode $r(k+1)=\beta$ at time $k+1$, the vector $E[\zeta_{k+1}]$ will not belong to the set $\mathscr{D}(\beta)$, nor to the set $\mathscr{D}$. This contradicts the hypothesis, then the condition (5.28) holds. □

5.2.3 Design of the Controller

In this section, we apply the previous results to the problem of saturated regulator of linear discrete-time system with Markovian jumping parameters. Consider system (5.1)-(5.3) with the feedback control law given by (5.4). The closed-loop system is then given by (5.5). The control obtained by the feedback law is admissible for every mode $\alpha\in\mathscr{I}$ if and only if the corresponding state is also constrained to evolve in a specified region given as

$$\mathscr{F}(\alpha) = \{x_k \in \mathbb{R}^n | -\bar{u}(\alpha) \leq F(\alpha)x_k \leq \bar{u}(\alpha), \bar{u}(\alpha) > 0, \}, \forall \alpha \in \mathscr{I} \tag{5.35}$$

where vectors $\bar{u}(\alpha)$ represent the constraint on the system input. The set $\mathscr{F}(\alpha)$ is an unbounded symmetrical polyhedral domain for $m < n$. Define the following set:

$$\mathscr{F} = \bigcap_{\alpha=1}^{N} \mathscr{F}(\alpha) . \tag{5.36}$$

Consider now the expression of the control at time $k+1$:

$$\begin{aligned} u_{k+1} &= F(r(k+1))x_{k+1} \\ &= F(r(k+1))\left[A(r(k)=\alpha) + B(r(k)=\alpha)F(r(k)=\alpha)\right]x_k . \end{aligned}$$

This equation shows that the control at time $k+1$ depends on the mode taken by the feedback control law at this time. In fact, at time k, when the state is x_k and the mode is at $r(k) = \alpha$, the gain matrix at time $k+1$ can only be known by means of its expectation . That is,

$$E\left[u_{k+1} | r(k), u_k\right] = \left\{ \sum_{\beta=1}^{N} p_{r(k)\beta} F(\beta) \right\} \left[A(r(k) + B(r(k)F(r(k)\right] x_k .$$

Define matrices $G(\alpha)$ as:

$$G(\alpha) = \sum_{\beta=1}^{N} p_{\alpha\beta} F(\beta), \forall \alpha \in \mathscr{I} . \tag{5.37}$$

If there exist matrices $H(\alpha)$ such that

$$G(\alpha)\left[A(\alpha) + B(\alpha)F(\alpha)\right] = H(\alpha)F(\alpha), \forall \alpha \in \mathscr{I} \tag{5.38}$$

in this case, the control will evolve as a dynamic system given as

$$E[u_{k+1} | r(k), u_k] = H(r(k))u_k, \forall \alpha \in \mathscr{I} . \tag{5.39}$$

Recall that this dynamic system has to be constrained into domain $\mathscr{U}(\alpha)$ given by (5.3), which is a polyhedral symmetric set. At this level, one has to use the necessary and sufficient conditions of Corollary 5.1 which guarantees that the set $\mathscr{U}(\alpha)$ is stochastically positively invariant w.r.t system (5.39) for all $\alpha \in \mathscr{I}$. Let $\mathscr{U}$ be defined as

$$\begin{aligned} \mathscr{U} &= \bigcap_{\alpha=1}^{N} \mathscr{U}(\alpha) \\ &= \{u_k \in \mathbb{R}^m | -\bar{\varphi} \leq u_k \leq \bar{\varphi}\} \end{aligned} \tag{5.40}$$

with, $\bar{\varphi}_i = \min_{1 \leq \alpha \leq N} \{\bar{u}_i(\alpha)\}$.

Therefore, the set $\mathscr{U}$ will be stochastically positively invariant w.r.t the system.

At this step, the result of Corollary 5.1 can be successfully applied to system (5.5) together with domain (5.36) leading to the following result obtained by [51].

Theorem 5.3. *If there exist matrices $H(\alpha), \alpha = 1, \ldots, N$ solution of equations (5.38) such that $|H(\alpha)|\bar{u}(\alpha) \leq \bar{u}(\alpha), \forall \alpha \in \mathscr{I}$, then the set $\mathscr{F}$ given by (5.36) is stochastically positively invariant w.r.t the closed-loop system (5.5) if and only if*

$$|H(\alpha)|\bar{u}(\alpha) \leq \bar{u}(\beta), \forall \alpha, \beta \in \mathscr{I} \tag{5.41}$$

where vectors $\bar{u}(\alpha)$ present the constraint on the control.

Proof: If matrices $H(\alpha), \alpha = 1, \ldots, N$ solution of equation (5.38) exist such that $|H(\alpha)|\bar{u}(\alpha) \leq \bar{u}(\alpha), \forall \alpha \in \mathscr{I}$, then one can obtain the dynamic system of the control given by (5.39) which is stochastically asymptotically stable for every mode by virtue of (5.6). Note that for every mode, domain $\mathscr{U}(\alpha)$ is positively invariant w.r.t its corresponding system (5.39). The domain $\mathscr{F}$, defined by (5.36), becomes domain $\mathscr{U}$ given by (5.40) (using the transformation $u_k = F(r(k))x_k$). By virtue of Corollary 5.1, condition (5.41) is the necessary and sufficient condition for this set to be stochastically positively invariant w.r.t system (5.39). It is also obvious that the set $\mathscr{F}$ is stochastically positively invariant w.r.t the closed-loop system (5.5) if and only if the set $\mathscr{U}$ is positively invariant w.r.t system (5.39). □

This result allows us to announce the stochastic stability condition for the system.

Corollary 5.2. *If there exist matrices $H(\alpha)$, and positive vectors $\mu(\alpha)$ such that*

(i)

$$G(\alpha)\left[A(\alpha) + B(\alpha)F(\alpha)\right] = H(\alpha)F(\alpha), \forall \alpha \in \mathscr{I} \tag{5.42}$$

(ii)

$$|H(\alpha)|\bar{u}(\alpha) \leq \bar{u}(\beta), \forall \alpha \in \mathscr{I}, \forall \beta \in \mathscr{I} \tag{5.43}$$

(iii)

$$\sum_{\beta=1}^{N} p_{\alpha\beta} \max_{1 \leq i \leq n} \left\{ \frac{[|A_c(\alpha)|\mu(\alpha)]_i}{\mu_i(\beta)} \right\} < 1, \forall \alpha \in \mathscr{I} \tag{5.44}$$

where vectors $\bar{u}(.)$ represents the constraint on the control, and matrix $G(.)$ is defined by (5.37); then, the closed-loop system (5.5) is stochastically stable for all $x_0 \in \mathscr{F}$ and all $\alpha \in \mathscr{I}$.

Proof: Condition (5.42)-(5.43) allow the closed-loop system (5.5) to have a linear behavior, despite the existence of constraint on the control, inside the set $\mathscr{F}$, which has the positive invariance property w.r.t the system. Indeed, condition (5.42) enables us to transform the dynamical system (5.5) to the dynamical system of the control (5.39) and according to Corollary 5.1, condition (5.43) ensures to the domain $\mathscr{F}$ the stochastic positive invariance. Further, (5.44) presents the sufficient condition of stochastic stability for the closed-loop system (5.5).

Remark 5.1. *It is worth noting that Corollary 5.2 presents in addition to Eq. (5.42), a condition on matrices $H(\alpha)$ and another on the matrices in the closed loop. However, one cannot find a relation between these two conditions that returns the resolution of Eq. (5.42), while respecting (5.43)-(5.44), rather difficult. Condition (5.44) can be given with vector $\bar{u}$ instead of vector μ.*

5.2.4 Resolution of Equation $G(\alpha)A_c(\alpha) = H(\alpha)F(\alpha)$

In this section, we present an algorithm to compute matrices $H(\alpha), \forall\alpha \in \mathscr{I}$ by using the result of [16] developed in Chapter *II*. This technique is based on the augmentation of the system entries to make the system square, i.e. the number of inputs equals the number of states, $m = n$, by introducing $(n-m)$ null columns in matrix B corresponding to $n-m$ fictitious entries, v_k, with its saturations, $\delta(\alpha)$ of the following form:

$$-\delta(\alpha) \leq v_k \leq \delta(\alpha), \forall\alpha \in \mathscr{I} .$$

The first step in our algorithm, which we present later, will consist of making a pole assignment to compute the gain matrices, $F(\alpha)$, $\alpha = 1,\dots,N$ by using the following equation:

$$F(\alpha)\left[A(\alpha) + B(\alpha)F(\alpha)\right] = H_1(\alpha)F(\alpha) \tag{5.45}$$

where the matrices $H_1(\alpha)$ are chosen in the following way:

$$H_1(\alpha)\theta_i(\alpha) = \lambda_i(\alpha)\theta_i(\alpha), i = 1,\dots,n \tag{5.46}$$

with $\{\theta_1(\alpha),\dots,\theta_n(\alpha)\}$ are the eigenvectors of $H_1(\alpha)$ which are assumed to be linearly independent and such that

$$B(\alpha)\theta_i(\alpha) \neq 0, \text{ and } \sigma(A(\alpha)) \bigcap \sigma(H_1(\alpha)) = \emptyset . \tag{5.47}$$

Based on the eigenvalues of $H_1(\alpha)$ one can compute the corresponding eigenvectors of the closed-loop matrices, $A_c(\alpha) = A(\alpha) + B(\alpha)F(\alpha)$ as follows:

$$\xi_i(\alpha) = \left[\lambda_i(\alpha)I_n - A(\alpha)\right]^{-1} B(\alpha)\theta_i(\alpha); i = 1,\dots,n . \tag{5.48}$$

Recall that the existence of a unique full rank solution, $F(\alpha)$, of Eq. (5.45) requires that $\{\xi_1(\alpha),\dots,\xi_n(\alpha)\}$ must be linearly independent. In this case, the gain matrices are given as

$$F(\alpha) = \left[\theta_1(\alpha),\dots,\theta_n(\alpha)\right]\left[\xi_1(\alpha),\dots,\xi_n(\alpha)\right]^{-1} . \tag{5.49}$$

At this level, matrices $G(\alpha), \alpha = 1,\ldots,N$ can be computed according to Eq.(5.37). It follows from above that the following relations hold for each i and each α as

$$F(\alpha)\xi_i(\alpha) = \theta_i(\alpha); \tag{5.50}$$
$$A_c(\alpha)\xi_i(\alpha) = \lambda_i(\alpha)\xi_i(\alpha) . \tag{5.51}$$

The second step of our algorithm consists in computing matrices $H(\alpha)$, $\alpha = 1,\ldots,N$ solution of Eq. (5.38). In fact, by post-multiplying Eq. (5.38) by vectors $\xi_i(\alpha)$, we get

$$G(\alpha)A_c(\alpha)\xi_i(\alpha) = H(\alpha)F(\alpha)\xi_i(\alpha) .$$

Taking account of (5.50)-(5.51), it follows that

$$\lambda_i(\alpha)G(\alpha)\xi_i(\alpha) = H(\alpha)\theta_i(\alpha) . \tag{5.52}$$

Now, by decomposing matrices $H(\alpha)$ as $H(\alpha) = H_2(\alpha)H_1(\alpha)$, with both matrices $H_1(\alpha)$ and $H_2(\alpha)$ $n \times n$ square matrices, using (5.46) and substituting in (5.52), one can obtain

$$\lambda_i(\alpha)G(\alpha)\xi_i(\alpha) = \lambda_i(\alpha)H_2(\alpha)\theta_i(\alpha), i = 1,\ldots,n . \tag{5.53}$$

Since $\lambda_i(\alpha) \neq 0$ by construction, one can easily deduce matrices $H_2(\alpha)$ from Eq. (5.53) and then obtain matrices $H(\alpha)$ i.e.,

$$\begin{aligned} H_2(\alpha) &= G(\alpha)\left[\xi_1(\alpha),\ldots,\xi_n(\alpha)\right]\left[\theta_1(\alpha),\ldots,\theta_n(\alpha)\right]^{-1} \\ &= G(\alpha)F^{-1}(\alpha) \\ H(\alpha) &= H_2(\alpha)H_1(\alpha) . \end{aligned} \tag{5.54}$$

Verify now that these matrices satisfy Eq. (5.38). For that, compute $H(\alpha)F(\alpha)$ with Eq. (5.45) kept in mind,

$$H(\alpha)F(\alpha) = G(\alpha)F^{-1}(\alpha)H_1(\alpha)F(\alpha) = G(\alpha)F^{-1}(\alpha)F(\alpha)A_c(\alpha) = G(\alpha)A_c(\alpha) .$$

Remark 5.2. *An interesting case in this resolution is the possibility to take matrices $H_1(\alpha) = H_0$, independent of α. The corresponding assigned spectrum of the closed-loop system will be independent of the modes and the computation becomes easier. Notice that the effective gain matrix for each α is obtained by extraction of the m lines of each computed matrix $F(\alpha)$.*

Further, if we choose matrices $H_1(\alpha)$ given by (5.46) such that

$$|H_1(\alpha)|\bar{q}(\alpha) \leq \phi, \bar{q}(\alpha) = \begin{bmatrix} \bar{u}(\alpha) \\ \delta(\alpha) \end{bmatrix} \in \mathbb{R}^n \tag{5.55}$$
$$\phi_i = \min_{\alpha}\{\bar{q}_i(\alpha)\} \tag{5.56}$$

the sufficient condition of Corollary 5.1 will be satisfied by the following:

$$|H_2(\alpha)|\phi < \phi. \tag{5.57}$$

To show that, one can write

$$|H(\alpha)|\bar{q}(\alpha) \leq |H_2(\alpha)||H_1(\alpha)|\bar{q}(\alpha)\ .$$

Taking account of (5.55) and (5.57), it follows that

$$|H(\alpha)|\bar{q}(\alpha) \leq |H_2(\alpha)|\phi < \phi \leq \bar{q}(\beta), \forall\alpha, \forall\beta \in \mathscr{I}\ . \tag{5.58}$$

Algorithm 5.1.

Step 1 Initialization: augment for each α *the matrix* $B(\alpha)$ *with* $n-m$ *null columns and choose* N *matrices* $H_1(\alpha)$ *according to (5.46) satisfying (5.55), one can use only one matrix* $H_1(\alpha) = H_0$ *independent of* α.

Step 2 Computation: get N *stabilizing gain matrices* $F(\alpha)$ *using (5.49) and* N *matrices* $A_c(\alpha)$.

Step 3 Test: if there exist positive real vectors $\mu(\alpha)$ *satisfying condition (5.44), compute matrices* $H_2(\alpha)$ *according to Eq. (5.54), otherwise choose new* N *matrices* $H_1(\alpha)$ *and go to Step 2. One can use one vector* $\mu = \mu(\alpha)$ *independent of* α.

Step 4 Test: if condition (5.57) is not satisfied choose new matrices $H_1(\alpha)$ *and go to Step 2, otherwise stop.*

In the case of augmented system entries, matrices $F(\alpha)$ become square. The set$\mathscr{F}(\alpha)$ can then be written as follows:

$$\begin{aligned}\mathscr{F}(\alpha) &= \{x_k \in \mathbb{R}^n| - \bar{q}(\alpha) \leq F(\alpha)x_k \leq \bar{q}(\alpha), \bar{q}(\alpha) > 0, \},\\ &= \left\{x_k \in \mathbb{R}^n | V(x_k, r(k)) = \max_{1\leq i\leq n} \left\{\frac{|F(\alpha)x_k|_i}{\bar{q}(\alpha)_i}\right\} < 1\right\}, \forall\alpha \in \mathscr{I}\end{aligned} \tag{5.59}$$

where vectors $\bar{q}(\alpha)$ represent the augmented constraint on the system input given by (5.55). The set $\mathscr{F}(\alpha)$ is now a bounded symmetrical polyhedral domain for $m = n$. Define the following set:

$$\mathscr{F}_u = \bigcup_{\alpha=1}^{N} \mathscr{F}(\alpha)\ . \tag{5.60}$$

The following result can replace Corollary 5.2 in the case of using the augmentation technique .

Corollary 5.3. *If there exist matrices* $H(\alpha) \in \mathbb{R}^{n\times n}$ *such that:*

(i)

$$G(\alpha)\left[A(\alpha) + B(\alpha)F(\alpha)\right] = H(\alpha)F(\alpha), \forall\alpha \in \mathscr{I} \tag{5.61}$$

(ii)

$$|H(\alpha)|\bar{q}(\alpha) \leq \bar{q}(\beta), \forall \alpha \in \mathscr{I}, \forall \beta \in \mathscr{I} \tag{5.62}$$

(iii)

$$\sum_{\beta=1}^{N} p_{\alpha\beta} \max_{1\leq i\leq n} \left\{ \frac{[|A_c(\alpha)|\bar{q}(\alpha)]_i}{\bar{q}_i(\beta)} \right\} < 1, \forall \alpha \in \mathscr{I} \tag{5.63}$$

where vectors $\bar{q}(.)$ represent the augmented constraint on the control, and matrix $G(.)$ is defined by (5.37); then, the closed-loop augmented system (5.5) is stochastically stable for all $x_0 \in \mathscr{F}_u$ and all $\alpha \in \mathscr{I}$.

Proof: Follows readily by using Theorem 5.2. □

To show the usefulness of our theoretical results, let us consider the following numerical example:

Example 5.1. *Let matrices $A(\alpha)$ and $B(\alpha)$ for $\alpha = 1,\ldots,3$ be given as*

- *mode 1:* $A(1) = \begin{bmatrix} 1 & 1 \\ -1 & 0.5 \end{bmatrix}, B(1) = \begin{bmatrix} 1 \\ 1 \end{bmatrix}, -3 \leq u_k \leq 3$
- *mode 2:* $A(2) = \begin{bmatrix} -1.5 & 1 \\ -1 & 0.32 \end{bmatrix}, B(2) = \begin{bmatrix} 1 \\ 0.5 \end{bmatrix}, -10 \leq u_k \leq 10$
- *mode 3:* $A(3) = \begin{bmatrix} 1.5 & 2 \\ 2 & 1 \end{bmatrix}, B(3) = \begin{bmatrix} 1 \\ 0.5 \end{bmatrix}, -8 \leq u_k \leq 8$.

The Markov chain is described by the following transition matrix Π:

$$\Pi = \begin{bmatrix} 0.67 & 0.1 & 0.23 \\ 0.35 & 0.47 & 0.18 \\ 0.06 & 0.3 & 0.64 \end{bmatrix} .$$

The augmentation of the system entries leads to the following:

$$B(1) = \begin{bmatrix} 1 & 0 \\ 1 & 0 \end{bmatrix}, B(2) = \begin{bmatrix} 1 & 0 \\ 0.5 & 0 \end{bmatrix}, B(3) = \begin{bmatrix} 1 & 0 \\ 0.5 & 0 \end{bmatrix} .$$

For each mode, fictitious control and constraint are introduced as follows:

$$-\delta(\alpha) \leq v_k \leq \delta(\alpha), \alpha = 1,2,3.$$

The parameters $\delta(\alpha), \alpha = 1,2,3$ of the fictitious constraint will be computed such that the required conditions will be satisfied. Let matrix $H_1(\alpha) = H_0$, $\forall \alpha \in \mathscr{I}$ with

$$H_0 = \begin{bmatrix} -0.1 & 0.1 \\ 0 & 0.2 \end{bmatrix}; \lambda_1 = -0.1, \lambda_2 = 0.2 .$$

For sake of simplicity, we will only give the computed matrices $H_2(\alpha)$, the gain matrices $F(\alpha)$ and the matrices $A_c(\alpha)$.

$$H_2(1) = \begin{bmatrix} 0.827 & 0.0244 \\ -0.0677 & 0.7341 \end{bmatrix}, H_2(2) = \begin{bmatrix} 0.6615 & -0.0728 \\ -0.2023 & 0.9391 \end{bmatrix},$$
$$H_2(3) = \begin{bmatrix} 0.968 & -0.027 \\ -0.2428 & -0.0841 \end{bmatrix}$$

$$F(1) = \begin{bmatrix} -0.512 & -0.8880 \\ 10.416 & 1.984 \end{bmatrix}, F(2) = \begin{bmatrix} 1.7353 & -0.9106 \\ 7.0235 & 1.8729 \end{bmatrix},$$
$$F(3) = \begin{bmatrix} -1.408 & -1.984 \\ -28.416 & -2.368 \end{bmatrix} .$$

The effective gain matrices are extracted from the computing ones as follows

$$F_1 = \begin{bmatrix} -0.512 & -0.888 \end{bmatrix}, F_2 = \begin{bmatrix} 1.7353 & -0.9106 \end{bmatrix},$$
$$F_3 = \begin{bmatrix} -1.408 & 0.0244 \end{bmatrix} .$$

Therefore, the closed-loop matrices are given as

$$A_c(1) = \begin{bmatrix} 0.4888 & 0.112 \\ -1.512 & -0.388 \end{bmatrix}, A_c(2) = \begin{bmatrix} 0.2353 & 0.0894 \\ -0.1324 & -0.1353 \end{bmatrix},$$
$$A_c(3) = \begin{bmatrix} 0.092 & 0.016 \\ 1.296 & 0.008 \end{bmatrix} .$$

The necessary and sufficient condition (5.43) is satisfied for $\delta_1 = 10, \delta_2 = 15, \delta_3 =$ 10. One can easily check that all the closed-loop matrices $A_c(\alpha)$, are assigned the same spectrum $\sigma(H_0)$.

On the other hand, the sufficient condition (5.44) of stochastic stability is also satisfied for $\mu = \begin{pmatrix} 2 \\ 5 \end{pmatrix}$.

In this section, necessary and sufficient conditions for domain $\mathscr{F}$ to be positively invariant w.r.t the system in the closed-loop (5.5) are established for linear discrete-time systems with Markovian jumping parameters and symmetrical constrained control. A new sufficient condition of stochastic stability is then deduced. These results are obtained by using non quadratic Lyapunov function as is usually the case in the problems with constraints of inequality type. An Algorithm is also presented to compute matrices $H(\alpha)$ together with a simple sufficient condition of stochastic positive invariance which is independent of the probability transition matrix Π. The case of obtaining the set $\bigcup_{\alpha=1}^{N} \mathscr{F}(\alpha)$ as a set of stochastic positive invariance and stochastic stability is deduced.

5.3 Continuous-Time Systems

In the previous section, the notion of stochastic positive invariance is introduced and applied on the common set between all the modes. However, this procedure in not followed in this section since the deterministic positive invariance is still applied for this common set. This solution is based on the simple fact that, if the common set is positively invariant w.r.t. each mode, then all the trajectories of the system with Markovian jumping parameters, emanating from this set, cannot leave it at any random jump. This technique enables us to deal with asymmetric constraints. These results can be found in [20, 71, 72, 73].

The aim of this section is to study the regulator problem for continuous-time systems with Markovian jumping parameters and constrained control by using the positive invariance approach. In this work, asymmetric constraints are considered. A sufficient condition for exponential stability of the system is obtained.

5.3.1 Problem Formulation

We consider the continuous-time linear system with Markovian jumping parameters defined as

$$(\Sigma) : \begin{cases} \dot{x}(t) = A(r(t))x(t) + B(r(t))u(t) \\ x(0) = x_0, r(0) = r_0 \end{cases} \tag{5.64}$$

where $x \in \mathbb{R}^n$ is the state vector, $u \in \mathbb{R}^m$ is the control vector. The system state is a function of the discrete parameter $r(t)$ which is a continuous-time homogeneous Markov process with finite discrete state space $\mathscr{I} = \{1, 2, \ldots, N\}$.

The previous equation can be regarded as the result of the following N equations:

$$\dot{x}(t) = A_\alpha x(t) + B_\alpha u(t), \qquad 1 \le \alpha \le N \tag{5.65}$$

switching from one to the others according to the evolution of the Markov process. The transition probability of $r(t)$ is defined by

$$P\left[r(t+h) = \beta | r(t) = \alpha\right] = \begin{cases} \lambda_{\alpha\beta} h + o(h) & \text{if, } \alpha \neq \beta \\ 1 + \lambda_{\alpha\alpha} h + o(h) & \text{if, } \alpha = \beta \end{cases} \tag{5.66}$$

$$\sum_\beta \lambda_{\alpha\beta} = 0; \lambda_{\alpha\beta} \ge 0 \text{ if, } \alpha \neq \beta \ .$$

The $\lambda_{\alpha\beta}$'s are the transition probability rates from the αth mode to the βth mode. The set of constraints for the control $u(t)$ is given for each mode α by

$$\Omega(\alpha) = \left\{u \in \mathbb{R}^m / - q_2(\alpha) \le u \le q_1(\alpha); q_1(\alpha), q_2(\alpha) \in \mathbb{R}^m_+\right\} \ . \tag{5.67}$$

In this section, we consider the problem of finding a controller which respects the constraints and exponentially stabilizes the system Σ by using a control law of the form

$$u(t) = F(r(t))x(t) . \tag{5.68}$$

The closed-loop system is given as

$$\dot{x}(t) = (A(r(t)) + B(r(t))F(r(t))x(t) = A_c(r(t))x(t) . \tag{5.69}$$

5.3.2 Preliminary Results

In this section, a new exponential stability condition of the system (5.69) is given in the unconstrained case. We first begin by recalling the definition of exponential stability as known in the literature.

Definition 5.3. *The system (5.69) is said:*

1. *Stochastically stable if there exists a positive scalar μ such that:* $E\left[\int_0^\infty \|x(t)\|^2 dt\right] \leq \mu(r_0, x_0)$.
2. *Exponentially stable if there exists two scalars $\mu > 0, \sigma > 0$ such that:* $E[\|x(t)\|] \leq \mu e^{-\sigma t}$.
3. *Mean square stable if* $\lim_{t\to\infty} E[\|x(t)\|^2] = 0$.

The following result of stochastic stability is now well known in the literature [53, 52].

Theorem 5.4. *The unsaturated system (5.69) is stochastically stable if there exist symmetric and positive definite matrices $P(\alpha)$, $\alpha \in \mathscr{I}$ $(P(\alpha) \succ 0)$, such that the following holds:*

$$A_c^T(\alpha)P(\alpha) + P(\alpha)A_c(\alpha) + \sum_{\beta\in\mathscr{I}} \lambda_{\alpha\beta}P(\beta) < 0, \forall \alpha \in \mathscr{I} . \tag{5.70}$$

Note that this result was obtained by using the quadratic Lyapunov function and is a necessary and sufficient condition of mean square stability. A second result can also be obtained by using a non quadratic Lyapunov function.

Theorem 5.5. *The unsaturated system (5.69) is exponentially stable if for all $\alpha \in \mathscr{I}$ there exist positive vectors $w(\alpha) \in \mathbb{R}^{n+}; \alpha = 1, \ldots, N$ such that*

$$\hat{A}_c(\alpha)w(\alpha) + \left[\sum_{\beta=1}^{N} \lambda_{\alpha\beta} \max_{1\leq i\leq n}\left(\frac{w_i(\alpha)}{w_i(\beta)}\right)\right] w(\alpha) < 0; \forall \alpha \in \mathscr{I} \tag{5.71}$$

with

$$\hat{A}_c(\alpha)(i,j) = \begin{cases} |A_c(i,j)(\alpha)| & \text{if, } i \neq j \\ A_c(i,i)(\alpha) & \text{if, } i = j \end{cases} . \tag{5.72}$$

Proof: Consider the following Lyapunov function candidate:

$$V(x(t),r(t)) = max_i \frac{|x_i(t)|}{w_i(r(t))} . \tag{5.73}$$

The infinitesimal operator is computed by

$$\begin{aligned} \mathscr{A}V(x(t),\alpha) &= \lim_{h\to 0}\frac{1}{h}\{E\left[V(x(t+h),r(t+h))|r(t)=\alpha\right] - V(x(t),\alpha)\} \quad (5.74)\\ &= \lim_{h\to 0}\frac{1}{h}\left\{E\left[max_i\frac{|x_i(t+h)|}{w_i(r(t+h))}|r(t)=\alpha\right] - V(x(t),\alpha)\right\} . \end{aligned}$$

We use the fact that

$$\begin{aligned} x_i(t+h) &= x_i(t) + h\left[A_c(r(t))x(t)\right]_i + o_i(h)\\ &= \left[L(r(t),h)x(t)\right]_i + o_i(h) \end{aligned}$$

where $L(r(t),h) = I_n + hA_c(r(t))$, I_n is the $n\times n$ identity matrix and $o_i(h)$ a scalar which satisfies $\lim_{h\to 0}\frac{o_i(h)}{h} = 0$. It follows that one can write

$$\mathscr{A}V(x(t),\alpha) \le \lim_{h\to 0}\frac{1}{h}\left\{E\left[max_i\frac{|\left[L(\alpha,h)x(t)\right]_i|}{w_i(r(t+h))}|r(t)=\alpha\right] - V(x(t),\alpha)\right\} .$$

Taking account of (5.66), one obtains

$$\mathscr{A}V(x(t),\alpha) \le \lim_{h\to 0}\frac{1}{h}\left\{\sum_\beta h\lambda_{\alpha\beta}max_i\frac{|\left[L(\alpha,h)x(t)\right]_i|}{w_i(\beta)} + max_i\frac{|\left[L(\alpha,h)x(t)\right]_i|}{w_i(\alpha)} - V(x(t),\alpha)\right\} .$$

Let $l_{ij}(\alpha,h), i = 1,\ldots,n, j = 1,\ldots,n$ be the components of matrix $L(\alpha,h)$, that is,

$$l_{ij}(\alpha,h) = \begin{cases} 1 + hA_c(i,j)(\alpha) & \text{if, } i = j\\ hA_c(i,j)(\alpha) & \text{if, } i \ne j \end{cases} . \tag{5.75}$$

It follows that

$$\begin{aligned} \mathscr{A}V(x(t),\alpha) &\le \sum_{\beta\in\mathscr{I}}\lambda_{\alpha\beta}max_i\frac{|x_i(t)|}{w_i(\beta)} + lim_{h\to 0}\frac{1}{h}\left\{max_i\frac{\sum_j|l_{ij}(\alpha,h)||x_j(t)|}{w_i(\alpha)} - V(x(t),\alpha)\right\}\\ &\le \sum_{\beta\in\mathscr{I}}\lambda_{\alpha\beta}max_i\frac{w_i(\alpha)}{w_i(\beta)}V(x(t),\alpha) + lim_{h\to 0}\frac{1}{h}\left\{max_i\frac{\sum_j|l_{ij}(\alpha,h)|w_j(\alpha)}{w_i(\alpha)} - 1\right\}V(x(t),\alpha) . \end{aligned}$$

Using (5.75),

$$\sum_j \frac{|l_{ij}(\alpha,h)|w_j(\alpha)}{w_i(\alpha)} = \sum_{j\neq i} h\frac{|A_c(i,j)(\alpha)|}{w_i(\alpha)}w_j(\alpha) + |1+hA_c(i,i)(\alpha)|$$
$$= \sum_{j\neq i} h\frac{|A_c(i,j)(\alpha)|}{w_i(\alpha)}w_j(\alpha) + 1+hA_c(i,i)(\alpha),$$

for a small h.

According to (5.72), this latter enables one to have

$$\mathscr{A}V(x(t),\alpha) \leq \left\{\sum_{\beta\in\mathscr{I}} \lambda_{\alpha\beta} max_i\left(\frac{w_i(\alpha)}{w_i(\beta)}\right) + max_i\frac{\left(\hat{A_c}(\alpha)w(\alpha)\right)_i}{w_i(\alpha)}\right\} V(x(t),\alpha)$$

Pose,

$$\mu(\alpha) = -\left\{\sum_{\beta\in\mathscr{I}} \lambda_{\alpha\beta} max_i\left(\frac{w_i(\alpha)}{w_i(\beta)}\right) + max_i\frac{\left(\hat{A_c}(\alpha)w(\alpha)\right)_i}{w_i(\alpha)}\right\}. \tag{5.76}$$

Note that if condition (5.71) holds, $\mu(\alpha) > 0$. This leads to

$$\mathscr{A}V(x(t),\alpha) < -\mu(\alpha)V(x(t),\alpha) . \tag{5.77}$$

That is,

$$E\left[\mathscr{A}V(x(t),\alpha)\right] < -\mu E\left[V(x(t),\alpha)\right],$$

with

$$\mu = \min_{\alpha\in\mathscr{I}} \mu(\alpha).$$

Applying the Dynkin's formula,

$$V(x(t),r(t)) = V(x_0,r_0) + \int_0^t \mathscr{A}V(x(t),r(s))ds,$$

one obtains

$$E\left[V(x(t),r(t)\right] \leq E\left[V(x_0,r_0\right] + \int_0^t E\left[\mathscr{A}V(x(s),r(s))\right]ds$$
$$< E\left[V(x_0,r_0\right] - \mu\int_0^t E\left[V(x(s),r(s))\right]ds .$$

By virtue of Gronwall lemma, we have

$$E\left[V(x(t),r(t)\right] < e^{-\mu t}E\left[V(x_0,r_0)\right] . \tag{5.78}$$

Finally, (5.78) implies that $E[\|x(t)\|] < \sigma e^{-\mu t}$ with:
$\|x(t)\| = \max_i |x_i(t)|$ and $\sigma = \max_{\alpha \in \mathscr{I}} \min_i w_i(\alpha) E[V(x_0, r_0)]$. □

Remark 5.3. *One can note the similarity between the results of Theorems 5.4 and 5.5. In case of* $n = 1$, *condition (5.71) becomes* $a_c(\alpha) + \sum_\beta \lambda_{\alpha\beta} \frac{w(\alpha)}{w(\beta)} < 0$, *where the term* $\sum_\beta \lambda_{\alpha\beta} \frac{w(\alpha)}{w(\beta)}$ *is not necessarily positive and condition (5.70),* $a_c(\alpha) + \frac{1}{2} \sum_\beta \lambda_{\alpha\beta} \frac{p(\beta)}{p(\alpha)} < 0$. *That is, condition (5.71) is more restrictive than (5.70), however, it needs the computation of only* $N\ n$ *parameters instead of* $N\ n^2$ *parameters for (5.70). Besides, if* $N = 1$, *condition (5.71) becomes* $\hat{A}_c w < 0$ *which is a sufficient condition of asymptotic stability.*

An important consequence of Theorem 5.5 is to obtain the union of all the level sets, associated to each mode, as a set of stochastic stability.

For this define the following level set:

$$\mathscr{S}_{r(t)} = \left\{ x \in \mathbb{R}^{n \times n} | V(x(t), r(t)) = \max_{1 \le i \le n} \left\{ \frac{|x(t)|_i}{w_i(r(t))} \right\} < 1 \right\}. \tag{5.79}$$

Theorem 5.6. *If condition (5.71) holds, then the set* $\bigcup_{\alpha=1}^{N} \mathscr{S}_\alpha$ *given by (5.79) is a set of stochastic stability of system (5.69).*

Proof: Assume that at time t, $x(t) \in \mathscr{S}_\alpha$. Recall that one can only compute $E[V(x(t+h), r(t+h))]$ for a small h, and at time $t+h$ the system may jump to any mode $r(t+h) = \beta$. It follows from (5.74) and (5.77) that

$$E[V(x(t+h), r(t+h)) | x(t), r(t)] \le (1 - h\mu) V(x(t), r(t)), \tag{5.80}$$

with $\mu = \min_i \mu_i$. If condition (5.71) holds, $\mu > 0$, then $1 - h\mu < 1$ for any small h which means that $E[V(x(t+h), r(t+h)) | x(t), r(t)] \le V(x(t), r(t))$. Since $x(t) \in \mathscr{S}_\alpha$, $V(x(t), r(t)) < 1$. It follows that $E[V(x(t+h), r(t+h)) | x(t), r(t)] < 1$. Consequently, $x(t+h) \in \mathscr{S}_\beta$. Finally, the set $\bigcup_{\alpha=1}^{N} \mathscr{S}_\alpha$ is a set of stochastic stability of system (5.69). □

5.3.3 Controller Design

In this section, the obtained results in the previous section allow us to deal with the problem of continuous-time systems with Markovian jumping parameters and saturated control.

Recall that the control law is given by (5.68) and each domain $\Omega(\alpha)$ generates by this feedback law a polyhedral domain in the state space $\mathscr{K}(\alpha)$ defined as

$$\mathscr{K}(\alpha) = \left\{ x(t) \in \mathbb{R}^n / -q_2(\alpha) \le F(\alpha) x \le q_1(\alpha); q_1(\alpha), q_2(\alpha) \in \mathbb{R}_+^m \right\}. \tag{5.81}$$

Let $\mathscr{K}_c$ be the common set of all the modes

$$\mathscr{K}_c = \bigcap_{\alpha \in \mathscr{I}} \mathscr{K}(\alpha) = \{x \in \mathbb{R}^n | -\omega_2 \leq F(\alpha)x \leq \omega_1\} \tag{5.82}$$

with,

$$\omega_j = \min_{\alpha \in \mathscr{I}} q_j(\alpha)\, ; j = 1,2\, ;\ \omega = \begin{bmatrix} \omega_1 \\ \omega_2 \end{bmatrix} . \tag{5.83}$$

The closed-loop system is given by (5.69).

Let us introduce the following change of variable,

$$z(t) = F(\alpha)x(t)\ ,\ \ \alpha \in \mathscr{I}\ . \tag{5.84}$$

In this case, for each $\alpha \in \mathscr{I}$ the domain $\mathscr{K}(\alpha)$ is transformed into the following domain:

$$\mathscr{D}(\alpha) = \{z \in \mathbb{R}^m | -q_2(\alpha) \leq z \leq q_1(\alpha), q_1(\alpha), q_2(\alpha) \in \mathbb{R}^m_+\}$$

Let

$$\begin{aligned} \mathscr{D}_c &= \bigcap_{\alpha \in \mathscr{I}} \mathscr{D}(\alpha) \\ &= \{z \in \mathbb{R}^m | -\omega_2 \leq z \leq \omega_1\} \end{aligned}$$

where ω_1 and ω_2 are defined by (5.83). According to (5.84), we obtain

$$\dot{z}(t) = F(\alpha)\,[A(\alpha) + B(\alpha)F(\alpha)]\,x(t)\ . \tag{5.85}$$

Then, if there exists matrix $H(\alpha) \in \mathbb{R}^{m\times m}$, for each $\alpha \in \mathscr{I}$ such that

$$F(\alpha)\,[A(\alpha) + B(\alpha)F(\alpha)] = H(\alpha)F(\alpha) \tag{5.86}$$

the dynamical system (5.69), for each mode $r(t) = \alpha$, $(\alpha \in \mathscr{I})$, is transformed by the use of (5.84) to the following dynamical system:

$$\dot{z}(t) = H(\alpha)z(t)\ . \tag{5.87}$$

Note that the variable $z(t)$ is identical to the control $u(t)$ while the set $\mathscr{D}(\alpha)$ is the same as the set of constraints $\Omega(\alpha)$. At this step, if domain $\mathscr{D}_c$ is positively invariant w.r.t system (5.87) for each mode $\alpha \in \mathscr{I}$, then the control (5.68) will be always admissible, i.e., $u(t) \in \Omega, \forall t \geq 0$ and therefore, the linear behavior of the system in the closed-loop (5.69) remains valid.

We now state the main stability result of this section.

Theorem 5.7. *If, for each $\alpha \in \mathscr{I}$, there exist matrices $H(\alpha) \in \mathbb{R}^{m\times m}$ and positive definite matrices $P(\alpha) \in \mathbb{R}^{n\times n}$, such that:*

(i)

$$F(\alpha)\left[A(\alpha)+B(\alpha)F(\alpha)\right] = H(\alpha)F(\alpha), \tag{5.88}$$

(ii)

$$\tilde{H}_c(\alpha)\omega \leq 0, \tag{5.89}$$

(iii)

$$A_c^T(\alpha)P(\alpha)+P(\alpha)A_c(\alpha)+\sum_{\beta\in\mathscr{I}}\lambda_{\alpha\beta}P(\beta) < 0 \tag{5.90}$$

where ω is given by (5.83).

Then, system (5.64) with (5.68) is stochastically stable while the control is admissible $\forall x_0 \in \mathscr{K}_c$.

Proof: Equation (5.88) allows to transform system (5.69) for each mode $r(t) = \alpha$, $\alpha \in \mathscr{I}$ to system (5.87). According to Theorem 1.6, condition (5.89) guarantees the positive invariance of domain $\mathscr{D}_c$ w.r.t system (5.87). That is, the positive invariance of domain $\mathscr{K}_c$ w.r.t system (5.69) is also guaranteed for each mode $r(t) = \alpha$, $\alpha \in \mathscr{I}$. Further, by virtue of Theorem 5.1, condition (5.90) ensures the stochastic stability of the system in the closed-loop (5.69) which is always valid since the control by state feedback is always admissible. □

Corollary 5.4. *If, for each $\alpha \in \mathscr{I}$, there exist matrices $H(\alpha) \in \mathbb{R}^{m\times m}$ and vectors $w(\alpha) \in \mathbb{R}^n_+$, such that conditions (5.88), (5.89) and (5.71) hold, then, system (5.64) with (5.68) is exponentially stable while the control is admissible $\forall x_0 \in \mathscr{K}_c$.*

Proof: Obvious.

Comment 5.1.

- *It is worth noting that these results concern the positive invariance property of the common polyhedral set $\mathscr{K}_c$ to all the modes $\alpha \in \mathscr{I}$, in the deterministic sense, using the fact that if this common set $\mathscr{K}_c$ is positively invariant for each mode, then all the trajectories of the system with Markovian jumping parameters cannot leave this set at any random jump for all $x_0 \in \mathscr{K}_c$. Further, these results guarantee the exponential stability and the stochastic stability respectively of the system with Markovian jumping parameters. Thus, the stability condition (5.90) and (5.71) are to be satisfied with matrices in the closed-loop $A_c(\alpha)$ where the interdependence between all the modes is explicit.*
- *It is also worth noting that two possibilities can be followed to use the results of Theorem 5.7 and Corollary 5.4. The first consists in computing feedback controls dependent on the jumping process by using the LMI (5.90), computing matrices*

$H(\alpha)$ according to (5.88) and testing (5.89) for each mode. The second consists in giving N matrices $H(\alpha)$ according to (5.89), computing the feedback controls by solving Eq. (5.88) which leads to gain feedback initially independent of the jumping process. The obtained matrices in closed loop are then used to compute N definite positive matrices $P(\alpha)$ for (5.90) (i.e., N n^2 parameters) or only N positive vectors $w(\alpha)$ for (5.71) (i.e., N n parameters). The resolution of the LMI (5.90) is available in MATLAB.

- *Note that no assumption on the stabilizability of each mode is needed even in the presence of saturation. This means that if the stochastic stability is ensured for the system with Markovian jumping parameters, one or more mode can always be unstable with admissible controls ensured by (5.89). For the unstable mode, the resolution of Eq. (5.88) leads to a spectrum in closed loop composed with m stable eigenvalues of matrix H and $n-m$ non necessary all stable eigenvalues of matrix A.*

The steps of constructing such controllers are summarized in the following algorithm.

Algorithm 5.2. *Step 1: Compute matrices $H_1(\alpha)$ and $H_2(\alpha)$ satisfying (5.89) by solving the following linear programming for each $\alpha \in \mathscr{I}$:*

$$(LP1)\begin{cases} \min \varepsilon / \\ \begin{bmatrix} H_1 \; H_2 \\ H_2 \; H_1 \end{bmatrix} \begin{bmatrix} g_1 \\ g_2 \end{bmatrix} \leq -\varepsilon \begin{bmatrix} \varphi_1 \\ \varphi_2 \end{bmatrix} \\ \varepsilon > 0 \\ H_1(i,i) < 0, H_2(i,i) = 0; H_1(i,j) > 0, H_2(i,j) > 0, i \neq j \end{cases} \tag{5.91}$$

where φ_1 and φ_2 are design positive vectors. If matrices $H(\alpha) = H_1(\alpha) - H_2(\alpha)$ are obtained according to the required assumptions of the resolution of Eq. (5.88) (i.e., $\sigma(H)(\alpha) \bigcap \sigma(A(\alpha)) = \emptyset$) continue, else change φ_1 and φ_2 .

Step 2: Compute the gain matrices $F(\alpha)$ solutions of Eqs. (5.88), by using the resolution given in [16] and recalled in Chapter II, and matrices $A_c(\alpha)$ of the system in closed loop. Go to Step 3 *or Step* 4.

Step 3: Compute matrices $P(\alpha)$ solutions of the LMIs (5.90). If the LMIs are not feasible go to Step 1 to modify the vectors $\varphi(\alpha)$.

Step 4: Compute vectors $w(\alpha)$ solution of the following linear programming for each $\alpha \in \mathscr{I}$:

$$(LP2)\begin{cases} \min \mu / \\ \hat{A}_c(\alpha)w(\alpha) + \left(\sum_{\beta=1}^{N} \lambda_{\alpha\beta}\Gamma(\alpha,\beta)\right) w(\alpha) < -\mu\, \psi(\alpha) \\ w(\alpha) \leq \Gamma(\alpha,\beta)\, w(\beta); \forall \beta \in \mathscr{I} \\ \mu > 0 \end{cases} \tag{5.92}$$

where vectors $\psi(\alpha)$ and matrix Γ are design positive parameters with $\Gamma(i,i) =$ 1; $\Gamma(i,j)\Gamma(j,i) > 1; i = 1,\dots N; j = 1,\dots N$. If the problem is feasible stop, if not go to Step 1 *to obtain a new matrix H.*

Theorem 5.7 can be improved, in the case of using the augmentation technique. This technique is based on the augmentation of the system entries to make the system square, *i.e.*, the number of inputs equals the number of states, $m = n$, by introducing $(n-m)$ null columns in matrix B corresponding to $n-m$ fictitious entries, $v(t)$, with its saturations, $\delta(\alpha)$. This technique is described in Chapters $IandII$. For this define

$$\begin{aligned} g_1(\alpha) &= \begin{bmatrix} q_1(\alpha) \\ \delta_1(\alpha) \end{bmatrix}, g_2(\alpha) = \begin{bmatrix} q_2(\alpha) \\ \delta_2(\alpha) \end{bmatrix}, g(\alpha) = \begin{bmatrix} g_1(\alpha) \\ g_2(\alpha) \end{bmatrix}, \\ \bar{g} &= \begin{bmatrix} \bar{g}_1 \\ \bar{g}_2 \end{bmatrix}, \bar{g}_j = \min_{\alpha \in \mathscr{I}} g_j(\alpha)\,; j = 1,2 \end{aligned} \tag{5.93}$$

with δ_1, δ_2 some fictitious saturation on the fictitious control used as parameters design. In this case, domain $\mathscr{K}(\alpha)$ can be rewritten equivalently as

$$\begin{aligned} \mathscr{K}(\alpha) &= \left\{x(t) \in \mathbb{R}^n / -g_2(\alpha) \leq F(\alpha)x \leq g_1(\alpha); g_1(\alpha), g_2(\alpha) \in \mathbb{R}^n_+\right\}, \\ &= \left\{x(t) \in \mathbb{R}^n | V(x(t), r(t)) = \max_{1 \leq i \leq n} \left\{ \frac{|F(\alpha)x(t)|_i}{g_i(\alpha)} \right\} < 1 \right\}. \end{aligned} \tag{5.94}$$

The following result can be announced.

Corollary 5.5. *If, for each $\alpha \in \mathscr{I}$, there exist matrices $H(\alpha) \in \mathbb{R}^{n\times n}$ such that:*

(i)

$$F(\alpha)\left[A(\alpha) + B(\alpha)F(\alpha)\right] = H(\alpha)F(\alpha), \tag{5.95}$$

(ii)

$$\tilde{H}_c(\alpha)\bar{g} \leq 0, \tag{5.96}$$

(iii)

$$\hat{A}_c(\alpha)g(\alpha) + \left[\sum_{\beta=1}^{N} \lambda_{\alpha\beta} \max_{1 \leq i \leq n} \left(\frac{g_i(\alpha)}{g_i(\beta)}\right)\right] g(\alpha) < 0; \forall \alpha \in \mathscr{I}\,. \tag{5.97}$$

Then, the augmented system (5.64) with (5.68) is stochastically stable while the control is admissible $\forall x_0 \in \bigcup_{\alpha=1}^{N} \mathscr{K}(\alpha)$.

Proof: Follows readily by using Theorem 5.6. □

Remark 5.4. *It is worth noting that when the system is augmented, matrix $F(\alpha)$ becomes square and non singular enabling function*

$$\vartheta(x(t), r(t)) = \max_{1 \leq i \leq n} \left\{ \frac{|F(\alpha)x(t)|_i}{g_i(\alpha)} \right\},$$

to become Lyapunov function candidate (i.e. positive definite), for both discrete-time and continuous-time systems.

Example 5.2. *Consider the following linear continuous-time system with Markovian jumping parameters with 2 modes:*

- *mode 1:* $A(1) = \begin{bmatrix} 2 & 1.5 \\ 0 & -2.5 \end{bmatrix}, B(1) = \begin{bmatrix} 1 \\ -1 \end{bmatrix}; q_1(1) = 5, q_2(1) = 10$
- *mode 2:* $A(2) = \begin{bmatrix} 3 & 0 \\ 0.1 & -2.1 \end{bmatrix}, B(2) = \begin{bmatrix} 0.5 \\ 1 \end{bmatrix}; q_1(2) = 10, q_2(2) = 20$

The Markovian process is described by its matrix transition given as

$$\Pi = \begin{bmatrix} -2 & 2 \\ 3 & -3 \end{bmatrix}.$$

For this example, we have $n - m = 1$. The resolution of the linear programming (LP1) leads to the following scalars $H(\alpha)$, $\alpha = 1,2$; $\varphi_1 = [200 \;\; 2.4]^T$ and $\varphi_2 = [155 \;\; 22]^T$:

$$H(1) = -0.495 \,,\; H(2) = -1.97 \,.$$

The resolution of the algebraic equation $XA + XBX = HX$ gives the following gain matrices $F(\alpha)$ and $A_c(\alpha)$:

$$F(1) = \begin{bmatrix} -3.7425 & -1.2475 \end{bmatrix}, F(2) = \begin{bmatrix} -9.94 & 0 \end{bmatrix}$$
$$A_c(1) = \begin{bmatrix} -1.7425 & 0.2525 \\ 3.7425 & -1.2525 \end{bmatrix}, A_c(2) = \begin{bmatrix} -1.97 & 0 \\ -9.84 & -2.1 \end{bmatrix}.$$

The resolution of the LMIs (5.90) gives the following matrices:

$$P(1) = \begin{bmatrix} 1.6502 & 0.3849 \\ 0.3849 & 0.6423 \end{bmatrix}, P(2) = \begin{bmatrix} 2.2398 & -0.455 \\ -0.455 & 0.4630 \end{bmatrix}.$$

Further, the exponential stability condition (5.71) of Corollary 5.4 is also satisfied with

$$w(1) = \begin{bmatrix} 0.585 & 3 \end{bmatrix}^T, w(2) = \begin{bmatrix} 0.571 & 2.67 \end{bmatrix}^T.$$

That is,

$$max_i \left(\frac{[\hat{A}_c(1)w(1)]_i}{w_i(1)} \right) + \sum_{\beta \in \mathscr{I}} \lambda_{1\beta} max_i \left(\frac{w_i(1)}{w_i(\beta)} \right) = -0.297$$
$$max_i \left(\frac{[\hat{A}_c(2)w(2)]_i}{w_i(2)} \right) + \sum_{\beta \in \mathscr{I}} \lambda_{2\beta} max_i \left(\frac{w_i(2)}{w_i(\beta)} \right) = -0.171 \,.$$

The results of simulation are given as follows: the first figure of Fig. 5.1 presents the common set of positive invariance and stochastic stability $\mathscr{K}_c$ with a trajectory of the system, the second concerns the corresponding jumping Markovian process while the third deals with the admissible control.

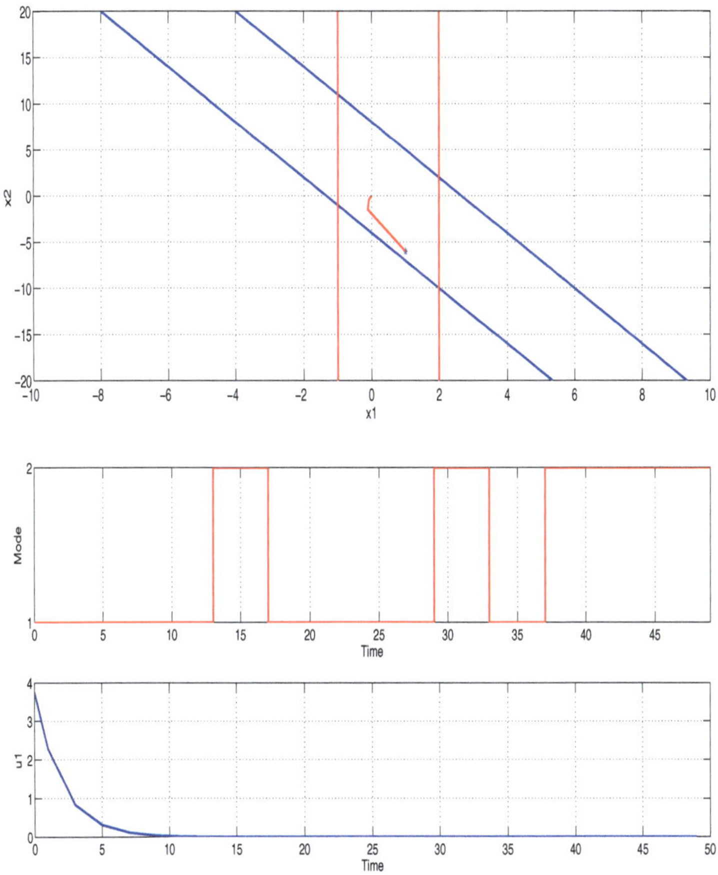

Fig. 5.1 Set of positive invariance and stochastic stability $\mathscr{K}_c$ in the state space (x_1, x_2), a jumping Markovian process in (α, t) and the admissible control

5.4 Complement to the Discrete-Time Case

In this section, we apply the same previous idea to discrete-time system class studied in the first section with non symmetrical constrained control. This extension was presented in [73]. The constraint control set, $\Omega(\alpha)$ for each mode is now given as

$$\Omega(\alpha) = \left\{ u \in \mathbb{R}^m / -q_2(\alpha) \leq u \leq q_1(\alpha); q_1(\alpha), q_2(\alpha) \in \mathbb{R}^{m+} \right\} . \quad (5.98)$$

5.4.1 Preliminary Result

The following result of stochastic stability is now well known in the literature [80, 52].

Theorem 5.8. *The unsaturated system (5.5) is stochastically stable if there exist symmetric and positive definite matrices $P(\alpha)$, $\alpha \in \mathscr{I}$ ($P(\alpha) \succ 0$), such that the following holds:*

$$A_c(\alpha) \sum_{\beta \in \mathscr{I}} \lambda_{\alpha\beta} P(\beta) A_c^T(\alpha) - P(\alpha) < 0, \forall \alpha \in \mathscr{I} \,. \tag{5.99}$$

5.4.2 Controller Design

The application of this result to the closed-loop system with non symmetrical constrained control and Markovian jumping parameters leads to the following result which is applied to the common set defined exactly as for continuous-time systems.

Theorem 5.9. *If, for each $\alpha \in \mathscr{I}$, there exist matrices $H(\alpha) \in \mathbb{R}^{m\times m}$ and positive definite matrices $P(\alpha) \in \mathbb{R}^{n\times n}$, such that:*

(i)

$$F(\alpha)\left[A(\alpha) + B(\alpha)F(\alpha)\right] = H(\alpha)F(\alpha), \tag{5.100}$$

(ii)

$$\tilde{H}_d(\alpha)\omega \leq \omega, \tag{5.101}$$

(iii)

$$A_c(\alpha) \sum_{\beta \in \mathscr{I}} \lambda_{\alpha\beta} P(\beta) A_c^T(\alpha) - P(\alpha) < 0, \forall \alpha \in \mathscr{I} \tag{5.102}$$

where ω is given by (5.83).

Then, system (5.5) with (5.98) is stochastically stable while the control is admissible $\forall x_0 \in \mathscr{K}_c$.

Proof: Follows readily from above and by using the same reasoning as for Theorem 1.1 in the proof of Corollary 5.1 to lead to condition 5.101. □

In the same way, one can also use the following corollary.

Corollary 5.6. *If, for each $\alpha \in \mathscr{I}$, there exist matrices $H(\alpha) \in \mathbb{R}^{m\times m}$ and vectors $w(\alpha) \in \mathbb{R}^{n+}$, such that conditions (5.100), (5.101) and (5.44) hold, then, system (5.5) with (5.98) is exponentially stable while the control is admissible $\forall x_0 \in \mathscr{K}_c$.*

Proof: Obvious. □

The results of Theorem 5.9 and Corollary 5.6 can be applied to square systems by using the augmentation technique. In this case, the set $\mathscr{F}_u$ given by (5.60) is a set of positive invariance and stochastic stability as shown by the following example.

Example 5.3. *Consider the following linear continuous-time system with Markovian jumping parameters with 2 modes:*

- *mode 1:* $A(1) = \begin{bmatrix} 1 & 1 \\ -1 & 0.5 \end{bmatrix}, B(1) = \begin{bmatrix} 1 \\ 1 \end{bmatrix}; q_1(1) = 3, q_2(1) = 6$
- *mode 2:* $A(2) = \begin{bmatrix} -1.5 & 1 \\ -1 & 0.33 \end{bmatrix}, B(2) = \begin{bmatrix} 0.5 \\ 1 \end{bmatrix}; q_1(2) = 10, q_2(2) = 5$.

The Markovian process is described by its matrix transition given as

$$\Pi = \begin{bmatrix} 0.67 & 0.33 \\ 0.35 & 0.65 \end{bmatrix} .$$

The introduced fictitious saturations are $\delta(1) = [10 \;\; 1]^T$ *for mode* 1 *and* $\delta(2) = [5 \;\; 10]^T$ *for mode* 2*. Matrices* H_1 *and* H_2 *are chosen as*

$$H(1) = \begin{bmatrix} -0.08 & 0.084 \\ 0.07 & 0.133 \end{bmatrix} H(2) = \begin{bmatrix} -0.066 & 0.052 \\ 0.02 & 0.16 \end{bmatrix} .$$

The resolution of the algebraic equation $XA + XBX = HX$ *gives the following gain matrices* $F(\alpha)$ *and* $A_c(\alpha)$*:*

$$F(1) = \begin{bmatrix} -0.5510 & -0.8960 \\ 13.0741 & 2.6887 \end{bmatrix}, F(2) = \begin{bmatrix} 1.7185 & -0.8891 \\ 11.8095 & 4.6671 \end{bmatrix}.$$

Further, the exponential stability condition (5.44) of Corollary 5.6 is also satisfied with

$$w(1) = \begin{bmatrix} 0.8 & 4.2 \end{bmatrix}^T, w(2) = \begin{bmatrix} 1 & 5 \end{bmatrix}^T .$$

The results of simulation are given as follows: the first figure of Fig. 5.2 presents the jumping Markovian process while the second deals with the admissible control. The third concerns the union set of positive invariance and stochastic stability $\mathscr{F}_u$ *given by (5.60) with a trajectory of the system.*

In this section, a sufficient condition of exponential stability is obtained by using the non quadratic Lyapunov function as is usually the case in the problems with constraints of inequality type. The result obtained in Chapters $I - II$ is used to guarantee the positive invariance of the common domain $\mathscr{K}_c$ w.r.t all the modes of the system. However, the quadratic condition of stochastic stability can also be used leading to a useful algorithm based on the resolution of the LMIs (5.90) available in Matlab.

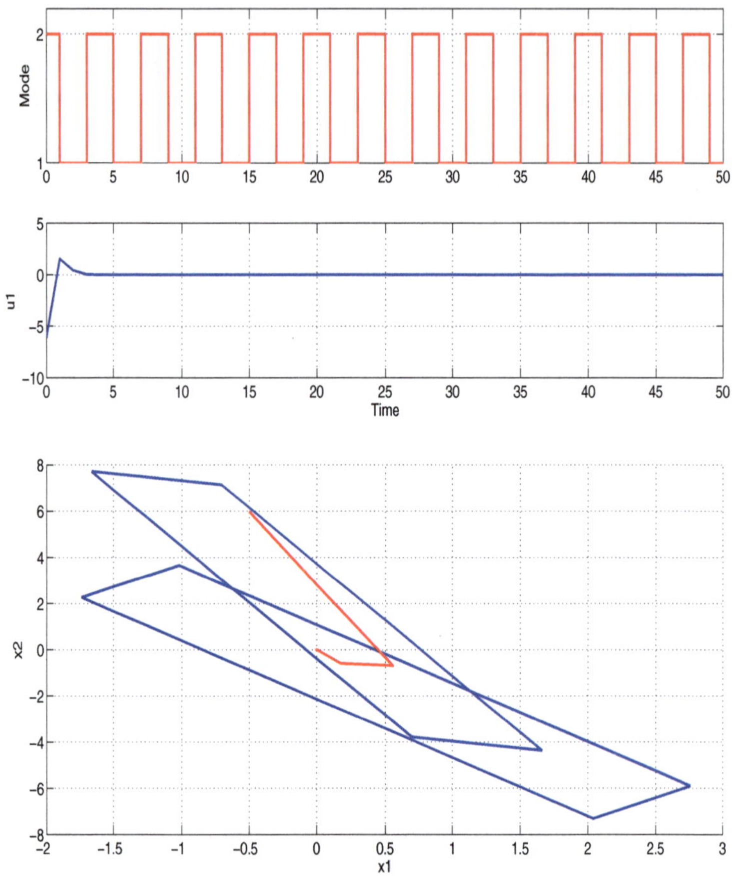

Fig. 5.2 A jumping Markovian process, the admissible control and the set of positive invariance and stochastic stability $\mathscr{F}_u$ given as (5.60) with a trajectory

5.5 Conclusion

In this chapter, necessary and sufficient conditions for domain $\mathscr{F}$(resp. $\mathscr{K}$) to be positively invariant w.r.t the system in the closed-loop (5.5) (resp. Σ) are established for discrete-time (resp. continuous-time) saturated Markovian switching systems. Sufficient conditions of stochastic stability are deduced. These results are obtained by using non quadratic Lyapunov functions as is usually the case in the problems with constraints of inequality type. Algorithms are also presented to compute matrices $H(\alpha)$. Illustrative examples are also presented. The case of obtaining the set $\bigcup_{\alpha=1}^{N} \mathscr{F}(\alpha)$ (resp. $\bigcup_{\alpha=1}^{N} \mathscr{K}(\alpha)$) as a set of stochastic stability is deduced for discrete-time (resp. continuous-time) systems.

Chapter 6
Stabilization of Saturated Switching Systems

6.1 Introduction

In this chapter, two directions are explored: the first concerns the synthesis of non saturating controllers, while the second direction deals with controllers tolerating saturations to take effect. The second method was firstly used in [22] with the use of a multiple Lyapunov function. However, only the intersection of all the corresponding level sets of the local functions was considered as a region of asymptotic stability of the switching system. This drawback is improved in [25, 35] by considering, for the first time, a large set of asymptotic stability composed by the union of all the level sets. In this context, different sufficient conditions of asymptotic stability are obtained for switching systems subject to actuator saturations. Furthermore, these conditions are presented in the form of LMIs for the state feedback control case.

The static output feedback control (SOFC) for dynamical systems plays a very important role in control theory and applications. The purpose is to design controllers such that the resulting closed-loop system is asymptotically stable without using any reconstruction method of the unavailable states. This type of control was already used for switching systems in [66] and extended to saturated switching systems in [25].

6.2 Problem Formulation

In this section, we give a more precise problem statement for the class of systems under consideration, namely, discrete-time switching linear systems with input saturation and state or output feedback. An equivalent description of such systems, based on the indicator function is also used in this work. The main results of this section are published in [35].

A. Benzaouia: Saturated Switching Systems, LNCIS 426, pp. 145–164.
springerlink.com

Thus, we consider systems described by:

$$x(t+1) = A_\alpha x(t) + B_\alpha sat(u(t)) \qquad (6.1)$$
$$y(t) = C_\alpha x(t)$$

where $x \in \mathbb{R}^n$ is the state, $u \in \mathbb{R}^m$ is the control, $y \in \mathbb{R}^p$ is the output, $sat(.)$ is the standard saturation function and α a switching rule which takes its values in the finite set $\mathscr{I} := \{1,\dots,N\}$, $t \in \mathbb{Z}_+$. The saturation function is assumed here to be normalized, *i. e.*, $(|sat(u)| = min\{1,|u|\})$. Each subsystem α is called a mode. It is assumed that:

- The switching system is stabilizable;
- Matrices C_α are of full rank;
- $t_{s+1} \geq t_s + 1, \forall s \in \mathscr{L}$;
- The switching rule is not known *a priori* but $\alpha(t)$ is available at each t.

The third assumption ensures that at each time only one subsystem is active. The fourth assumption corresponds to practical implementations where the switched system is supervised by a discrete-event system or operator allowing for $\alpha(t)$ to be known in real time.

In this work, we are interested by the synthesis of stabilizing controllers for this class of hybrid systems subject to actuator saturation. We use a feedback control law as:

$$u(t) = F_\alpha x(t) = K_\alpha y(t), \qquad (6.2)$$

and write the closed-loop system as

$$\begin{aligned} x(t+1) &= A_\alpha x(t) + B_\alpha sat(F_\alpha x(t)). \qquad (6.3)\\ &= A_\alpha x(t) + B_\alpha sat(K_\alpha y(t)) \end{aligned}$$

Upon introducing the indicator function:

$$\xi(t) = [\xi_1(t),\dots,\xi_N(t)]^T \qquad (6.4)$$

where $\xi_i(t) = 1$ if the switching system is in mode i and $\xi_i(t) = 0$ if it is in a different mode, one can write the closed-loop system (6.3) as follows:

$$\begin{aligned} x(t+1) &= \sum_{i=1}^{N} \xi_i(t)[A_i x(t) + B_i sat(F_i x(t))]. \qquad (6.5)\\ &= \sum_{i=1}^{N} \xi_i(t)[A_i x(t) + B_i sat(K_i C_i x(t))] \end{aligned}$$

6.3 Preliminary Results

Recall the following subsets of $\mathbb{R}^n$ defined in Chapter *1*:

$$\varepsilon(P,\rho) = \{x \in \mathbb{R}^n | x^T P x \leq \rho\}, \tag{6.6}$$

$$\mathscr{L}(F) = \{x \in \mathbb{R}^n || F_l x| \leq 1, l = 1, \ldots, m\}, \tag{6.7}$$

with P a positive definite matrix, $\rho > 0$ and F_l the lth row of the matrix $F \in \mathbb{R}^{m \times n}$. Thus $\varepsilon(P,\rho)$ is an ellipsoid while $\mathscr{L}(F)$ is a polyhedral consisting of states for which the saturation does not occur. Consider the following autonomous switching system:

$$x(t+1) = \sum_{i=1}^{N} \xi_i(t) A_i x(t) \tag{6.8}$$

A multiple Lyapunov function candidate for the system is given by:

$$V(t,x) = x^T(t)(\sum_{i=1}^{N} \xi_i(t) P_i) x(t) \tag{6.9}$$

Conditions of asymptotic stability of the switching system (6.5) without saturation are given by Theorem 3.3, while the controller synthesis are presented by LMIs (3.33) of Theorem 3.4 or LMIs (3.36) of Corollary 3.1.

Recall that a level set of the Lyapunov function $V(t,x)$ given by (6.9) and associated to the switching system (6.8) is given by the set $\varepsilon(P,\rho)$ defined by (6.6) with $P = \sum_{i=1}^{N} \xi_i(t) P_i$. This region of asymptotic stability is very difficult to construct. Nevertheless, an estimate can be obtained as large as possible as presented in the next section based on the union of the sub-level ellipsoid sets $\varepsilon(P_i, 1)$. Other type of level sets obtained with different Lyapunov functions for switched systems can be found in [105]. A useful lemma is also recalled.

Lemma 6.1. *Let R, S and Γ be matrices with appropriate dimension. Suppose $\Gamma^T \Gamma \leq \mathbb{I}$, then for any scalar $\lambda > 0$, we have:*

$$R\Gamma S + S^T \Gamma^T R^T \leq \lambda R R^T + \lambda^{-1} S S^T \tag{6.10}$$

6.4 Analysis and Synthesis of Stabilizability

In this section, the region of local asymptotic stability associated to the saturated switching system is firstly studied. The design of a stabilizing controller for the class of switching system with actuator saturation is then presented by following two ways, the first concerns controllers working inside the region of linear behavior where the saturations do not occur while the second is based on Lemma 2.1 tolerating saturations to occur.

6.4.1 Region of Asymptotic Stability

The link between the result of Theorem 3.3 applied to forced switching systems and the level set (6.6) is not mentioned in [128, 85] nor in [65, 66]. The absence of saturation on the control of the switching systems in these works does not necessitate to take care with the level sets since the asymptotic stability is global. We present here after an interesting result on this subject inspired from the general class studied in Chapter *IV*.

Theorem 6.1. *If there exist N symmetric positive definite matrices $P_1,\ldots,P_N$ such that the inequalities (3.23) are satisfied then the set given by,*

$$\Omega = \bigcup_{i=1}^{N} \varepsilon(P_i, 1) \tag{6.11}$$

is a level set of the multiple Lyapunov function $V(t,x)$ given by (6.9) with respect to the trajectories of the autonomous switching system (6.8).

Proof: Note that condition (3.23) for $i = j$ ensures that $\Delta V(t,x) < 0$, for $t \in I_s, s \in \mathscr{L}$, where the set $\mathscr{L}$ is defined in Chapter *III* [65, 66]. Now let $x(\bar{t}_{k-1}) \in \varepsilon(P_i, 1)$. What happens at the switching time $\bar{t}_{k-1}$?

For this, compute $x(t_k)$. With an identity reset function we have,

$$x(t_k) = x(\bar{t}_{k-1}) \tag{6.12}$$

The switch to the next region $\varepsilon(P_j, 1)$ will be seen an unit time after,

$$\begin{aligned} x(t_k + 1) &= x(\bar{t}_{k-1} + 1) \\ &= A_i x(\bar{t}_{k-1}) \end{aligned} \tag{6.13}$$

We have,

$$x^T(t_k + 1) P_j x(t_k + 1) = x^T(\bar{t}_{k-1}) \left(A_i^T P_j A_i\right) x(\bar{t}_{k-1}) \tag{6.14}$$

According to condition (3.23), one can obtain,

$$x^T(t_k + 1) P_j x(t_k + 1) < x^T(\bar{t}_{k-1}) P_i x(\bar{t}_{k-1}) \tag{6.15}$$

Since $x(\bar{t}_{k-1}) \in \varepsilon(P_i, 1)$, then, $x(t_k + 1) \in \varepsilon(P_j, 1), \forall i, j \in \mathscr{I} \times \mathscr{I}$. This proves that $V(t_k, x(t_k)) < V(t_{k-1}, x(t_{k-1}))$, $\forall k \in \mathscr{L}$. Consequently, function $V(t,x)$ is a multiple Lyapunov function of the switching system according to Definition 3.5 and the set Ω is a level set associated to this function. It is worth to note that the same reasoning holds for $\varepsilon(P_j, \rho)$ for any positive scalar ρ. □

Note that an important remark is to be done at this level: since the studied system is a switching one composed of N subsystems (modes), the switching system can be initialized inside any level set $\varepsilon(P_i, 1)$, however, the initial mode to be selected is the corresponding mode i.

6.4.2 State Feedback Control

We assume that the state is available, so a state feedback control can be performed. The first result we present is a simple extension of known result given by [97] for linear systems and concerns the synthesis of non saturating controllers ensuring that a large region of linear behavior is a region of asymptotic stability.

Theorem 6.2. *If there exists N symmetric matrices $X_1,\ldots,X_N$ and N matrices $Y_1,\ldots,Y_N$ solutions of the following LMIs:*

$$\begin{bmatrix} X_i & (A_iX_i+B_iY_i)^T \\ * & X_j \end{bmatrix} > 0, \tag{6.16}$$

$$\begin{bmatrix} 1 & Y_{li} \\ * & X_i \end{bmatrix} > 0, \tag{6.17}$$

$$\forall (i,j) \in \mathscr{I} \times \mathscr{I}, \forall l \in [1,m]$$

where Y_{li} is the lth row of matrix Y_i; then the switching system with saturations in closed-loop (6.3), with,

$$F_i = Y_iX_i^{-1}, P_i = X_i^{-1} \tag{6.18}$$

is asymptotically stable at the origin $\forall x_0 \in \Omega$ and for any sequences of switching $\alpha(t)$.

Proof: In order to guarantee that the control by state feedback is always admissible, each sub-level set of the Lyapunov function has to be contained inside the polyhedral set where the saturations do not occur $\varepsilon(P_i,1) \subset \mathscr{L}(F_i), \forall i \in \mathscr{I}$ [97]. Using [53], this inclusion condition can also be transformed to the equivalent LMI (6.17) by letting $X_i = P_i^{-1}$ and $F_iX_i = Y_i$. Let $x(t) \in \Omega, \forall t \in I_s$. According to condition (6.17), the system in closed-loop (6.5) can be equivalently written as,

$$x(t+1) = \sum_{i=1}^{N} \xi_i(t) A_{c_i} x(t). \tag{6.19}$$

with, $A_{c_i} = A_i + B_iF_i$. The proof follows from the asymptotic stability conditions of the switching system given by Theorem 3.4. Further, Theorem 6.1 ensures that the set Ω is a set of asymptotic stability of the switching system with saturations in closed-loop (6.3). □

To achieve a domain of attraction as large as possible, we can solve the following optimization problem:

$$(Pb.1): \begin{cases} \sup_{(X_i,Y_i)} Trace(X_i) \\ s.t.\ (6.16),(6.17), \\ \quad i = 1,\ldots,N \end{cases}$$

When this optimization problem is feasible, the obtained ellipsoid volumes are maximum with respect to the data of the system.

The obtained LMIs (6.16) of Theorem 6.2 are similar to those obtained in [128, 85, 65, 66] for non saturated switching systems. The presence of saturation on the control in our problem leads to the additional LMIs (6.17), which will obviously restrict the set of solutions. Nevertheless, the associated large region of asymptotic stability Ω for the saturated switching system enables one to conclude that these LMIs are not conservative.

The second result we present concerns the synthesis of saturating controllers tolerating saturations to take effect inside a large region of asymptotic stability.

Theorem 6.3. *If there exist symmetric positive definite matrices $P_1,\ldots,P_N \in \mathbb{R}^{n\times n}$ and matrices $H_1,\ldots,H_N \in \mathbb{R}^{m\times n}$ such that,*

$$\begin{bmatrix} P_i & [A_i + B_i(D_{is}F_i + D_{is}^- H_i)]^T P_j \\ * & P_j \end{bmatrix} > 0 \ , \qquad (6.20)$$
$$\forall (i,j) \in \mathscr{I} \times \mathscr{I}, \forall s \in [1,\eta]$$

and,

$$\varepsilon(P_i,1) \subset \mathscr{L}(H_i), \forall i \in \mathscr{I} \qquad (6.21)$$

then, the closed-loop switching system (6.3) is asymptotically stable at the origin $\forall x_0 \in \Omega$ and for all sequences of switching $\alpha(t)$.

Proof: Assume that there exist N matrices $H_1,\ldots,H_N$ and N symmetric matrices $P_1,\ldots,P_N$ such that condition (6.20) and (6.21) are satisfied. Using the expression in (2.98) and rewriting System (6.3) as in (6.5) yields that:

$$sat(F_i x(t)) = \sum_{s=1}^{\eta} \delta_{si}(t)[D_{is}F_i + D_{is}^- H_i]x(t) \ ; \qquad (6.22)$$

$$\delta_{si}(t) \geq 0, \ \sum_{s=1}^{\eta} \delta_{si}(t) = 1 \qquad (6.23)$$

and, subsequently:

$$x(t+1) = \sum_{s=1}^{\eta}\sum_{i=1}^{N} \xi_i(t)\delta_{si}(t)Ac_{is}x(t); \qquad (6.24)$$
$$Ac_{is} := A_i + B_i(D_{is}F_i + D_{is}^- H_i), \quad s \in [1,\eta]$$

The rate of change of the Lyapunov function candidate (6.9) along the trajectories of (6.24) is given by:

$$\Delta V(t,x(t)) = x^T(t+1)(\sum_{j=1}^{N} \xi_j(t+1)P_j)x(t+1) - x^T(t)(\sum_{i=1}^{N} \xi_i(t)P_i)x(t)$$

$$= x^T(t)\left[\left(\sum_{s=1}^{\eta}\sum_{i=1}^{N}\xi_i(t)\delta_{si}(t)Ac_{is}\right)^T\left(\sum_{j=1}^{N}\xi_j(t+1)P_j\right)\left(\sum_{s=1}^{\eta}\sum_{i=1}^{N}\xi_i(t)\delta_{si}(t)Ac_{is}\right)\right.$$
$$\left. - \sum_{i=1}^{N}\xi_i(t)P_i\right]x(t)$$

Let condition (6.20) be satisfied. At this level, for each i, multiply the $j = 1,\ldots,N$ inequalities (6.20) by $\xi_j(t+1)$ and sum. Multiply the resulting $i = 1,\ldots,N$ inequalities by $\xi_i(t)$ and sum. Multiply again the resulting $s = 1,\ldots,\eta$ inequalities by $\delta_{si}(t)$ and sum. As $\sum_{i=1}^{N}\xi_i(t) = \sum_{j=1}^{N}\xi_j(t+1) = \sum_{s=1}^{\eta}\delta_{si}(t) = 1$, one gets:

$$\begin{bmatrix} \sum_{i=1}^{N}\xi_i(t)P_i & \Gamma\sum_{j=1}^{N}\xi_j(t+1)P_j \\ * & \sum_{j=1}^{N}\xi_j(t+1)P_j \end{bmatrix} > 0, \tag{6.25}$$

with, $\Gamma = \left[\sum_{s=1}^{\eta}\sum_{i=1}^{N}\xi_i(t)\delta_{si}(t)Ac_{is}\right]^T$. The use of Schur complement allows us to write condition (6.25) under the equivalent form,

$$\left(\sum_{s=1}^{\eta}\sum_{i=1}^{N}\xi_i(t)\delta_{si}(t)Ac_{is}\right)^T\left(\sum_{j=1}^{N}\xi_j(t+1)P_j\right)\left(\sum_{s=1}^{\eta}\sum_{i=1}^{N}\xi_i(t)\delta_{si}(t)Ac_{is}\right)$$
$$- \sum_{i=1}^{N}\xi_i(t)P_i < 0, \forall (i,j) \in \mathscr{I}\times\mathscr{I}, \forall s \in [1,\eta]$$

Letting λ be the largest eigenvalue among all the above matrices, we obtain that

$$\Delta V(t,x(t)) \le \lambda x^T(t)x(t) < 0. \tag{6.26}$$

which ensures the desired result. Noting that condition (6.20) is also satisfied for $i = j$, this implies that each set $\varepsilon(P_i,1)$ is a contractively invariant set for the corresponding subsystem. Further, by taking account of condition (6.21), one can guarantee that for every $x_o \in \varepsilon(P_i,1) \subset \mathscr{L}(H_i)$, each subsystem is asymptotically stable at the origin. Besides, following Theorem 6.1, conditions (6.20)-(6.21) also allow for a state belonging, before the switch, to a set $\varepsilon(P_i,1) \subset \mathscr{L}(H_i)$, if a switch occurs at any time t_k, that the switch will handle the state to the desired set $\varepsilon(P_j,1) \subset \mathscr{L}(H_j)$. That means that the set Ω is a set of asymptotic stability of the switching system. □

This result of stabilizability can be used for control synthesis as follows.

Theorem 6.4. *If there exists N symmetric matrices $X_1,\ldots,X_N$ and $2N$ matrices $Y_1,\ldots,Y_N$ and $Z_1,\ldots,Z_N$ solutions of the following LMIs :*

$$\begin{bmatrix} X_i & (A_iX_i + B_iD_{is}Y_i + B_iD_{is}^-Z_i)^T \\ * & X_j \end{bmatrix} > 0, \tag{6.27}$$

$$\begin{bmatrix} 1 & Z_{li} \\ * & X_i \end{bmatrix} > 0, \qquad (6.28)$$

$$\forall (i,j) \in \mathscr{I} \times \mathscr{I}, \forall s \in [1,\eta], \forall l \in [1,m],$$

then the switching system with saturations in closed-loop (6.3), with,

$$F_i = Y_i X_i^{-1}, H_i = Z_i X_i^{-1}, P_i = X_i^{-1}, \qquad (6.29)$$

is asymptotically stable at the origin $\forall x_0 \in \Omega$ *and for any sequences of switching* $\alpha(t)$.

Proof: The Inequalities (6.20) can be transformed equivalently by Schur complement to the following:

$$P_j \left[A_i + B_i(D_{is}F_i + D_{is}^- H_i)\right] P_i^{-1} \left[A_i + B_i(D_{is}F_i + D_{is}^- H_i)\right]^T P_j - P_j < 0 \qquad (6.30)$$

By pre and post-multiplying the latter by P_j^{-1}, it follows:

$$\left[A_i + B_i(D_{is}F_i + D_{is}^- H_i)\right] P_i^{-1} \left[A_i + B_i(D_{is}F_i + D_{is}^- H_i)\right]^T - P_j^{-1} < 0 \qquad (6.31)$$

Pose $X_i = P_i^{-1}$, inequality (6.31) can then be rewritten as,

$$\left[A_i + B_i(D_{is}F_i + D_{is}^- H_i)\right] X_i \left[A_i + B_i(D_{is}F_i + D_{is}^- H_i)\right]^T - X_j < 0,$$

that is

$$\left[A_i X_i + B_i(D_{is}F_i X_i + D_{is}^- H_i X_i)\right] X_i^{-1} \left[A_i X_i + B_i(D_{is}F_i X_i + D_{is}^- H_i X_i)\right]^T - X_j < 0.$$

The use of the Schur complement a second time leads to:

$$\begin{bmatrix} X_i & (A_i X_i + B_i D_{is} F_i X_i + B_i D_{is}^- H_i X_i)^T \\ * & X_j \end{bmatrix} > 0 \qquad (6.32)$$

By letting $F_i X_i = Y_i$ and $H_i X_i = Z_i$, the LMI (6.27) follows readily. Using [53], the inclusion condition (6.21) can also be transformed to the equivalent LMI (6.28). □

To obtain larger ellipsoid domains $\varepsilon(P_i, 1)$, we can apply the following optimization problem:

$$(Pb.2): \begin{cases} \sup_{(X_i,Y_i,Z_i)} Trace(X_i) \\ s.t. \ (6.27),(6.28), \\ i = 1,\ldots,N \end{cases}$$

When this optimization problem is feasible, the obtained ellipsoid volumes are maximum with respect to the data of the system.

Remark 6.1. *The results presented in this section can be extended easily to level sets $\varepsilon(P_i,\rho_i)$ where the scalars $\rho_i, i=1,\dots,N$ can be a priori fixed, by using the variables $X_i = P_i/\rho_i$ in the LMIs (6.16)- (6.17) and (6.27)-(6.28).*

Comment 6.1. *The resolution of the LMIs (6.27) for $s \in [1,\eta]$, may be very restrictive. One can relax this resolution by accepting controllers working in a region of linear behavior where the saturations do not occur. The obtained results in this work, where the switching system is with saturated control, are a set of $(2^m \times N^2 + Nm)$ LMIs for computing stabilizing controllers working inside the saturated regions $\mathcal{L}(H_i)$ or only a set of (N^2+Nm) LMIs for computing stabilizing controllers working inside regions of linear behavior $\mathcal{L}(F_i)$. Note also that the LMIs of Theorem 6.2 can be obtained from LMIs of Theorem 6.4 by letting $D_{is} = \mathbb{I}$ and $D_{is}^+ = 0$.*

Example 6.1. *Consider a numerical switching discrete-time system with saturation specified by the two subsystems:*

$$A_1 = \begin{bmatrix} -0.7 & 1 \\ -0.5 & -1.5 \end{bmatrix}; B_1 = \begin{bmatrix} 1 \\ 0 \end{bmatrix}; A_2 = \begin{bmatrix} 0.9 & -1 \\ 1.7 & -1.5 \end{bmatrix}; B_2 = \begin{bmatrix} 1 \\ 1 \end{bmatrix}.$$

We have to solve 5 LMIs with 4 variables to compute controllers working inside a region of linear behavior. The use of the Matlab LMI Toolbox to check our conditions leads to the following results for $\rho_1 = \rho_2 = 1$ and without using the optimization problem $(Pb.1)$.

$$\begin{aligned} P_1 &= \begin{bmatrix} 1.0569 & 1.4232 \\ 1.4232 & 2.7574 \end{bmatrix}; P_2 = \begin{bmatrix} 0.0288 & 0.0183 \\ 0.0183 & 0.0498 \end{bmatrix}; \\ F_1 &= \begin{bmatrix} 0.7014 & 0.5009 \end{bmatrix}; F_2 = \begin{bmatrix} -0.1127 & 0.0695 \end{bmatrix}; \\ Ac_1 &= \begin{bmatrix} 0.0014 & 1.5009 \\ -0.5000 & -1.5000 \end{bmatrix}; Ac_2 = \begin{bmatrix} 0.7873 & -0.9305 \\ 1.5873 & -1.4305 \end{bmatrix}. \end{aligned}$$

Figure 6.1 presents the two ellipsoid sets of invariance and contractivity $\varepsilon(P_i,1)$ together with the sets of linear behavior $\mathcal{L}(F_i)$ computed with LMIs (6.16)-(6.17).

For the same example, the use of the Matlab LMI Toolbox to check conditions (6.27)-(6.28) leads to the following results without using the optimization problem $(Pb.2)$ with 10 LMIs and 6 variables.

$$\begin{aligned} P_1 &= \begin{bmatrix} 1.9771 & 2.7897 \\ 2.7897 & 5.7254 \end{bmatrix}; H_1 = \begin{bmatrix} 0.9235 & 0.8069 \end{bmatrix}; \\ F_1 &= \begin{bmatrix} 1.4618 & 1.2283 \end{bmatrix}; Ac_1 = \begin{bmatrix} 0.7618 & 2.2283 \\ -0.5000 & -1.5000 \end{bmatrix}; \\ P_2 &= \begin{bmatrix} 0.1172 & 0.0783 \\ 0.0783 & 0.2114 \end{bmatrix}; H_2 = \begin{bmatrix} -0.2353 & 0.1266 \end{bmatrix}; \end{aligned}$$

$$F2 = \begin{bmatrix} -0.2451 & 0.1310 \end{bmatrix}; Ac_2 = \begin{bmatrix} 0.6549 & -0.8690 \\ 1.4549 & -1.3690 \end{bmatrix}.$$

Figure 6.2 presents the two ellipsoid sets of invariance and contractivity $\varepsilon(P_i, 1)$ *together with the set of saturations* $\mathscr{L}(H_i)$ *with the same color in solid lines and the sets* $\mathscr{L}(F_i)$ *in dotted lines.*

Comparing the results obtained by LMIs (6.27)-(6.28) and LMIs (6.16)-(6.17), one can note that the matrices H_i are very closed up to matrices F_i obtained with LMIs (6.27)-(6.28), that is, $\mathscr{L}(H_i) \cong \mathscr{L}(F_i)$. This means that in the case of this example, the saturations allowed by this technique are not very important. Recall that inside

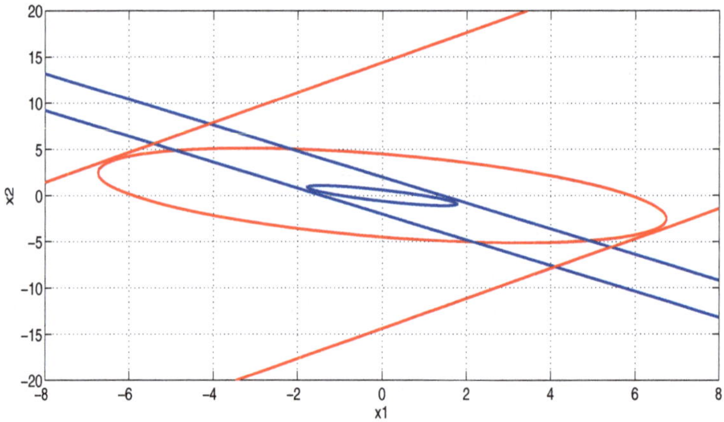

Fig. 6.1 The ellipsoids sets of invariance and contractivity for the switching discrete-time linear system computed with LMIs (6.16)-(6.17)

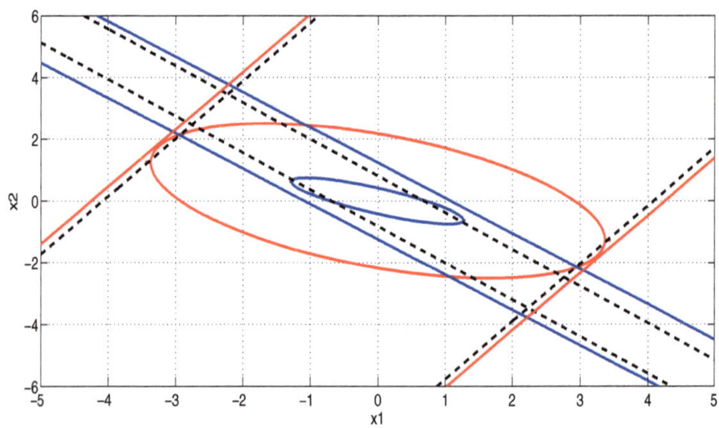

Fig. 6.2 The ellipsoids sets of invariance and contractivity for the switching discrete-time linear system computed with LMIs (6.27)-(6.28)

the sets $\mathscr{L}(F_i)$ no saturations occur. Furthermore, the results obtained with LMIs (6.16)-(6.17) are less conservative. This is due to the fact that among the number of LMIs to be solved in (6.27)-(6.28), one find the LMIs (6.16)-(6.17).

6.4.3 Output Feedback Control

Static output-feedback control plays a very important role in control applications: The purpose is to design controllers such that the resulting closed-loop system is asymptotically stable without using any reconstruction method of the unavailable states. In this section, we begin by the synthesis of the saturating controllers since the non saturating ones can be derived as a particular case. The synthesis of the stabilizing controller by output feedback control for the class of switching system with actuator saturation is presented by applying the results of Theorem 6.3.

Theorem 6.5. *If there exist symmetric matrices X_i, matrices V_i, Y_i and Z_i solutions of the following LMIs:*

$$\begin{bmatrix} X_i & (A_iX_i + B_iD_{is}Y_iC_i + B_iD_{is}^{-}Z_i)^T \\ * & X_j \end{bmatrix} > 0, \tag{6.33}$$

$$\begin{bmatrix} 1 & Z_{il} \\ * & X_i \end{bmatrix} > 0, \tag{6.34}$$

$$V_iC_i = C_iX_i, \tag{6.35}$$

$$\forall (i,j) \in \mathscr{I}^2, \ \forall s \in [1,\eta], \ \forall l \in [1,m]$$

then the closed-loop saturated switching system (6.3) with

$$K_i = Y_iV_i^{-1}, H_i = Z_iX_i^{-1}, P_i = X_i^{-1}, \tag{6.36}$$

is asymptotically stable $\forall x_0 \in \Omega$ and for all switching sequences $\alpha(k)$.

Proof: The same development as (6.30)-(6.32) while replacing F_i by K_iC_i, can be followed to obtain,

$$\begin{bmatrix} X_i & (A_iX_i + B_iD_{is}K_iC_iX_i + B_iD_{is}^{-}H_iX_i)^T \\ * & X_j \end{bmatrix} > 0.$$

According to Equation (6.35), we have $K_iC_iX_i = K_iV_iC_i$. By letting $K_iV_i = Y_i$ and $H_iX_i = Z_i$, the LMI (6.33) follows together with relations (6.36). Similarly, the inclusion condition (6.21) can also be transformed to the equivalent LMI (6.34). Finally, the inequality (6.33) ensures that the obtained solutions X_i are positive definite while equalities (6.35) guarantee that matrices V_i are nonsingular. □

It is worth noting that the state feedback control follows readily from Theorem 6.5 by letting $C_i = \mathbb{I}_m$. In this case, $V_i = X_i$.

The LMI (6.33), relating matrices C_i to matrix X_i by means of Equation (6.35), can be relaxed by using the LMI (3.39), where the new variables G_i are related to matrices C_i instead of the matrix X_i.

Corollary 6.1. *If there exist symmetric matrices X_i, matrices G_i, V_i, Y_i and Z_i solutions of the following LMIs:*

$$\begin{bmatrix} G_i + G_i^T - X_i & \Psi_{is}^T \\ * & X_j \end{bmatrix} > 0, \tag{6.37}$$

$$\begin{bmatrix} 1, & Z_{il} \\ * & G_i + G_i^T - X_i \end{bmatrix} > 0, \tag{6.38}$$

$$V_i C_i = C_i G_i, \tag{6.39}$$

$$\forall (i,j) \in \mathscr{I}^2, \;\; \forall s \in [1,\eta], \;\; \forall l \in [1,m],$$

where $\Psi_{is} = (A_i G_i + B_i D_{is} Y_i C_i + B_i D_{is}^- Z_i)$, then the closed-loop saturated switching system (6.3) with

$$K_i = Y_i V_i^{-1}, H_i = Z_i G_i^{-1}, P_i = X_i^{-1}, \tag{6.40}$$

is asymptotically stable $\forall x_0 \in \Omega$ and all switching sequences $\alpha(k)$.

Proof: The proof uses the equivalent LMI (3.39) and is similar to that of Theorem 6.5. However, the inclusion condition $\varepsilon(P_i,1) \subset \mathscr{L}(H_i) \quad \forall i \in \mathscr{I}$ holds if $1 - H_{il} X_i H_{il}^T > 0, \forall l \in [1,m]$, which is equivalent to,
$1 - (H_i G_i)_l (G_i^T X_i^{-1} G_i)^{-1} (H_i G_i)_l^T > 0$. That is, by virtue of (6.40)
$1 - (Z_{il})(G_i^T X_i^{-1} G_i)^{-1} (Z_{il})^T > 0$. Since, $(G_i - X_i)^T X_i^{-1} (G_i - X_i) \geq 0$, then $G_i X_i^{-1} G_i^T \geq G_i + G_i^T - X_i$. It follows that, $1 - Z_{il}(G_i + G_i^T - X_i) Z_{il}^T > 0$ is sufficient to have $1 - H_{il} X_i H_{il}^T > 0$. By Schur complement, the LMI (6.38) is obtained. □

To achieve a domain of attraction as large as possible, we can solve similar optimization problem as for state feedback control.

Another way to deal with the problem by relaxing the more constraining relations (6.35) of Theorem 6.5, especially in MIMO case, is presented by the following result which uses the same idea as in [82] and [61].

Theorem 6.6. *If there exist symmetric matrices P_i and S_i, matrices K_i and H_i solutions of the following LMIs:*

$$\begin{bmatrix} P_i & (A_i + B_i D_{is} K_i C_i + B_i D_{is}^- H_i)^T \\ * & S_j \end{bmatrix} > 0, \tag{6.41}$$

$$\begin{bmatrix} 1 & H_{il} \\ * & P_i \end{bmatrix} > 0, \tag{6.42}$$

$$\begin{bmatrix} P_i & \mathbb{I} \\ * & S_i \end{bmatrix} \geq 0, \tag{6.43}$$

$$\forall (i,j) \in \mathscr{I}^2, \;\; \forall s \in [1,\eta], \;\; \forall l \in [1,m],$$

such that $Trace(P_iS_i) = n$ *, then the closed-loop saturated switching system (6.3) is asymptotically stable* $\forall x_0 \in \Omega$ *and for all switching sequences* $\alpha(k)$.

Proof: The inequality (6.20) can be equivalently transformed via the Schur complement to the following:

$$\left[A_i + B_i(D_{is}K_iC_i + D_{is}^- H_i)\right]^T P_j \left[A_i + B_i(D_{is}K_iC_i + D_{is}^- H_i)\right] - P_i < 0 \quad (6.44)$$

The use of the Schur complement a second time leads to

$$\begin{bmatrix} P_i & (A_i + B_iD_{is}K_iC_i + B_iD_{is}^- H_i)^T \\ * & P_j^{-1} \end{bmatrix} > 0 \quad (6.45)$$

By letting $S_j = P_j^{-1}$, LMI (6.41) follows. The inclusion condition (6.21) can also be transformed to the equivalent LMI (6.42) [53]: $H_{il}P_i^{-1}H_{il}^T \leq 1$ can be rewritten by Schur complement, as (6.42). Note that the LMI (6.41) ensures that the obtained solutions X_i and S_i are positive definite. Finally, the LMI (6.43) is equivalent to $S_iP_i \geq \mathbb{I}$. □

This result presents the advantage of computing directly the matrices K_i and H_i. Nevertheless, an optimization problem must be solved to achieve $P_iS_i \simeq \mathbb{I}$ by minimizing iteratively the trace of matrix P_iS_i using an algorithm presented in [82]. This heuristic is based on a linear approximation of $Tr(P_iS_i)$ by $Tr(P_0S_i + S_0P_i)$ where P_0 and S_0 are particular solutions of the LMI constraints (6.41), (6.42) and (6.43).

$$(Pb.3): \begin{cases} \min_{(S_j,P_i,K_i,H_i)} Trace(P_iS_i) \\ s.t. \ (6.41), (6.42), (6.43) \\ \quad i,j = 1,\ldots,N \end{cases}$$

The following result presents a new way of applying cone complementarity technique without any use of any optimization problem of type $(Pb.3)$.

Corollary 6.2. *If there exist symmetric matrices* P_i, R_i*, matrices* G_i, N_i, K_i *and* H_i *solutions of the following LMIs:*

$$\begin{bmatrix} P_i & (A_i + B_iD_{is}K_iC_i + B_iD_{is}^- H_i)^T \\ * & R_j \end{bmatrix} > 0, \quad (6.46)$$

$$\begin{bmatrix} 1 & H_{il} \\ * & P_i \end{bmatrix} > 0, \quad (6.47)$$

$$\begin{bmatrix} G_i + G_i^T - R_i & G_i + G_i^T - \mathbb{I} \\ * & G_i + G_i^T - P_i \end{bmatrix} > 0, \quad (6.48)$$

$$\forall (i,j) \in \mathscr{I}^2, \ \forall s \in [1,\eta], \ \forall l \in [1,m],$$

then the closed-loop saturated switching system (6.3) is asymptotically stable $\forall x_0 \in \Omega$ and for all switching sequences $\alpha(k)$.

Proof: The idea is to compute symmetric definite positive matrices R_i realizing $P_i^{-1} > R_i$. This inequality is satisfied by inequality $P_i < R_i^{-1}$ and is equivalent by Schur complement to:

$$\begin{bmatrix} R_i^{-1} & \mathbb{I} \\ * & P_i^{-1} \end{bmatrix} > 0, \tag{6.49}$$

$$\tag{6.50}$$

By pre multiplying inequality (6.49) by $diag\{G_i^T, G_i^T\}$ and post multiplying by $diag\{G_i, G_i\}$, where G_i is any non singular matrix, it follows equivalently:

$$\begin{bmatrix} G_i^T R_i^{-1} G_i & G_i^T G_i \\ * & G_i^T P_i^{-1} G_i \end{bmatrix} > 0. \tag{6.51}$$

The following development

$$(G_i^T - R_i) R_i^{-1} (G_i - R_i) = G_i^T R_i^{-1} G_i - G_i - G_i^T + R_i > 0, \tag{6.52}$$

leads to

$$G_i^T R_i^{-1} G_i > Gi + G_i^T - R_i, \tag{6.53}$$

$$G_i^T P_i^{-1} G_i > G_i + G_i^T - P_i, \tag{6.54}$$

$$G_i^T G_i > G_i + G_i^T - \mathbb{I}. \tag{6.55}$$

Replacing (6.53)-(6.54) one obtains (6.48). Inequality $P_i^{-1} > R_i$ is then used to bound Inequality (6.41) to obtain LMIs (6.46)-(6.48). □

An output feedback unsaturating controller can be obtained, as noted in Comment 6.1, as a particular case of the saturating controller results by letting $D_{is} = \mathbb{I}$ and $D_{is}^+ = 0$. It is also obvious that the inclusion condition becomes $\varepsilon(P_i, 1) \subset \mathscr{L}(K_i C_i)$. The following results are then directly obtained.

Corollary 6.3. *If there exist symmetric matrices X_i, matrices V_i and Y_i solutions of the following LMIs:*

$$\begin{bmatrix} X_i & (A_i X_i + B_i Y_i C_i)^T \\ * & X_j \end{bmatrix} > 0, \tag{6.56}$$

$$\begin{bmatrix} 1 & (Y_i C_i)_l \\ * & X_i \end{bmatrix} > 0, \tag{6.57}$$

$$V_i C_i = C_i X_i, \tag{6.58}$$

$$\forall (i,j) \in \mathscr{I}^2, \forall l \in [1,m],$$

then the closed-loop saturated switching system (6.3) with

$$K_i = Y_i V_i^{-1}, P_i = X_i^{-1}, \tag{6.59}$$

is asymptotically stable $\forall x_0 \in \Omega$ *and for all switching sequences* $\alpha(k)$.

Proof: This proof can be obtained easily by letting $D_{is} = \mathbb{I}$ and $D_{is}^{+} = 0$ in Theorem 6.5. However, the inclusion condition $\varepsilon(P_i, 1) \subset \mathscr{L}(K_iC_i)$ leads to

$$(K_iC_i)_l P_i^{-1} (K_iC_i)_l^T \leq 1$$

which is equivalent to,

$$(K_iC_iP_i^{-1})_l P_i (K_iC_iP_i^{-1})_l^T \leq 1,$$

by using (6.58), $(K_iV_iC_i)_l X_i^{-1} (K_iV_iC_i)_l^T \leq 1$, can be rewritten by Schur complement, while using $Y_i = K_iV_i$ as (6.57). □

In a similar way, the result of Theorem 6.6 can be applied for designing non saturating controls by letting $D_{is} = \mathbb{I}$ and $D_{is}^{+} = 0$.

The last result can be derived from Corollary 6.1 by using slack variables for unsaturating controllers.

Corollary 6.4. *If there exist symmetric matrices* X_i, *matrices* G_i, V_i *and* Y_i *solutions of the following LMIs:*

$$\begin{bmatrix} G_i + G_i^T - X_i & (A_iG_i + B_iY_iC_i)^T \\ * & X_j \end{bmatrix} > 0, \tag{6.60}$$

$$\begin{bmatrix} 1, & (Y_iC_i)_l \\ * & G_i + G_i^T - X_i \end{bmatrix} > 0, \tag{6.61}$$

$$V_iC_i = C_iG_i, \tag{6.62}$$

$$\forall (i,j) \in \mathscr{I}^2, \;\; \forall s \in [1,\eta], \;\; \forall l \in [1,m],$$

then the closed-loop saturated switching system (6.3) with

$$K_i = Y_iV_i^{-1}, P_i = X_i^{-1}, \tag{6.63}$$

is asymptotically stable $\forall x_0 \in \Omega$ *and all switching sequences* $\alpha(k)$.

Proof: *Obvious.*

Example 6.2. *In order to illustrate these results, consider a SISO saturated switching discrete-time system with two modes given by the following matrices:*

$$A_1 = \begin{bmatrix} 1 & 1 \\ 0 & 1 \end{bmatrix}; B_1 = \begin{bmatrix} 10 \\ 5 \end{bmatrix}; C_1 = \begin{bmatrix} 1 & 1 \end{bmatrix}$$

$$A_2 = \begin{bmatrix} 0 & -1 \\ 0.0001 & 1 \end{bmatrix}; B_2 = \begin{bmatrix} 0.5 \\ -2 \end{bmatrix}; C_2 = \begin{bmatrix} 2 & 3 \end{bmatrix}$$

For this example with $n = 2$, $m = p = 1$ and $N = 2$, we have to solve 13 LMIs with 9 variables by using the LMIs of Corollary 6.1. Let the scalar $\rho = 1$. The use of the Matlab LMI Toolbox yields the following results:

$$P_1 = \begin{bmatrix} 0.0016 & 0.0006 \\ 0.0006 & 0.0025 \end{bmatrix}; H_1 = \begin{bmatrix} -0.0196 & -0.0479 \end{bmatrix};$$
$$K_1 = -0.0743; Ac_1 = \begin{bmatrix} 0.2574 & 0.2574 \\ -0.3713 & 0.6287 \end{bmatrix};$$
$$P_2 = \begin{bmatrix} 0.0010 & 0.0004 \\ 0.0004 & 0.0027 \end{bmatrix}; H_2 = \begin{bmatrix} 0.000007 & 0.0483 \end{bmatrix};$$
$$K_2 = 0.0347; Ac_2 = \begin{bmatrix} 0.0615 & -0.9077 \\ -0.2460 & 0.6309 \end{bmatrix}.$$

The main conditions (3.23) are also satisfied:

$$Ac_1^T P_2 Ac_1 - P_1 = 10^{-3} \begin{bmatrix} -0.0012 & -0.0011 \\ -0.0011 & -0.0013 \end{bmatrix};$$
$$Ac_2^T P_1 Ac_2 - P_2 = 10^{-3} \begin{bmatrix} -0.0008 & -0.0007 \\ -0.0007 & -0.0011 \end{bmatrix}.$$

Figure 6.3 shows the level set as the union of two ellipsoid sets of invariance and contractivity for the switching discrete-time linear system together with three trajectories and a switching sequence, obtained by Corollary 6.1. Each ellipsoid set is contained inside the set of admissible saturations $\mathscr{L}(H_i)$ in dotted lines and same color. While the sets $\mathscr{L}(K_iC)$ are plotted in solid lines in the same color. Inside this set, the asymptotic stability of the saturated switching system is guaranteed for any arbitrary switch from any linear subsystem to another. This SISO example is studied only in order to illustrate the results of this work by plotting the level set in the plane.

The application of Corollary 6.4 leads to the following results:

$$P_1 = \begin{bmatrix} 0.0061 & 0.0028 \\ 0.0028 & 0.0079 \end{bmatrix}; G_1 = \begin{bmatrix} 177.1128 & -66.1787 \\ -87.9906 & 155.3009 \end{bmatrix};$$
$$K_1 = -0.0684; Ac_1 = \begin{bmatrix} 0.3158 & 0.3158 \\ -0.3421 & 0.6579 \end{bmatrix};$$
$$P_2 = \begin{bmatrix} 0.0032 & 0.0014 \\ 0.0014 & 0.0081 \end{bmatrix}; G_2 = \begin{bmatrix} 291.3878 & -57.1530 \\ -130.9738 & 133.0291 \end{bmatrix};$$
$$K_2 = 0.0198; Ac_2 = \begin{bmatrix} 0.0198 & -0.9704 \\ -0.0789 & 0.8815 \end{bmatrix}.$$

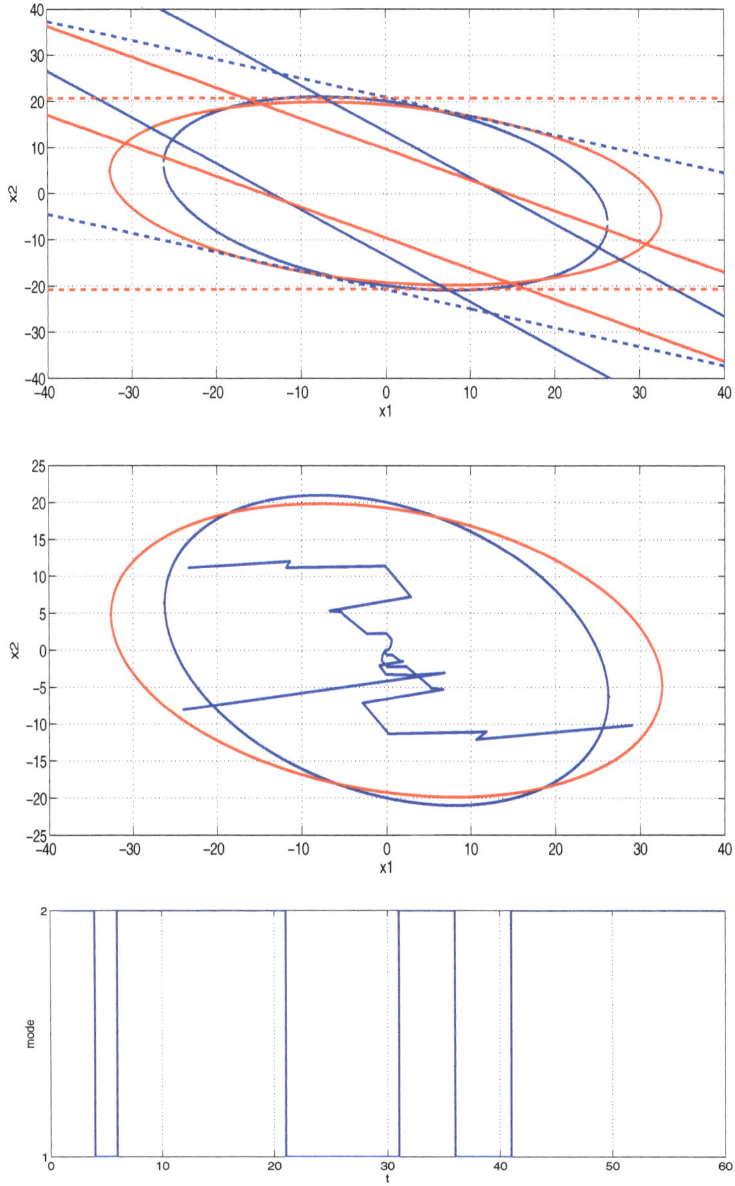

Fig. 6.3 The level set as the union of two ellipsoid sets of invariance and contractivity for the switching discrete-time linear system obtained by Corollary 6.1 together with three trajectories and a switching sequence

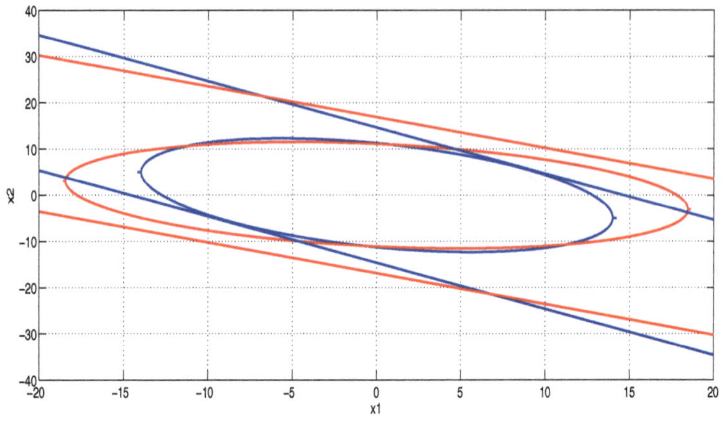

Fig. 6.4 The level set as the union of two ellipsoid sets of invariance and contractivity for the switching discrete-time linear system obtained by Corollary 6.4

Figure 6.4 shows the level set as the union of two ellipsoid sets of invariance and contractivity for the switching discrete-time linear system obtained by Corollary 6.4. Each ellipsoid set is contained inside the set of admissible saturations $\mathscr{L}(K_iC_i)$ in the same color.

The application of Corollary 6.2 with $D_i = \mathbb{I}, D_i^- = 0$ leads to the following results:

$$P_1 = \begin{bmatrix} 0.4074 & -0.5589 \\ -0.5589 & 1.0070 \end{bmatrix}; K_1 = -0.0855; Ac_1 = \begin{bmatrix} 0.1452 & 0.1452 \\ -0.4274 & 0.5726 \end{bmatrix};$$

$$P_2 = \begin{bmatrix} 0.2169 & -0.4276 \\ -0.4276 & 1.2596 \end{bmatrix}; K_2 = 0.0862; Ac_2 = \begin{bmatrix} 0.0862 & -0.8706 \\ -0.3449 & 0.4826 \end{bmatrix}.$$

The initial four inequalities of Theorem 6.6 are all checked and are well satisfied as shown there after for $i = j$.

$$\begin{bmatrix} P_1 & (A_1 + B_1K_1C_1)^T \\ * & P_1^{-1} \end{bmatrix} = \begin{bmatrix} 0.4074 & -0.5589 & 0.1452 & -0.4274 \\ -0.5589 & 1.0070 & 0.1452 & 0.5726 \\ 0.1452 & 0.1452 & 13.9431 & 4.7334 \\ -0.4274 & 0.5726 & 4.7334 & 2.4008 \end{bmatrix} > 0,$$

$$\begin{bmatrix} P_2 & (A_2 + B_2K_2C_2)^T \\ * & P_2^{-1} \end{bmatrix} = \begin{bmatrix} 0.2169 & -0.4276 & 0.0862 & -0.3449 \\ -0.4276 & 1.2596 & -0.8706 & 0.4826 \\ 0.0862 & -0.8706 & 10.2836 & 5.7072 \\ -0.3449 & 0.4826 & 5.7072 & 4.1604 \end{bmatrix} > 0.$$

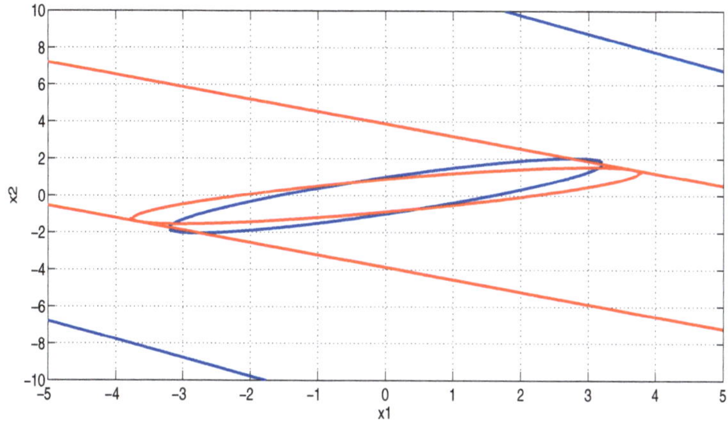

Fig. 6.5 The level set as the union of two ellipsoid sets of invariance and contractivity for the switching discrete-time linear system obtained by Corollary 6.2

Nevertheless, these results lead to more conservative regions of stability, even obtained by maximizing the trace of matrices P_1 and P_2. The corresponding level sets are plotted in Figure 6.5.

Closed results are also obtained for the saturated controller case given directly by Corollary 6.2.

Figure 6.5 shows the level set as the union of two ellipsoid sets of invariance and contractivity for the switching discrete-time linear system obtained by Corollary 6.2. Each ellipsoid set is contained inside the set of admissible saturations $\mathscr{L}(K_iC_i)$ in the same color.

One can notice that the obtained level sets with Corollary 6.1 are the most larger. One can conclude, via this example, that the result of Corollary 6.1 is the less conservative among all the other results due to the introduced slack variables G_i. Combining the comparisons made for state feedback control and output feedback control, one can expect that the introduction of slack variables in the Corollary 6.3 can lead to the more less conservative results. However, as shown by Figure 6.4, this is not the case.

6.5 Conclusion

In this chapter, two main different sufficient conditions of asymptotic stability are obtained for switching discrete-time linear systems subject to actuator saturations for each case: state feedback and output feedback control. The first allows the synthesis of stabilizing controllers inside a large region of linear behavior while the second applies the idea of Lemma 2.1 which rewrites the saturation function under a combination of 2^m elements to obtain stabilizing controllers tolerating saturations to take effect. A particular attention is given to the output feedback case which has additive complexity due to the output equation. The main results of this work are given under LMIs formulation leading to the design of the stabilizing state feedback and output feedback controllers for the system. Even the dynamical system is a switching system, it is shown that the set Ω formed by the union of all the ellipsoid level sets associated to each subsystem, constitutes a set of asymptotic stability. It was the first time that this important result was established for saturated switching systems in [25, 27]. Illustrative examples are studied by using the solution of the proposed LMIs. A comparison of the obtained solutions is also given.

Chapter 7
Stabilization of Saturated Switching Systems with Uncertainties

7.1 Stabilization of Saturated Switching Systems with Polytopic Uncertainties

The objective of this section is to extend the results of Chapter *VI* to uncertain switching systems subject to actuator saturations . The uncertainty type considered in this work is the polytopic one. This type of uncertainty was also studied, without saturation, in [98]. Other works were interested in different problems for uncertain switched systems in [117, 122]. Thus, in this work, two directions are explored: the first concerns the synthesis of non-saturating controllers, while the second direction deals with controllers tolerating saturations to take effect under polytopic uncertainties. The main results of this section are published in [30, 32].

7.1.1 Problem Presentation

Let us consider the uncertain saturated switching discrete-time linear system described by

$$\begin{cases} x_{k+1} = A_\alpha(q_\alpha(k))x_k + B_\alpha(q_\alpha(k))sat(u_k) \\ y_k = C_\alpha(q_\alpha(k))x_k \end{cases} \tag{7.1}$$

where $x_k \in \mathbb{R}^n$, $u_k \in \mathbb{R}^m$, $y_k \in \mathbb{R}^p$ are the state, the input and the output, respectively, $sat(.)$ is the standard saturation (assumed here to be normalized, *i.e.*, $|sat(u_k)| = min(1,|u_k|)$), function $\alpha(k) : \mathbb{N} \longmapsto \mathscr{I}$ is a switching rule taking its values $\alpha(k) = i$ in the finite set $\mathscr{I} = \{1,...,N\}$ and $q_\alpha(k) \in \Gamma_\alpha \subset \mathbb{R}^{d_\alpha}$ are the bounded uncertainties that affect the system parameters in such a way that

$$A_\alpha(q_\alpha(k)) = A_\alpha + \sum_{h=1}^{d_\alpha} A_{\alpha h} q_{\alpha h}(k) \tag{7.2}$$

$$B_\alpha(q_\alpha(k)) = B_\alpha + \sum_{h=1}^{d_\alpha} B_{\alpha h} q_{\alpha h}(k) \tag{7.3}$$

A. Benzaouia: Saturated Switching Systems, LNCIS 426, pp. 165–193.
springerlink.com

$$C_\alpha(q_\alpha(k)) = C_\alpha + \sum_{h=1}^{d_\alpha} C_{\alpha h} q_{\alpha h}(k) \tag{7.4}$$

where matrices $A_\alpha, B_\alpha, C_\alpha$ represent the nominal matrices and $q_{\alpha h}(k)$ the hth component of vector $q_\alpha(k)$:

$$q_\alpha(k) = [q_{\alpha 1}(k) \;\; q_{\alpha 2}(k) \;\; ... \;\; q_{\alpha h}(k) \;\; ... \;\; q_{\alpha d_\alpha}(k)]^T.$$

The following assumptions are required:

- Γ_α are compact convex sets.
- The switching system is stabilizable.
- Matrices C_α are of full rank.
- Each time only one subsystem is active.

Let the control be obtained by an output feedback control law:

$$\begin{aligned} u_k &= K_\alpha y_k = F_\alpha x_k, \\ F_\alpha &= K_\alpha C_\alpha. \end{aligned} \tag{7.5}$$

Thus, the closed-loop system is given by

$$x_{k+1} = A_\alpha(q_\alpha(k))x_k + B_\alpha(q_\alpha(k))sat(K_\alpha C_\alpha(q_\alpha(k))x_k). \tag{7.6}$$

Defining the indicator function:

$$\xi(k) := [\xi_1(k), ..., \xi_N(k)]^T, \tag{7.7}$$

where $\xi_i(k) = 1$ if the switching system is in mode i and 0 otherwise, yields the following representation for the closed-loop system:

$$x_{k+1} = \sum_{i=1}^{N} \xi_i(k)[A_i(q_i(k))x_k + B_i(q_i(k))sat(K_i C_i(q_i(k))x_k)]. \tag{7.8}$$

Let each convex set Γ_i has μ_i vertices $\nu_{i\kappa}, \kappa = 1, \ldots, \mu_i$ so that for every $q_i \in \Gamma_i$, one can write $q_i = \sum_{\kappa=1}^{\mu_i} \beta_{i\kappa} \nu_{i\kappa}$ with $\sum_{\kappa=1}^{\mu_i} \beta_{i\kappa} = 1, 0 \leq \beta_{i\kappa} \leq 1$. The consequence of this, is that each matrix $A_i(q_i(k))$, $B_i(q_i(k))$ and $C_i(q_i(k))$ can be expressed as a convex combination of the corresponding vertices of the compact set Γ_i as follows:

$$\begin{aligned} M(q_i) :&= M_i + \sum_{\kappa=1}^{\mu_i} \beta_{i\kappa} M(\nu_{i\kappa}) = \sum_{\kappa=1}^{\mu_i} \beta_{i\kappa} M_{i\kappa}, \\ M(\nu_{i\kappa}) &= \sum_{h=1}^{d_i} M_{ih} \nu_{i\kappa h}, \\ M_{i\kappa} &= M_i + M(\nu_{i\kappa}), \\ \sum_{\kappa=1}^{\mu_i} \beta_{i\kappa} &= 1, 0 \leq \beta_{i\kappa} \leq 1 \end{aligned} \tag{7.9}$$

where M_i represents the nominal matrix. Matrix M can be taken differently as A, B or C. Note that the system without uncertainties can be obtained as a particular case of this representation by letting the vertices $v_{i\kappa} = 0, \forall i, \forall \kappa$. Besides, Equations (7.9) are directly related to the dimension d_i of the convex compact set Γ_i. The saturated uncertain switching system given by (7.8) can be rewritten as

$$x_{k+1} = \sum_{i=1}^{N} \sum_{\kappa=1}^{\mu_i} \xi_i(k)\beta_{i\kappa}(k)[A_{i\kappa}x_k + B_{i\kappa}sat(K_iC_{i\kappa}x_k)] \ . \tag{7.10}$$

The nominal matrices will be represented by A_i, B_i and C_i. The nominal system in closed loop is then given by

$$x_{k+1} = \sum_{i=1}^{N} \xi_i(k)[A_ix_k + B_isat(K_iC_ix_k)] \ . \tag{7.11}$$

7.1.2 Analysis and Synthesis of Stabilizability

This section presents sufficient conditions of asymptotic stability of the saturated uncertain switching system given by (7.10). The synthesis of the controller follows two different approaches, the first one deals first with the nominal system and then uses a test to check the asymptotic stability in presence of uncertainties, while the second considers the global representation of the uncertain system (7.10).

Theorem 7.1. *If there exist symmetric positive definite matrices $P_1, \dots, P_N \in \mathbb{R}^{n \times n}$ and matrices $H_1, \dots, H_N \in \mathbb{R}^{m \times n}$ such that*

$$\begin{bmatrix} P_i & [A_{i\kappa} + B_{i\kappa}(D_{is}K_iC_{i\kappa} + D_{is}^{-}H_i)]^T P_j \\ * & P_j \end{bmatrix} > 0,$$
$$\forall \kappa = 1, \dots, \mu_i, \forall (i,j) \in \mathscr{I}^2, \forall s \in [1,\ \eta], \tag{7.12}$$

and

$$\varepsilon(P_i,\ 1) \subset \mathscr{L}(H_i), \tag{7.13}$$

then the closed-loop uncertain saturated switching system (7.10) is asymptotically stable $\forall\ x_0 \in \Omega := \bigcup_{i=1}^{N} \varepsilon(P_i,\ 1)$ and for all switching sequences $\alpha(k)$.

Proof: By using Lemma 2.1, for all $H_i \in \mathbb{R}^{m \times n}$ with $|H_{is}x_k| < 1$, $s \in [1, m]$, where H_{is} denotes the sth row of matrix H_i, there exist $\delta_{i1\kappa} \geq 0$,..., $\delta_{i\eta\kappa} \geq 0$ such that $sat(K_iC_{i\kappa}x_k) = \sum_{s=1}^{\eta} \delta_{is\kappa}(k)[D_{is}K_iC_{i\kappa} + D_{is}^{-}H_i]x_k$, $\delta_{i\kappa s}(k) \geq 0$, $\sum_{s=1}^{\eta} \delta_{i\kappa s}(k) = 1$.

Then the closed-loop system (7.10) can be rewritten as

$$x_{k+1} = \sum_{s=1}^{\eta} \sum_{i=1}^{N} \sum_{\kappa=1}^{\mu_i} \xi_i(k)\beta_{i\kappa}(k)\delta_{i\kappa s}(k)Ac_{i\kappa s}x_k \tag{7.14}$$

$$Ac_{i\kappa s} := A_{i\kappa} + B_{i\kappa}(D_{is}K_iC_{i\kappa} + D_{is}^{-}H_i).$$

Consider the Lyapunov function candidate $V(x) = x_k^T(\sum_{i=1}^{N}\xi_i(k)P_i)x_k$. Computing its rate of increase along the trajectories of system (7.10) yields

$$\begin{aligned}\Delta V(x_k) &= x_{k+1}^T(\sum_{j=1}^{N}\xi_j(k+1)P_j)x_{k+1} - x_k^T(\sum_{i=1}^{N}\xi_i(k)P_i)x_k \\ &= x_k^T\left\{\Sigma^T(\sum_{j=1}^{N}\xi_j(k+1)P_j)\Sigma - \sum_{i=1}^{N}\xi_i(k)P_i\right\}x_k\end{aligned}$$

where,

$$\Sigma = \sum_{s=1}^{\eta}\sum_{i=1}^{N}\sum_{\kappa=1}^{\mu_i}\xi_i(k)\beta_{i\kappa}(k)\delta_{s\kappa i}(k)Ac_{i\kappa s}$$

Let condition (7.12) be satisfied. For each i and j multiply successively by $\xi_i(k)$, $\xi_j(k+1)$, $\beta_{i\kappa}(k)$ and $\delta_{i\kappa s}(k)$ and sum. As $\sum_{i=1}^{N}\xi_i(k) = \sum_{j=1}^{N}\xi_j(k+1) = \sum_{s=1}^{\eta}\delta_{i\kappa s}(k)$ $= \sum_{\kappa=1}^{\mu_i}\beta_{i\kappa}(k) = 1$, one gets

$$\begin{bmatrix}\sum_{i=1}^{N}\xi_i(k)P_i & \Pi \\ * & \sum_{j=1}^{N}\xi_j(k+1)P_j\end{bmatrix} > 0, \tag{7.15}$$

where

$$\Pi = \Sigma^T(\sum_{j=1}^{N}\xi_j(k+1)P_j).$$

Inequality (7.15) is equivalent, by Schur complement, to

$$\Sigma^T(\sum_{j=1}^{N}\xi_j(k+1)P_j)\Sigma - \sum_{i=1}^{N}\xi_i(k)P_i < 0\,.$$

Letting λ be the largest eigenvalue among all the above matrices, we obtain that

$$\Delta V(x_k) \le \lambda x_k^T x_k < 0, \tag{7.16}$$

which ensures the desired result. Besides, following Theorem 6.3, conditions (7.12)-(7.13) also allow for a state belonging to a set $\varepsilon(P_i,1) \subset \mathscr{L}(H_i)$, before that the switch occurs at time k, the switch will drive the state to the set $\varepsilon(P_j,1) \subset \mathscr{L}(H_j)$. That means that the set Ω is a set of asymptotic stability of the uncertain saturated switching system. □

Remark 7.1. *It is worth to note that the result of Theorem 6.3 can be obtained as a particular case of Theorem 7.1.*

This stability result is now used for control synthesis in two ways: the first consists in computing the controllers only with the nominal system and to test their robustness in a second step; while the second consists in computing in a single step the robust controllers. At this end, the result of Theorem 6.5 can be used to compute matrices K_i, H_i and P_i for the nominal switching system (7.11). At this step, the stabilizing controllers K_i and H_i of the nominal system are assumed to be known. Then, the following test has to be performed.

Corollary 7.1. *If there exist symmetric positive definite matrices X_i such that*

$$\begin{bmatrix} X_i & (A_{i\kappa}X_i + B_{i\kappa}D_{is}K_iC_{i\kappa}X_i + B_{i\kappa}D_{is}^{-}H_iX_i)^T \\ * & X_j \end{bmatrix} > 0, \tag{7.17}$$

$$\begin{bmatrix} 1 & (H_iX_i)_l \\ * & X_i \end{bmatrix} > 0, \tag{7.18}$$

$\forall (i,j) \in \mathscr{I}^2, \quad \forall s \in [1,\ \eta], \forall l \in [1,m],\ \forall \kappa \in [1,\ \mu_i],$
with $P_i = X_i^{-1}$, then the closed-loop uncertain switching system (7.10) is asymptotically stable $\forall\ x_0 \in \bigcup_{i=1}^{N} \varepsilon(P_i,\ 1)$ and for all switching sequences $\alpha(k)$.

Proof: The proof is similar to that given for Theorem 6.5. □
The second way to deal with robust controller design is to run a global set of LMIs leading directly, if it is feasible, to the robust controllers. However, one can note that this method is computationally more intensive.

Theorem 7.2. *If there exist symmetric positive definite matrices X_i, matrices, Y_i, V_i and Z_i such that*

$$\begin{bmatrix} X_i & (A_{i\kappa}X_i + B_{i\kappa}D_{is}Y_iC_{i\kappa} + B_{i\kappa}D_{is}^{-}Z_i)^T \\ * & X_j \end{bmatrix} > 0, \tag{7.19}$$

$$\begin{bmatrix} 1 & Z_{il} \\ * & X_i \end{bmatrix} > 0, \tag{7.20}$$

$$V_iC_{i\kappa} = C_{i\kappa}X_i, \tag{7.21}$$

$\forall (i,j) \in \mathscr{I}^2, \quad \forall s \in [1,\ \eta], \forall l \in [1,m], \forall \kappa \in [1,\ \mu_i]$
with

$$H_i = Z_iX_i^{-1}, K_i = Y_iV_i^{-1}, P_i = X_i^{-1}, \tag{7.22}$$

then the closed-loop uncertain saturated switching system (7.10) is asymptotically stable $\forall\ x_0 \in \Omega$, and for all switching sequences $\alpha(k)$.

Proof: The proof is also similar to that given for Theorem 6.5. □

In order to relax the previous LMIs , one can introduce some slack variables as in [66, 25], as it is now shown:

Theorem 7.3. *If there exist symmetric positive definite matrices X_i, matrices, Y_i, V_i, G_i and Z_i such that*

$$\begin{bmatrix} G_i + G_i^T - X_i & \Psi \\ * & X_j \end{bmatrix} > 0, \tag{7.23}$$

with $\Psi = (A_{i\kappa}G_i + B_{i\kappa}D_{is}Y_iC_{i\kappa} + B_{i\kappa}D_{is}^{-}Z_i)^T$,

$$\begin{bmatrix} 1 & Z_{il} \\ * & G_i + G_i^T - X_i \end{bmatrix} > 0, \tag{7.24}$$

$$V_iC_{i\kappa} = C_{i\kappa}G_i, \tag{7.25}$$

$\forall\kappa = 1,\ldots,\mu_i, \forall(i,j) \in \mathscr{I}^2, \forall s \in [1,\eta], \forall l \in [1,m]$, *with*

$$H_i = Z_iG_i^{-1}, K_i = Y_iV_i^{-1}, P_i = X_i^{-1}; \tag{7.26}$$

then the closed-loop uncertain saturated switching system (7.10) is asymptotically stable $\forall\ x_0 \in \Omega$ *and for all switching sequences* $\alpha(k)$.

Proof: The proof is similar to that given for Corollary 6.1. Another way to deal with the problem by relaxing the more constraining relations (7.21) of Theorem 7.2, especially in MIMO case, is presented by the following result which uses the same idea as in [61] and [82].

Theorem 7.4. *If there exist symmetric matrices* P_i *and* S_i, *matrices* K_i *and* H_i *solutions of the following LMIs:*

$$\begin{bmatrix} P_i & (A_{i\kappa} + B_{i\kappa}D_{is}K_iC_{i\kappa} + B_{i\kappa}D_{is}^{-}H_i)^T \\ * & S_j \end{bmatrix} > 0, \tag{7.27}$$

$$\begin{bmatrix} 1 & H_{il} \\ * & P_i \end{bmatrix} > 0, \tag{7.28}$$

$$\begin{bmatrix} P_i & \mathbb{I} \\ * & S_i \end{bmatrix} \geq 0, \tag{7.29}$$

$$\forall(i,j) \in \mathscr{I}^2,\ \ \forall s \in [1,\eta],\ \ \forall l \in [1,m],$$

such that $Trace(P_iS_i) = n$, *then the closed-loop saturated switching system (7.10) is asymptotically stable* $\forall x_0 \in \Omega$ *and for all switching sequences* $\alpha(k)$.

Proof: The proof is similar to that given for Corollary 6.6. □

As explained in Chapter VI, this result needs to solve a linear programming of type (6.41). These results can be illustrated with the following example.

Example 7.1. *Consider a SISO saturated switching discrete system with two modes given by the following matrices:*

$$\begin{aligned} A_1(q_1(k)) &= \begin{bmatrix} 1 & 1 \\ 0 & 1+q_{11} \end{bmatrix}; B_1(q_1(k)) = \begin{bmatrix} 10 \\ 5 \end{bmatrix}; \\ C_1(q_1(k)) &= \begin{bmatrix} 1+q_{12} & 1 \end{bmatrix}; \end{aligned}$$

$$A_2(q_2(k)) = \begin{bmatrix} 0+q_{21} & -1 \\ 0.0001 & 1 \end{bmatrix}; B_2(q_2(k)) = \begin{bmatrix} 0.5 \\ -2+q_{22} \end{bmatrix};$$
$$C_2(q_2(k)) = \begin{bmatrix} 2 \; 3 \end{bmatrix}.$$

The vertices of the domain of uncertainties that affect the first mode are:

$$v_{11} = (-0.1, -0.2), \;\; v_{12} = (-0.1,\, 0.2)$$
$$v_{13} = (0.1, -0.2), \;\; v_{14} = (0.1, -0.2).$$

The vertices of the domain of uncertainties that affect the second mode are:

$$v_{21} = (-0.2, 0.5), \;\; v_{22} = (-0.2,\, -0.1)$$
$$v_{23} = (0.3, 0.5), \;\; v_{24} = (0.3, -0.1).$$

Using Theorem 6.5, a stabilizing controller for the nominal system is

$$K_1 = -0.1000, \quad K_2 = 0.1397.$$

$$H_1 = \begin{bmatrix} -0.0282613 \; -0.069 \end{bmatrix}; H_2 = \begin{bmatrix} 0.0000045 \; 0.0811986 \end{bmatrix}$$

To test the robustness, we can use Corollary 7.1 which leads to the following results:

$$P_1 = \begin{bmatrix} 0.01027 & -0.0017225 \\ -0.0017225 & 0.0104868 \end{bmatrix}; P_2 = \begin{bmatrix} 0.0093467 \; 0.0000525 \\ 0.0000525 \; 0.0165149 \end{bmatrix}$$

Figures 7.1 and 7.2 concern the first method. In Figure 7.1, the switching signals $\alpha(k)$ and the evolution of uncertainties used for simulation, are plotted. Figure 7.2 shows the obtained level set of stability $\bigcup_{i=1}^{N} \varepsilon(P_i,\, 1)$ which is well contained inside the sets of saturations together with three motions evolving inside the level set starting from different initial states.

On the other hand, the use of Theorem 7.2 leads to the following results:

$$K_1 = -0.0902, \quad K_2 = 0.1858.$$

$$P_1 = \begin{bmatrix} 0.0615 & 0 \\ 0 & 0.0615 \end{bmatrix}; P_2 = \begin{bmatrix} 0.0698 \; 0.0125 \\ 0.0125 \; 0.0799 \end{bmatrix};$$

$$H_1 = \begin{bmatrix} -0.0755 \; -0.1227 \end{bmatrix}; H_2 = \begin{bmatrix} -0.0461 \; 0.2586 \end{bmatrix}.$$

Figure 7.3 concerns the second method and shows the level set of stability $\bigcup_{i=1}^{N} \varepsilon(P_i,\, 1)$ using the second method of Theorem 7.2 which is well contained inside the sets of saturations. The uncertainties and sequence of switching evolutions are taken similar to Figure 7.1.

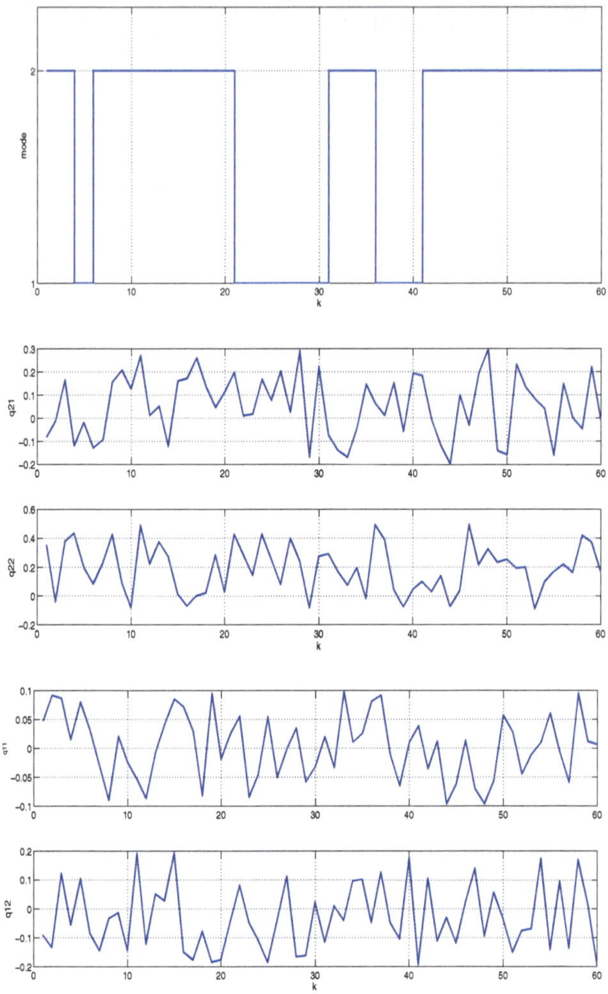

Fig. 7.1 Switching signals $\alpha(k)$ and uncertainties evolution

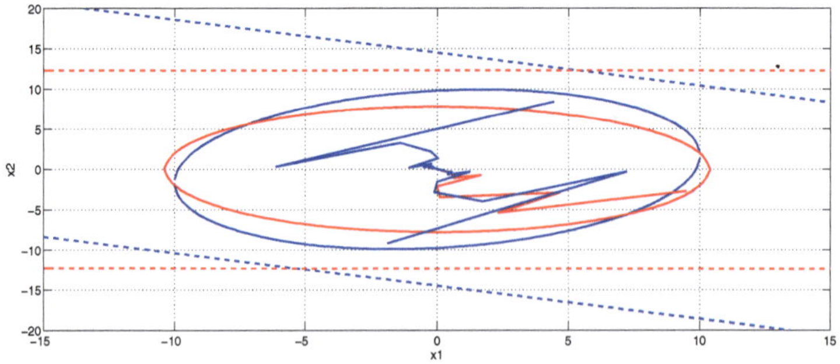

Fig. 7.2 Motion of the system with controllers obtained with Theorem 6.5 and Corollary 7.1

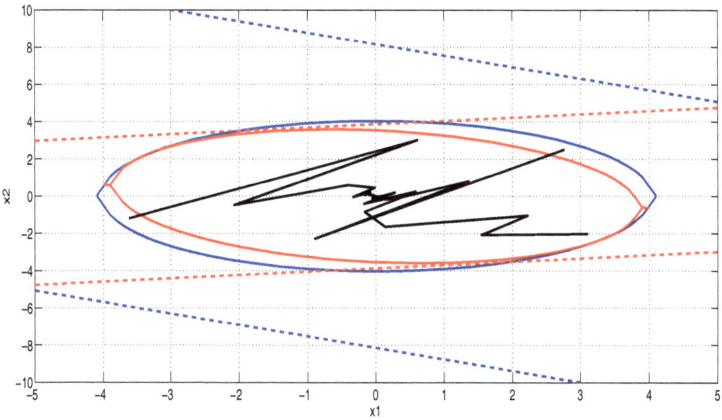

Fig. 7.3 Motion of the system with controllers obtained with Theorem7.2

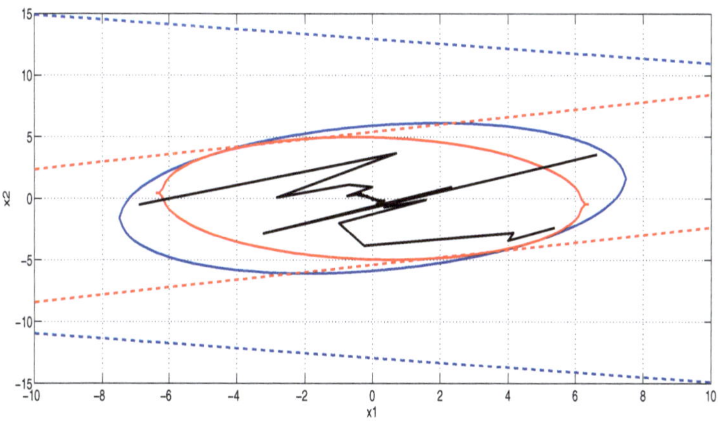

Fig. 7.4 Inclusion of the ellipsoids inside the polyhedral sets obtained with Theorem 7.3

The use of Theorem 7.3 leads to the following results:

$$K_1 = -0.0752, \quad K_2 = 0.1386;$$

$$P_1 = \begin{bmatrix} 0.0193 & -0.0062 \\ -0.0062 & 0.0288 \end{bmatrix}; P_2 = \begin{bmatrix} 0.0261 & 0.0028 \\ 0.0028 & 0.0406 \end{bmatrix};$$

$$H_1 = \begin{bmatrix} -0.0152 & -0.0773 \end{bmatrix}; H_2 = \begin{bmatrix} -0.0559 & 0.1854 \end{bmatrix}.$$

Figure 7.4 shows the level set of stability $\bigcup_{i=1}^{N} \varepsilon(P_i, 1)$ *obtained with Theorem 7.3, which is also well contained inside the sets of saturations. The uncertainties and sequence of switching evolutions are taken similar to Figure 7.1.*

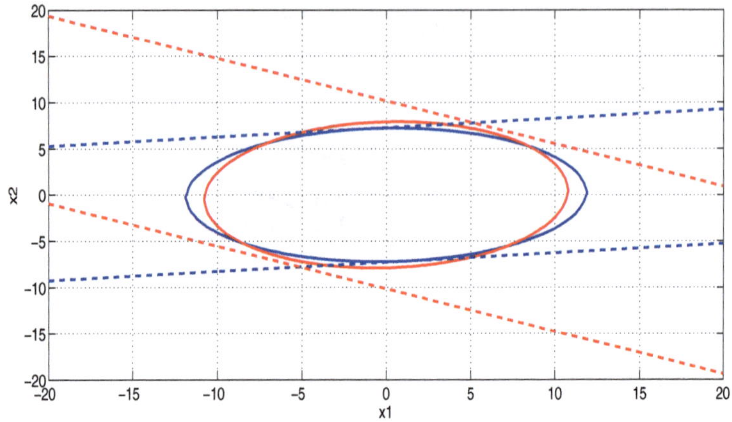

Fig. 7.5 Level sets and the corresponding polyhedral sets of saturations obtained with Theorem 7.4

Applying the results of Theorem 7.4 to the same example, the following solutions are obtained:

$$P_1 = \begin{bmatrix} 0.0071576 & -0.0004465 \\ -0.0004465 & 0.0192320 \end{bmatrix}; P_2 = \begin{bmatrix} 0.0086832 - 0.0007213 \\ -0.00072130.0160737 \end{bmatrix};$$

$$P_1 S_1 = \begin{bmatrix} 1 & 2.031E-11 \\ 1.35E-11 & 1 \end{bmatrix}; S_2 P_2 = \begin{bmatrix} 1 & -7.189E-13 \\ -2.892E-13 & 1 \end{bmatrix};$$

$$\begin{aligned} K_1 &= -0.0641862; \quad K_2 = 0.0947991, \\ H_1 &= \begin{bmatrix} -0.0136709\ 0.137607 \end{bmatrix}; H_2 = \begin{bmatrix} -0.0454111\ -0.098550 \end{bmatrix}. \end{aligned}$$

In Figure 7.5, the level set $\bigcup_{i=1}^{N} \varepsilon(P_i,\ 1)$ *is also well contained inside the regions of saturations.*

7.1.3 Synthesis of Unsaturating Controllers

The non-saturating controllers which work inside a region of linear behavior can be obtained from the previous results by replacing $D_{is} = \mathbb{I}$ and $D_{is}^{-} = 0$. The following result presents the synthesis of such controllers.

Theorem 7.5. *If there exist symmetric matrices X_i and matrices Y_i such that*

$$\begin{bmatrix} X_i & (A_{i\kappa} X_i + B_{i\kappa} Y_i C_{i\kappa})^T \\ * & X_j \end{bmatrix} > 0, \tag{7.30}$$

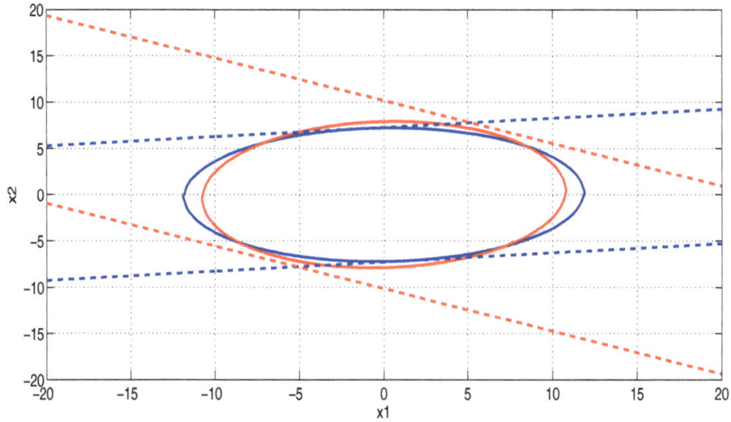

Fig. 7.6 Level sets and the corresponding polyhedral sets of linear behavior obtained with Theorem 7.5

$$\begin{bmatrix} 1 & Y_{il}C_{i\kappa} \\ * & X_i \end{bmatrix} > 0, \tag{7.31}$$

$$V_i C_{i\kappa} = C_{i\kappa} X_i, \tag{7.32}$$

$\forall\ \kappa = 1,\ldots,\mu_i,\quad \forall (i,j) \in \mathscr{I}^2, \forall l \in [1,m]$, with, $K_i = Y_i V_i^{-1}, P_i = X_i^{-1}$, then the uncertain closed-loop switching system (7.10) is asymptotically stable $\forall\ x_0 \in \Omega$ and for all switching sequences $\alpha(k)$.

To illustrate this result, the same system as Example 7.1 is used.

Theorem 7.5 leads to the following results:

$$P_1 = \begin{bmatrix} 0.2574 & 0 \\ 0 & 0.2574 \end{bmatrix}; P_2 = \begin{bmatrix} 0.2930 & 0.0535 \\ 0.0535 & 0.3376 \end{bmatrix};$$

$$K_1 = -0.0902; \quad K_2 = 0.1694.$$

In Figure 7.6, the level set $\bigcup_{i=1}^{N} \varepsilon(P_i,\ 1)$ is also well contained inside the regions of linear behavior.

Comment 7.1. *The application of all the proposed results to the same example, shows that the result applied in two steps (Theorem 6.5 and Corollary 7.1), the result using the cone complementarity technique (Theorem 6.6) and the result using unsaturating controllers (Theorem 7.5) lead to similar level sets which are less conservative than the ones of Theorems 7.2 and 7.3.*

In this section, different sufficient conditions of asymptotic stability are obtained for output feedback control of uncertain switching discrete-time linear systems subject to actuator saturations . These conditions allow the synthesis of stabilizing controllers inside a large region of saturation under LMIs formulation. The unsaturating controller result is easily deduced. The synthesis of controllers by state feedback control can also be obtained as a particular case of the study presented in this section. An illustrative example is studied by using the direct resolution of the proposed LMIs. A comparison of the obtained solutions is also given.

7.2 Stabilization of Saturated Switching Systems with Parametric Uncertainties

The objective of this section is to extend the results of Chapter VI to uncertain switching systems subject to actuator saturations by using output feedback control. This technique allows to design stabilizing controllers by output feedback for switching discrete-time systems despite the presence of actuator saturations and uncertainties on the system parameters. The case of state feedback control is derived as a particular case. It is also shown that the results obtained in this section with state feedback control are less conservative than those presented in [165] where only the state feedback control case is addressed. The main results of this section are published in [31, 32].

7.2.1 Problem Presentation

Let us consider the linear uncertain discrete-time switching system described by

$$\begin{cases} x_{k+1} = \mathscr{A}_\alpha(k)x_k + \mathscr{B}_\alpha(k)sat(u_k) \\ y_k = \mathscr{C}_\alpha(k)x_k \end{cases} \tag{7.33}$$

where $x_k \in \mathbb{R}^n$, $u_k \in \mathbb{R}^m$ are the state and the input respectively, $sat(.)$ is the standard saturation, $y_k \in \mathbb{R}^p$ the output. α is a switching rule taking its values in the finite set $\mathscr{I} = \{1,...,N\}$.
The saturation function is assumed here to be normalized:
i. e., $|sat(u_l| = min(1,|u_l|), l = 1,\dots m.$
The system matrices are assumed to be uncertain and satisfy

$$[\mathscr{A}_i(k) \quad \mathscr{B}_i(k)] = [A_i \quad B_i] + M_i\Gamma_i(k)\,[N_{1i} \quad N_{2i}]\ . \tag{7.34}$$

The following assumptions are required:

- The switching system is stabilizable.
- Matrices $C_\alpha$are of full rank.
- Each time only one subsystem is active.
- $\Gamma_i^T(k)\Gamma_i(k) \leq \mathbb{I}, \forall k \in \mathbb{N}$.

Let the control be obtained by an output feedback control law

$$u_k = K_\alpha y_k = K_\alpha C_\alpha x_k = F_\alpha x_k \ .$$

The closed-loop system is given by

$$x_{k+1} = \mathscr{A}_\alpha(k)x_k + \mathscr{B}_\alpha(k)sat(K_\alpha C_\alpha x_k) \tag{7.35}$$
$$x_{k+1} = \mathscr{A}_\alpha(k)x_k + \mathscr{B}_\alpha(k)sat(F_\alpha x_k) \ . \tag{7.36}$$

Defining the indicator function:

$$\xi(k) := [\xi_1(k), ..., \xi_N(k)]^T \tag{7.37}$$

where $\xi_i(k) = 1$ if the switching system is in mode i and 0 otherwise, yields the following representation for the closed-loop system:

$$x_{k+1} = \sum_{i=1}^{N} \xi_i(k)[\mathscr{A}_i x_k + \mathscr{B}_i sat(K_i C_i x_k)] \tag{7.38}$$

The following result of [165] extends the ones of [22, 25] to saturated switching systems with parametric uncertainties by using feedback control.

Theorem 7.6. *[165] If there exist symmetric matrices X_i, matrices Y_i, Z_i, $i = 1, \dots N$ and a real positive scalar β, such that*

$$\begin{bmatrix} X_i & W_{is}^T & S_{is}^T \\ * & X_j - \beta M_i M_i^T & 0 \\ * & * & \beta \mathbb{I} \end{bmatrix} > 0,$$

$$\forall (i,j) \in \mathscr{I} \times \mathscr{I}, \forall s \in [1, \dots \eta] \tag{7.39}$$
$$\begin{bmatrix} 1 & Z_{il} \\ * & X_i \end{bmatrix} > 0, \tag{7.40}$$
$$\forall i \in \mathscr{I}, \forall l \in [1, \dots m],$$

where $W_{is} = A_i X_i + B_i(D_{is}Y_i + D_{is}^- Z_i)$ and $S_{is} = N_{1i}X_i + N_{2i}(D_{is}Y_i + D_{is}^- Z_i)$. Then, the uncertain switching system with input saturation in closed-loop (7.36) with

$$F_i = Y_i X_i^{-1}, H_i = Z_i X_i^{-1}, P_i = X_i^{-1},$$

is asymptotically stable $\forall\ x_0 \in \Omega = \bigcup_{i=1}^{N} \varepsilon(X_i^{-1},\ 1)$ and for all switching sequences $\alpha(k)$.

7.2.2 Analysis and Synthesis of Stabilizability

Consider now the saturated uncertain switching system given by (7.38). The first result synthesizing stabilizing controllers of the uncertain saturated switching system by output feedback is now presented.

Theorem 7.7. *If there exist symmetric matrices X_i, matrices Y_i, Z_i, V_i, $i = 1,\dots N$ and a set of real positive scalars λ_{ijs}, such that*

$$\begin{bmatrix} X_i & \Theta_{is}^T & \Phi_{is}^T \\ * & X_j - \lambda_{ijs} M_i M_i^T & 0 \\ * & * & \lambda_{ijs}\mathbb{I} \end{bmatrix} > 0,$$

$$\forall (i,j) \in \mathscr{I} \times \mathscr{I}, \forall s \in [1,\dots \eta] \tag{7.41}$$

$$C_i X_i = V_i C_i \tag{7.42}$$

$$\begin{bmatrix} 1 & Z_{il} \\ * & X_i \end{bmatrix} > 0, \tag{7.43}$$

$$\forall i \in \mathscr{I}, \forall l \in [1,\dots m],$$

where $\Theta_{is} = A_i X_i + B_i(D_{is} Y_i C_i + D_{is}^{-} Z_i)$ and $\Phi_{is} = N_{1i} X_i + N_{2i}(D_{is} Y_i C_i + D_{is}^{-} Z_i)$. Then, the uncertain switching system with input saturation in closed-loop (7.38) with

$$K_i = Y_i V_i^{-1}, H_i = Z_i X_i^{-1}, P_i = X_i^{-1},$$

is asymptotically stable $\forall\, x_0 \in \Omega = \bigcup_{i=1}^{N} \varepsilon(X_i^{-1},\ 1)$ and for all switching sequences $\alpha(k)$.

Proof: By using Lemma 2.1, for all $H_i \in \mathbb{R}^{m\times n}$ with $|H_{il} x_k| < 1, l \in [1,m]$, there exist $\delta_{i1} \geq 0$,..., $\delta_{i\eta} \geq 0$ such that, $sat(F_i x_k) = \sum_{s=1}^{\eta} \delta_{is}(k)[D_{is} F_i + D_{is}^{-} H_i] x_k$, $\delta_{is}(k) \geq 0$, $\sum_{s=1}^{\eta} \delta_{is}(k) = 1$.

Consider the Lyapunov function candidate $V(x) = x_k^T(\sum_{i=1}^{N} \xi_i(k) P_i) x_k$. Computing its rate of increase along the trajectories of system (7.38) yields:

$$\Delta V(x_k) = x_{k+1}^T(\sum_{j=1}^{N} \xi_j(k+1) P_j) x_{k+1} - x_k^T(\sum_{i=1}^{N} \xi_i(k) P_i) x_k$$

$$= \sum_{s=1}^{\eta}\sum_{j=1}^{N} \xi_j(k+1)\delta_{is} x_k^T [\mathscr{A}_i + \mathscr{B}_i(D_{is}F_i + D_{is}^{-}H_i)]^T P_j [\mathscr{A}_i + \mathscr{B}_i(D_{is}F_i + D_{is}^{-}H_i)] x_k$$

$$- \sum_{i=1}^{N} \xi_i(k) x_k^T P_i x_k$$

Since, $\sum_{s=1}^{\eta} \delta_{is}(k) = \sum_{j=1}^{N} \xi_j(k+1) = \sum_{i=1}^{N} \xi_i(k) = 1$, one should obtain

$$\Delta V(x_k) = \sum_{j=1}^{N}\sum_{i=1}^{N}\sum_{s=1}^{\eta} \xi_i(k)\xi_j(k+1)\delta_{is}(k)$$

$$x_k^T \left(\left[\mathscr{A}_i + \mathscr{B}_i(D_{is}F_i + D_{is}^{-}H_i)\right]^T P_j \left[\mathscr{A}_i + \mathscr{B}_i(D_{is}F_i + D_{is}^{-}H_i)\right] - P_i \right) x_k \,.$$

A sufficient condition to obtain $\Delta V(x_k) < 0$ is that

$$-\Psi_{ijs} = \left[\mathscr{A}_i + \mathscr{B}_i(D_{is}F_i + D_{is}^- H_i)\right]^T P_j \left[\mathscr{A}_i + \mathscr{B}_i(D_{is}F_i + D_{is}^- H_i)\right] - P_i < 0 \quad (7.44)$$

By applying Schur complement to (7.44), the following equivalent inequality is obtained:

$$\begin{bmatrix} P_i & [\mathscr{A}_i + \mathscr{B}_i(D_{is}K_iC_i + D_{is}^- H_i)]^T P_i \\ * & P_j \end{bmatrix} > 0\,. \quad (7.45)$$

Letting $X_i = P_i^{-1}$, $Y_i = K_iV_i$, $C_iX_i = V_iC_i$, $Z_i = H_iX_i$ and multiplying the above inequality on both sides by $diag(X_i, X_j)$ we get

$$\begin{bmatrix} X_i & [\mathscr{A}_iX_i + \mathscr{B}_i(D_{is}K_iC_i + D_{is}^- H_i)X_i]^T \\ * & X_j \end{bmatrix} > 0\,. \quad (7.46)$$

Taking account of (7.34), inequality (7.46) can be developed as follows:

$$-\begin{bmatrix} X_i & [A_iX_i + B_i(D_{is}Y_iC_i + D_{is}^- Z_i)]^T \\ * & X_j \end{bmatrix} +$$
$$\begin{bmatrix} [N_{1i}X_i + N_{2i}(D_{is}Y_iC_i + D_{is}^- Z_i)]^T \\ 0 \end{bmatrix} \Gamma_i^T \begin{bmatrix} 0 & -M_i^T \end{bmatrix} +$$
$$\begin{bmatrix} 0 \\ -M_i \end{bmatrix} \Gamma_i \begin{bmatrix} [N_{1i}X_i + N_{2i}(D_{is}Y_iC_i + D_{is}^- Z_i)] & 0 \end{bmatrix} < 0\,.$$

By virtue of Lemma 6.1, this inequality holds if and only if there exist positive scalars λ_{ijs} such that

$$-\begin{bmatrix} X_i & \Theta_{is}^T \\ * & X_j \end{bmatrix} + \lambda_{ijs}\begin{bmatrix} 0 \\ -M_i \end{bmatrix}\begin{bmatrix} 0 & -M_i^T \end{bmatrix} + \frac{1}{\lambda_{ijs}}\begin{bmatrix} \Phi_{is} & 0 \end{bmatrix}\begin{bmatrix} \Phi_{is}^T \\ 0 \end{bmatrix} < 0,$$
$$\forall (i,j) \in \mathscr{I} \times \mathscr{I}, \forall s \in [1, \ldots \eta].$$

Or in a compact form,

$$\begin{bmatrix} X_i - \frac{1}{\lambda_{ijs}}\Phi_{is}\Phi_{is}^T & \Theta_{is}^T \\ * & X_j - \lambda_{ijs}M_iM_i^T \end{bmatrix} > 0, \quad (7.47)$$
$$\forall (i,j) \in \mathscr{I} \times \mathscr{I}, \forall s \in [1,\ \eta],$$

where Φ_{is} and Θ_{is} are defined before.

By Schur complement, inequality (7.47) is equivalent to (7.41). One can then bound the rate of increase as follows:

$$\Delta V(x_k) \leq -\gamma(\|x_k\|);$$
$$\gamma(\|x_k\|) = min_{ijs}\lambda_{min}(\Psi_{ijs})\|x_k\|^2.$$

Using [104], the inclusion condition (6.21) can also be transformed into the equivalent LMI (7.43) by virtue of the results of [53]. □

To obtain larger ellipsoid domains $\varepsilon(P_i,\ 1)$, we can use a shape reference set $\mathscr{X}_R \subset \mathbb{R}^n$, in terms of a polyhedron or ellipsoid to measure the size of the domain of attraction.

For a set $\mathscr{L} \subset \mathbb{R}^n$ which contains the origin, define

$$\mu(\mathscr{X}_R, \mathscr{L}) = sup\{\mu > 0,\ \ \mu\mathscr{X}_R \subset \mathscr{L}\}.$$

Here, we choose $\mathscr{X}_R$ to be a polyhedral defined as $\mathscr{X}_R = co\{\omega_1, \omega_2,\ \ldots\ , \omega_q\}$, where $\omega_1, \omega_2,\ \ldots\ , \omega_q$ are a prior given vectors in $\mathbb{R}^n$.

The problem can be formulated as the following constrained optimization problem:

$$(Pb.4): \begin{cases} \max_{X_i>0, Y_i, Z_i, \lambda_{ijs}} \mu_i \\ s.t.\ \ \mu\mathscr{X}_R \subset \varepsilon(P_i, 1) \\ (7.41)-(7.43), \\ i = 1, \ldots, N \end{cases}.$$

As explained in [103, 104], the constraint $\mu\mathscr{X}_R \subset \varepsilon(P_i, 1)$ is satisfied if the following matrix inequalities hold:

$$\begin{bmatrix} \mu_i^{-2} & \omega_l^T \\ \omega_l & X_i \end{bmatrix} \geq 0, \qquad (7.48)$$
$$\forall i \in \mathscr{I},\ \ \forall l \in [1,\ q].$$

The problem of enlarging the domain of attraction can be reduced to an LMI optimization problem defined as follows:

$$(Pb.5): \begin{cases} \min_{X_i>0, Y_i, Z_i, \lambda_{ijs}} \gamma_i \\ s.t.\ \ (7.41)-(7.43), (7.48) \\ i = 1, \ldots, N \end{cases}$$

where $\gamma_i = \mu_i^{-2}$.

Comment 7.2. *The results of Theorem 7.7 apply directly to switching systems with state feedback control by taking $C_i = \mathbb{I}$. In this case, these results can be compared to the one given by Theorem 7.6 of [165]. The fact that the scalars λ_{ijs} are all kept equal in Theorem 7.6, makes the result obviously more conservative. An example will show this conservatism.*

In order to more improve the result of Theorem 7.7 by introducing additional slack variables, the following corollary is presented:

Corollary 7.2. *If there exist symmetric matrices $X_i > 0$, matrices G_i, Y_i, V_i, Z_i and positive scalars λ_{ijs} such that*

$$\begin{bmatrix} G_i^T + G_i - X_i & \Upsilon_{is}^T & 0 & \Lambda_{is}^T \\ * & X_j & \lambda_{ijs} M_i & 0 \\ * & * & \lambda_{ijs}\mathbb{I} & 0 \\ * & * & * & \lambda_{ijs}\mathbb{I} \end{bmatrix} > 0, \tag{7.49}$$

$$\forall (i,j) \in \mathscr{I}^2, \forall s \in [1, \dots \eta]$$

$$C_i G_i = V_i C_i \tag{7.50}$$

$$\begin{bmatrix} 1 & Z_{il} \\ * & G_i^T + G_i - X_i \end{bmatrix} > 0, \tag{7.51}$$

$$\forall i \in \mathscr{I}, \forall s \in [1, \dots \eta], \forall l \in [1, \dots m]$$

where $\Upsilon_{is} = A_i G_i + B_i (D_{is} Y_i C_i + D_{is}^- Z_i)$ *and* $\Lambda_{is} = N_{1i} G_i + N_{2i} (D_{is} Y_i C_i + D_{is}^- Z_i)$.

Then, the uncertain switching system with input saturation in the closed-loop (7.38) with

$$K_i = Y_i V_i^{-1}, H_i = Z_i G_i^{-1}, P_i = X_i^{-1},$$

is asymptotically stable $\forall\ x_0 \in \Omega = \bigcup_{i=1}^{N} \varepsilon(X_i^{-1},\ 1)$ *and for all switching sequences* $\alpha(k)$.

Proof: It was proven in Chapter *III* that condition (7.45) is feasible if and only if there exists non-singular matrices G_i such that the following inequality holds:

$$\begin{bmatrix} G_i + G_i^T - X_i & G_i^T [\mathscr{A}_i + \mathscr{B}_i (D_{is} K_i C_i + D_{is}^- H_i)]^T \\ * & X_j \end{bmatrix} > 0, \tag{7.52}$$

$$\forall (i,j) \in \mathscr{I} \times \mathscr{I}, \forall s \in [1,\ \eta]$$

where $X_i = P_i^{-1}$. The same reasoning is then followed as in the proof of Theorem 7.7 leading to (7.49). Inequality (7.51) was also proven in [25] and recalled in Chapter *VI* by using [53]. □

These results can be illustrated with the following example.

Example 7.2. *Consider a SISO saturated switching discrete-time system with two modes given by the following matrices:*

$$A_1 = \begin{bmatrix} 1 & 1 \\ 0 & 1 \end{bmatrix}, B_1 = \begin{bmatrix} 10 \\ 5 \end{bmatrix}, M_1 = 0.1\mathbb{I}, N_{11} = N_{12} = 0.01[1\ \ 1]^T,$$

$$A_2 = \begin{bmatrix} 0 & -1 \\ 0.0001 & 1 \end{bmatrix}, B_2 = \begin{bmatrix} 0.5 \\ -2 \end{bmatrix}, M_2 = 0.1\mathbb{I}, N_{21} = N_{22} = 0.01[1\ \ 1]^T.$$

By solving the optimization problem (Pb.5) for the above system, we can obtain the following results:

$$P_1 = 1E-03\begin{bmatrix}4.3324 & 1.2516\\ 1.2516 & 4.3324\end{bmatrix}; P_2 = 1E-03\begin{bmatrix}4.3988 & 2.0934\\ 2.0934 & 6.1433\end{bmatrix},$$
$$H_1 = [-0.0261536 \ \ -0.0653823]; H_2 = [-0.0000192 \ \ 0.0717335]$$
$$K_1 = -0.1000089; K_2 = 0.1256683$$

The corresponding figures are given by Figure 7.7 while the sequence of switching and the uncertainties evolutions are given by Figures 7.8 and 7.9, respectively. Even the LMIs are feasible with uncertainty matrices $N_{ij} = 0.01[1 \ \ 1]^T$, *the simulations are plotted with* $N_{ij} = [1 \ \ 1]^T$.

By applying Corollary 7.2, the following results are obtained:

$$P_1 = 1E-03\begin{bmatrix}1.0047 & 0.1917\\ 0.1917 & 2.0626\end{bmatrix}; P_2 = 1E-04\begin{bmatrix}7.380 & 1.823\\ 1.823 & 23.530\end{bmatrix},$$
$$H_1 = [-0.0179753 \ \ -0.0405034]; H_2 = [-0.0000072 \ \ 0.0480397]$$
$$K_1 = -0.0597163; K_2 = 0.0365389.$$

The corresponding figures are given by Figure 7.10 while the sequence of switching and the uncertainties evolutions are given by Figures 7.8 and 7.9, respectively.

In order to compare the present results with Theorem 7.6 for state feedback control, the same example is considered. The results of Theorem 7.7 give:

$$P_1 = 1E-08\begin{bmatrix}8.656 & 60\\ 60 & 6900\end{bmatrix}, P_2 = 1E-08\begin{bmatrix}1.691 & 7.501\\ 7.501 & 6750\end{bmatrix},$$
$$H_1 = [-0.0002470 \ \ -0.0060812]; H_2 = [-0.0000081 \ \ 0.0081436]$$
$$F_1 = [-0.0003254 \ \ -0.0099873]; F_2 = [-0.0068339 \ \ 0.6156659],$$

while the results of [165] give:

$$P_1 = 1E-07\begin{bmatrix}2 & 8\\ 8 & 773\end{bmatrix}; P_2 = 1E-08\begin{bmatrix}4.894 & 20\\ 20 & 7630\end{bmatrix},$$
$$H_1 = [-0.0003203 \ \ -0.0066782]; H_2 = [-0.0000016 \ \ 0.0086701]$$
$$F_1 = [-0.0050262 \ \ -0.3821976]; F_2 = [-0.0112893 \ \ 0.3695061].$$

The corresponding level sets are depicted in Figures 7.11 and 7.12, where the conservatism of the results of [165] is obvious. The sequence of switching and the uncertainties evolutions are given in Figures 7.8 and 7.9, respectively.

This section studied uncertain switching systems with output feedback control which extends the results of Theorem 7.6 given with state feedback control. In order to compare these two results, a numerical study using only the particular case of the present work, as mentioned by the Comment 7.2, is also presented. The obtained improvements with the methods presented in this work are shown in Figures 7.11 and 7.12. A numerical example is used to illustrate all these techniques.

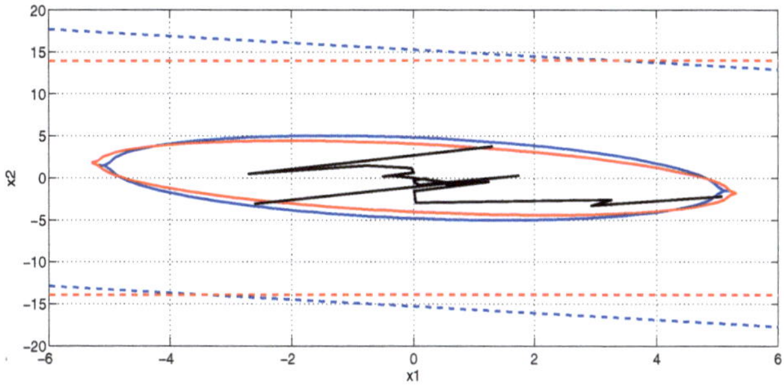

Fig. 7.7 Inclusion of the ellipsoids inside the polyhedral sets using Theorem 7.7

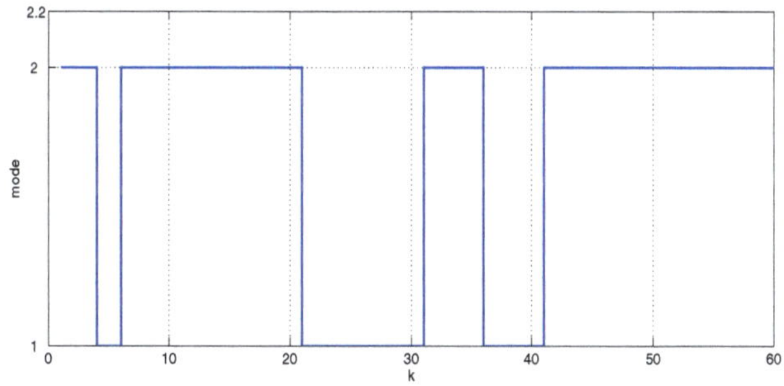

Fig. 7.8 The switching sequences

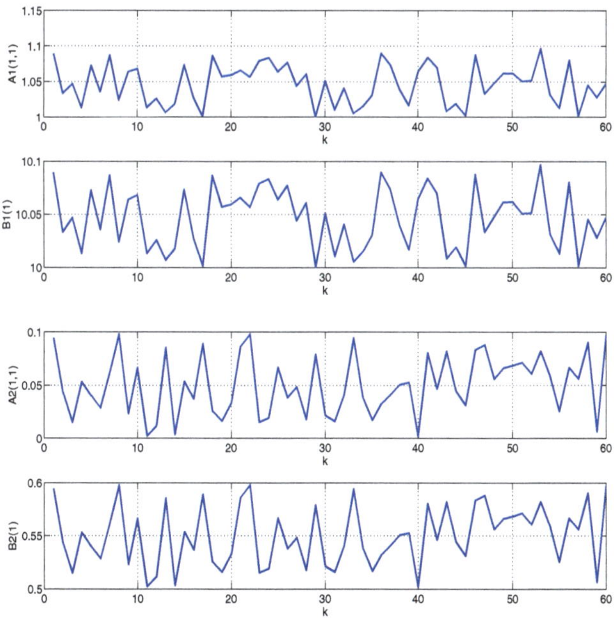

Fig. 7.9 The uncertainties evolution

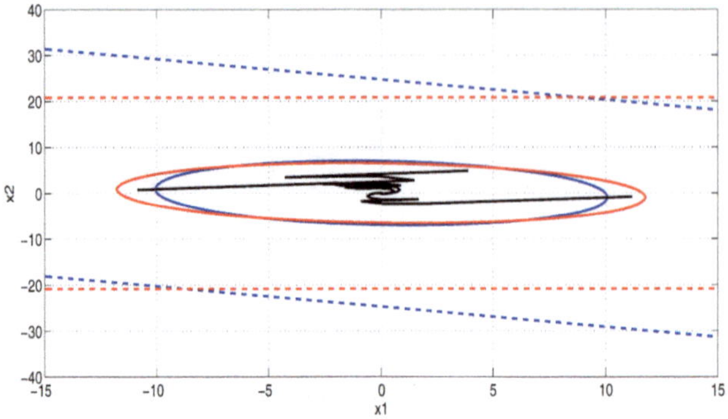

Fig. 7.10 Inclusion of the ellipsoids inside the polyhedral sets using Theorem 7.2

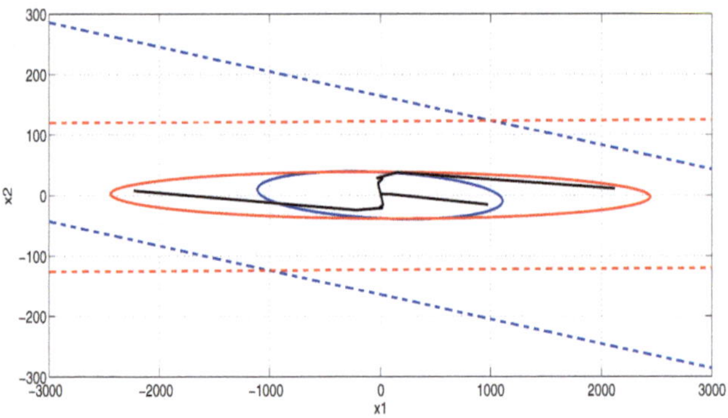

Fig. 7.11 Inclusion of the ellipsoids inside the polyhedral sets using Theorem 7.7

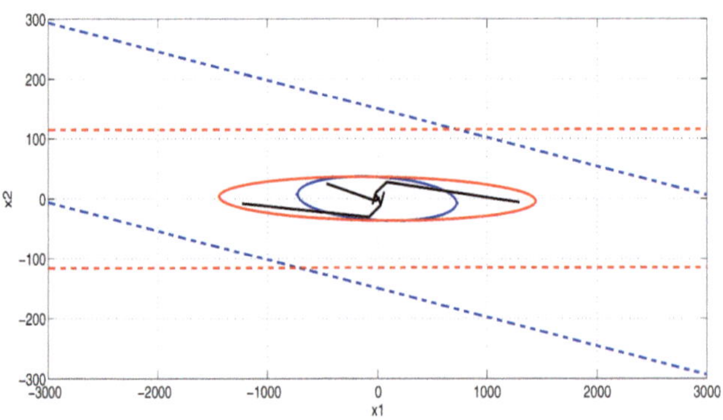

Fig. 7.12 Inclusion of the ellipsoids inside the polyhedral sets using Theorem 7.6

7.3 Stabilization of Uncertain State Saturated Discrete-Time Switched Systems

The objective of this section is to design a stabilizing switching law for uncertain switched system subject to state constraints. In this section, the switching sequence is not arbitrary as studied in the previous chapters while the subsystems are not necessarily all stable. In this case, the switching sequence is called the control law while the class of systems is called switched systems . Recall that when the switching sequence is arbitrary, the class of systems is known as switching systems. The type of uncertainty considered in this work is the polytopic one which is still of great interest even for linear systems [145]. This type of uncertainty was also studied, without saturation, in [98]. Thus, this work deals with controllers non tolerating saturations to take effect under polytopic uncertainties. The idea of including an ellipsoid set inside a polyhedral set is used. This work extends the result obtained in [92] for switched systems without uncertainties nor state constraints. These results were first published in [9].

7.3.1 Problem Presentation

Let us consider the uncertain saturated switched discrete-time linear system described by

$$x_{k+1} = A_\alpha(q_\alpha(k))x_k, \tag{7.53}$$

where $x_k \in \mathbb{R}^n$ is the state vector, the function $\alpha(k) : \mathbb{N} \longmapsto \mathscr{I}$ is a switching rule taking its values $\alpha(k)$ in the finite set $\mathscr{I} = \{1,...,N\}$ and $q_\alpha(k) \in \Gamma_\alpha \subset \mathbb{R}^{d_\alpha}$ are the bounded uncertainties that affect the system parameters in such a way that

$$A_\alpha(q_\alpha(k)) = A_\alpha + \sum_{h=1}^{d_\alpha} A_{\alpha h} q_{\alpha h}(k), \tag{7.54}$$

where matrix A_α represents the nominal matrix and $q_{\alpha h}(k)$ the hth component of vector $q_\alpha(k)$:

$$q_\alpha(k) = [q_{\alpha 1}(k),\ q_{\alpha 2}(k) \ldots q_{\alpha h}(k),\ \ldots q_{\alpha d_\alpha}(k)]^T.$$

Let convex sets Γ_i have p_i vertices $v_{il}, l = 1,\ldots,p_i$ so that for every $q_i \in \Gamma_i$, one can write $q_i = \sum_{l=1}^{p_i} \beta_{il} v_{il}$ with $\sum_{l=1}^{p_i} \beta_{il} = 1, 0 \le \beta_{il} \le 1$. The consequence is that each matrix $A_i(q_i(k))$ can be expressed as a convex combination of the corresponding vertices of the compact set Γ_i as follows:

$$A(q_i) := A_i + \sum_{l=1}^{p_i} \beta_{il} A(v_{il}) = \sum_{l=1}^{p_i} \beta_{il} A_{il},$$
$$A(v_{il}) = \sum_{h=1}^{d_i} A_{ih} v_{ilh},$$
$$A_{il} = A_i + A(v_{il}),$$
$$\sum_{l=1}^{p_i} \beta_{il} = 1, 0 \le \beta_{il} \le 1, \tag{7.55}$$

where A_i represents the nominal matrix. Note that the system without uncertainties can be obtained as a particular case of this representation by letting the vertices $v_{il} = 0, \forall i, \forall l$. Besides, Equations (7.55) are directly related to the dimension d_i of the convex compact set Γ_i. Hence, the uncertain saturated switched discrete-time linear system (7.53) can be described by

$$x_{k+1} = \sum_{l=1}^{p_i} \beta_{il} A_{il} x_k. \tag{7.56}$$

Consider the following system with uncertainties and constraints:

$$x_{k+1} = A_{\alpha(k)}(k) x_k, \tag{7.57}$$

with

$$\alpha(k) \in \{1, ..., N\}, \tag{7.58}$$

and

$$A_i(k) = A_i(q(k)) \in Co\{A_{i1}, A_{i2} ... A_{ip_i}\}. \tag{7.59}$$

Matrices A_i are not necessarily stable. Matrices A_{il} represent the vertices of the uncertainty polytope, $l = 1 .... p_i$ for the mode i.

The system is subject to the following state constraint

$$x_k \in \chi = \{x | C_x x < 1_v\}, \tag{7.60}$$

where $C_x \in \mathbb{R}^{p \times n}$, 1_v is a column vector of appropriate length containing only 1's. The goal is to determine the function $u(.) : \mathbb{R}^{\kappa} \to \{1, ..., N\}$, such that

$$\alpha(k) = u(x(k)) \tag{7.61}$$

makes the equilibrium point $x = 0$ of system (7.57) asymptotically stable without violation of the state constraint (7.60).

7.3.2 Preliminary Results

In this section, some preliminary results are recalled for their use in this work.

Consider the following system without uncertainties and constraints:

$$x_{k+1} = A_{\alpha(k)} x_k, \tag{7.62}$$

with

$$\alpha(k) \in \{1,...,N\}. \tag{7.63}$$

Define the simplex

$$\Lambda := \{\lambda \in \mathbb{R}^N : \sum_{i=1}^{N} \lambda_i = 1, \lambda_i \geq 0\}. \tag{7.64}$$

From the set of positive definite matrices $\{P_1 ..., P_N\}$, we define the following non-quadratic Lyapunov function

$$v(x) := \min_{i=1,...,N} x^T P_i x = \min_{\lambda \in \Lambda} (\sum_{i=1}^{N} \lambda_i x^T P_i x). \tag{7.65}$$

Denote the 1-level set of $v(x)$ as

$$L_v := \{x \in \mathbb{R}^n : v(x) \leq 1\}. \tag{7.66}$$

Let us recall from [92] the class of Metzler matrices for discrete-time systems denoted by $\mathscr{M}$ and constituted by all matrices $\Pi \in \mathbb{R}^{N \times N}$ with elements π_{ij} such that

$$\pi_{ij} \in [0,1], \sum_{i=1}^{N} \pi_{ij} = 1, \forall j. \tag{7.67}$$

For a positive definite matrix P_i, we define the following ellipsoids of $\mathbb{R}^n$

$$\xi(P_i, \rho) := \{x \in \mathbb{R}^n : x^T P_i x \leq \rho\}. \tag{7.68}$$

Define the union set of all the level sets associated to matrices $P_i, i = 1, \ldots, N$ as

$$\Omega = \bigcup_{i=1}^{N} \xi(P_i, 1) \tag{7.69}$$

Theorem 7.8. *[92] Assume that there exist $\Pi \in \mathscr{M}$ and a set of positive definite matrices $\{P_1, ..., P_N\}$ satisfying the Lyapunov-Metzler inequalities*

$$A_i^T (\sum_{j=1}^{N} \pi_{ji} P_j) A_i - P_i < 0, i = 1, ..., N, \tag{7.70}$$

the state switching control with

$$u(x(k)) = arg \min_{i=1,...,N} x^T(k) P_i x(k), \tag{7.71}$$

makes the equilibrium solution of (7.62) globally asymptotically stable.

The Lyapunov-Metzler inequalities(7.70) are nonlinear. The following result gives a simpler stability condition that can be expressed by means of LMIs .

Theorem 7.9. *[92] Let $Q_i \geq 0$ be given and assume that there exist P_i and a scalar $0 \leq \gamma \leq 1$ satisfying the following Lyapunov- Metzler inequalities*

$$A_i^T(\gamma P_i + (1-\gamma)P_j)A_i - P_i + Q_i < 0, j \neq i = 1,...,N, \tag{7.72}$$

then the state switching control with

$$u(x(k)) = arg \min_{i=1,...,N} x^T(k)P_i x(k) \tag{7.73}$$

makes the equilibrium solution of (7.62) globally asymptotically stable.

7.3.3 Stability Analysis

7.3.3.1 Unsaturated System

This section presents sufficient conditions of asymptotic stability of the uncertain switched system in term of a set of Lyapunov-Metzler inequalities. The point to be noticed is that these conditions do not require any stability property associated to each individual matrix of the set $\{A_1,\ldots,A_N\}$. However, the price to be paid is to deal with non-convex Lyapunov-Metzler inequalities, which are difficult to solve numerically.

Consider the system (7.56). The following result can be stated.

Theorem 7.10. *Assume that there exist $\Pi \in \mathcal{M}$ and a set of positive definite matrices $\{P_1,...,P_N\}$ satisfying the following Lyapunov-Metzler inequalities:*

$$A_{il}^T\left(\sum_{j=1}^{N} \pi_{ji} P_j\right)A_{il} - P_i < 0, i = 1,...,N; l = 1,...,p_i, \tag{7.74}$$

then the state switching control with

$$u(x(k)) = arg \min_{i=1,...,N} x^T(k)P_i x(k) \tag{7.75}$$

makes the equilibrium solution of system (7.56) asymptotically stable.

Proof: Assume that at an arbitrary time k, $\alpha(k) = u(x(k)) = i$ for $i \in \{1,...,N\}$. Hence, $v(x(k)) = x^T(k)P_i x(k)$ so from the system dynamic equation (7.56), we have the following developments:

$$\begin{aligned} v(x(k+1)) &= \min_{j=1,\ldots,N} x^T(k)A_i^T P_j A_i x(k), \\ &= \min_{\lambda\in\Lambda} x^T(k)A_i^T(\sum_{j=1}^{N} \lambda_j P_j)A_i x(k), \\ &\le x^T(k)A_i^T(\sum_{j=1}^{N} \pi_{ji} P_j)A_i x(k). \end{aligned}$$

This can be used to bound the rate of increase of the Lyapunov function candidate

$$\Delta v(x(k)) \le x^T(k)[A_i^T(\sum_{j=1}^{N} \pi_{ji} P_j)A_i - P_i]x(k).$$

It follows that the condition

$$A_i^T(\sum_{j=1}^{N} \pi_{ji} P_j)A_i - P_i < 0 \tag{7.76}$$

is sufficient to have $\Delta v(x(k)) < 0$. Let (7.76) be satisfied. By Schur complement, one can equivalently write

$$\begin{bmatrix} P_i & A_i \\ * & \Upsilon^{-1} \end{bmatrix} > 0, \tag{7.77}$$

with $\Upsilon = \sum_{j=1}^{N} \pi_{ji} P_j$.

As $\sum_{l=1}^{p_i} \beta_{il}(k) = 1$, one gets

$$\sum_{l=1}^{p_i} \beta_{il}(k) \begin{bmatrix} P_i & A_{il} \\ * & \Upsilon^{-1} \end{bmatrix} > 0. \tag{7.78}$$

It is obvious that the following inequality is sufficient to realize (7.78):

$$\begin{bmatrix} P_i & A_{il} \\ * & \Upsilon^{-1} \end{bmatrix} > 0. \tag{7.79}$$

Inequality (7.79) is equivalent, by Schur complement, to

$$A_{il}^T(\sum_{j=1}^{N} \pi_{ji} P_j)A_{il} - P_i) < 0, i = 1,\ldots,N; l = 1,\ldots,p_i,$$

which ensures the desired result. □

As used for Theorem 7.8, the result of Theorem 7.10 can be expressed in LMI form.

Theorem 7.11. *Let $Q_i \ge 0$ be given and assume that there exist matrices P_i and a scalar $0 \le \gamma \le 1$ satisfying the Lyapunov- Metzler inequalities*

$$A_{il}^T(\gamma P_i + (1-\gamma)P_j)A_{il} - P_i + Q_i < 0, \qquad (7.80)$$
$$j \neq i = 1,...,N, l = 1,...,p_i,$$

then the state switching control law given by

$$u(x(k)) = arg \min_{i=1,...,N} x^T(k) P_i x(k) \qquad (7.81)$$

makes the equilibrium solution of (7.56) globally asymptotically stable.

Proof: It follows readily from Theorem 7.9. □

In order to illustrate the obtained results, the following example is presented.

Example 7.3. *Consider a discrete-time system with two modes given by the following matrices:*

$$A_1(q_1(k)) = \begin{bmatrix} 1.0076 + q_1 & 0.0662 \\ 0.1323 & 0.4122 \end{bmatrix};$$
$$A_2(q_2(k)) = \begin{bmatrix} 0.9867 & 0.1527 + q_2 \\ -0.3054 & 2.2082 \end{bmatrix}.$$

The vertex matrices are:

$$A_{11} = \begin{bmatrix} 1.0096 & 0.0662 \\ 0.1323 & 0.4122 \end{bmatrix}; \ A_{12} = \begin{bmatrix} 1.0056 & 0.0662 \\ 0.1323 & 0.4122 \end{bmatrix};$$
$$A_{21} = \begin{bmatrix} 0.9867 & 0.1327 \\ -0.3054 & 2.2082 \end{bmatrix}; \ A_{22} = \begin{bmatrix} 0.9867 & 0.1727 \\ -0.3054 & 2.2082 \end{bmatrix}.$$

For $\gamma = 0.006$, the obtained optimal solutions are

$$P_1 = \begin{bmatrix} 670.48519 & 793.72622 \\ 793.72622 & 1009.213 \end{bmatrix};$$
$$P_2 = \begin{bmatrix} 269.64471 & 1111.5182 \\ 1111.5182 & 5567.0435 \end{bmatrix}.$$

7.3.3.2 Saturated System

In this subsection, the switched system with state constraint is studied. The results obtained so far are now extended to this class of systems. Consider now the system (7.57) with constraint (7.60).

Theorem 7.12. *Let $Q_i \geq 0$ be given and assume that there exist matrices P_i and scalars $0 \leq \gamma \leq 1$, $\rho > 0$ satisfying the following Lyapunov- Metzler inequalities*

$$A_{il}^T(\gamma P_i + (1-\gamma)P_j)A_{il} - P_i + Q_i < 0, \qquad (7.82)$$
$$j \neq i = 1,...,N, l = 1,...,p_i,$$

$$\begin{bmatrix} 1/\rho & C_x \\ * & P_i \end{bmatrix} > 0, \tag{7.83}$$

then the state switching control law

$$u(x(k)) = arg \min_{i=1,\ldots,N} x^T(k)P_ix(k) \tag{7.84}$$

makes the equilibrium solution of (7.57) with constraint (7.60) asymptotically stable for any $x_0 \in \xi(P_i, \rho), i = 1, \ldots, N$.

Proof: Condition (7.83) implies that $\xi(P_i, \rho) \subset \chi$ [53]. Using the same reasoning as in [106], the union set of all the ellipsoids $\xi(P_i, \rho)$ is equal to the level set L_v. It follows that $L_v \subset \chi$. This means that inside this level set, all the trajectories of the system satisfy the state constraint. Consequently, the system is asymptotically stable for any $x_0 \in \xi(P_i, \rho), i = 1, \ldots, N$. □.

This result is illustrated by the following example.

Example 7.4. *Consider a discrete-time system with two modes given by the following matrices:*

$$A_1(q_1(k)) = \begin{bmatrix} 0.9575 & 0.2185 \\ -0.4369 & 1.1759 + q_1 \end{bmatrix};$$

$$A_2(q_2(k)) = \begin{bmatrix} 1.0176 & 0.1659 + q_2 \\ 0.1659 & 0.6858 \end{bmatrix}.$$

The vertex matrices are:

$$A_{11} = \begin{bmatrix} 0.9575 & 0.2185 \\ -0.4369 & 1.1659 \end{bmatrix}; \; A_{12} = \begin{bmatrix} 0.9575 & 0.2185 \\ -0.4369 & 1.1859 \end{bmatrix};$$

$$A_{21} = \begin{bmatrix} 1.0176 & 0.1959 \\ 0.1659 & 0.6858 \end{bmatrix}; \; A_{22} = \begin{bmatrix} 1.0176 & 0.1359 \\ 0.1659 & 0.6858 \end{bmatrix},$$

with

$$C_x = [4\mathbb{I}_2; -4\mathbb{I}_2], \tag{7.85}$$

where $\mathbb{I}_2$ stands for the identity matrix 2 by 2. For $\gamma = 0.006$ and $\rho = 1$, the LMIs (7.82)-(7.83) are feasible. The obtained solutions are

$$P_1 = \begin{bmatrix} 16.1125 & 1.9559 \\ 1.9559 & 33.9938 \end{bmatrix}; \; P_2 = \begin{bmatrix} 19.4830 & 8.0680 \\ 8.0680 & 19.3410 \end{bmatrix}.$$

In Figure 7.13, the evolution of uncertainties used for simulation, is plotted. Figure 7.14 shows the switching sequence $\alpha(k)$.

Figure 7.15 illustrates the fact that the level set of stability $\bigcup_{i=1}^N \xi(P_i, \rho)$ is contained inside the sets of state constraints. Two trajectories are plotted to show the property of the set $\bigcup_{i=1}^N \xi(P_i, \rho)$.

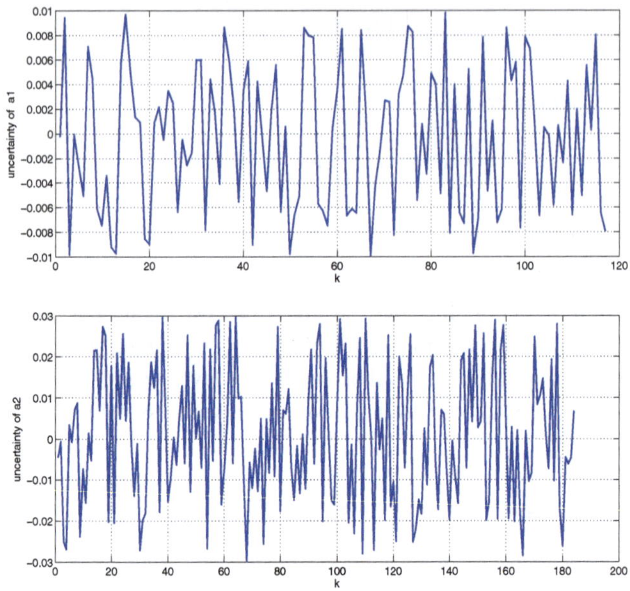

Fig. 7.13 Uncertainties evolution

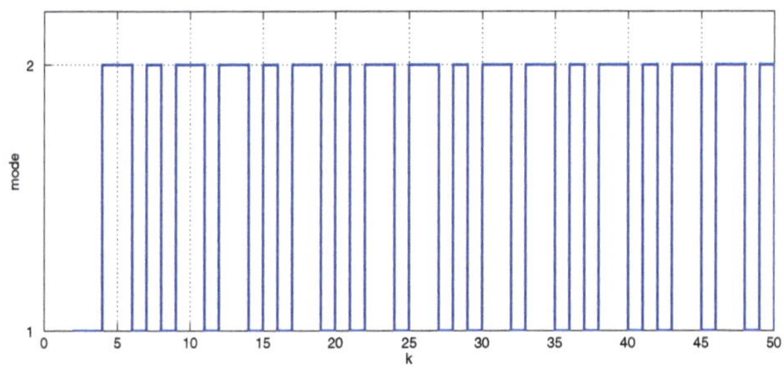

Fig. 7.14 Switching signals $\alpha(k)$

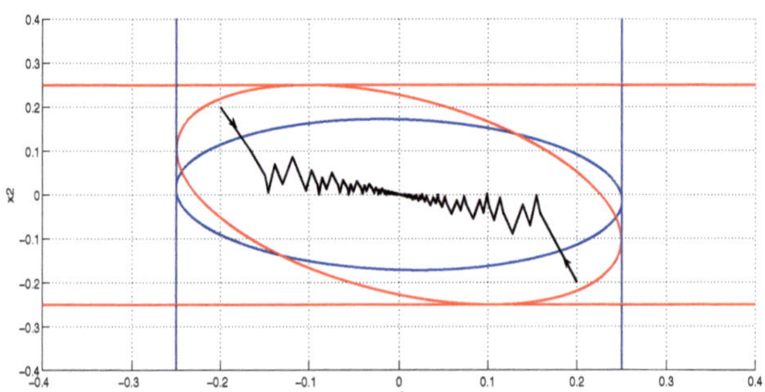

Fig. 7.15 Inclusion of the ellipsoids inside the polyhedral set of constraint together with two trajectories

In this section, sufficient conditions of asymptotic stability of discrete-time uncertain switched systems with state constraints are obtained. The main results of this section are given under LMI formulation. The proposed technique consists in designing switching sequences that render the switched system asymptotically stable, even in presence of state constraints and polytopic uncertainties. Illustrative examples are studied by using the proposed LMIs. As a perspective of this work, one can cite the extension of these techniques to deal with switching nonlinear systems by using the idea developed in [162].

7.4 Conclusion

In this chapter, sufficient conditions of stabilization of switching linear discrete-time systems with polytopic and parametric uncertainties are obtained. These conditions are given under LMIs form. Both the cases of state feedback control and output feedback control are studied for polytopic uncertainties. However, for parametric uncertainties, the output feedback control is presented extending the results of [165] given with state feedback control. A comparison study is given with a numerical particular case. The obtained improvements with the presented method are also shown. A numerical example is used to illustrate all these techniques. As an open subject, the work developed for switching systems with output feedback control and without saturation by [8], can be used with saturated controls. Besides, sufficient conditions of asymptotic stability of discrete-time uncertain switched systems with state constraints are presented.

Chapter 8
Stability and Stabilization of Positive Switching Linear Discrete-Time Systems

8.1 Introduction

In the literature, systems whose states are nonnegative, whenever the initial conditions are nonnegative, are referred to as positive [84]. The design of controllers for these positive systems has been studied (see, for example, [1], [99] and references therein). However, to the best of our knowledge, few works have directly considered positive switching linear systems as in [29]. We can also cite [146], which studies the stability of positive continuous-time switching systems composed of two subsystems. It must be pointed out that in [96], it is shown that the conjecture of [146], which says that the "Hurwitz stability of the convex hull of a set of Metzler matrices is a necessary and sufficient condition for the asymptotic stability of the associated switched linear system under arbitrary switching", is not true in general. Another interesting problem that has been studied in the literature is the reachability problem for positive discrete-time switching systems [143]. In particular, in [78], systems whose state variables remain nonnegative independently of the control and the external events are studied; then, control strategies are studied to maintain the nonnegative state evolution, against external events acting on the system. Although no synthesis method for building controllers ensuring this objective has been proposed, in [78], different positivity questions are addressed:

- Systems where state variables evolve while remaining nonnegative, independently of the chosen control, and the external events are analyzed.
- Systems where a control strategy can be designed to maintain the state evolution as nonnegative against external events acting on the system are also studied.

However, no synthesis method for building controllers to ensure this objective was proposed.

This chapter presents new results on the use of multiple Lyapunov functions for control positive switching systems, motivated by the studies of [146], which address the difficulty of constructing common Lyapunov quadratic functions for positive

A. Benzaouia: Saturated Switching Systems, LNCIS 426, pp. 195–216.
springerlink.com

switching systems. The idea is to extend the results for general switching discrete-time linear systems given in [66] and [128], in order to include the nonnegativity constraint on the states. To this end, a classical multiple Lyapunov function is used as in [66].

8.2 Stability Analysis for Positive Switching Linear Discrete-Time Systems

The goal of this section is to characterize the stability of positive switching discrete-time systems with multiple Lyapunov functions as it was done for positive linear systems. In this sense, this work completes the results of [34] by presenting the necessary condition for the positive switched system to admit a multiple Lyapunov function of the form $V(t,x)$. A linear program is also presented to construct such multiple Lyapunov functions. The results of this chapter are illustrated with an example composed of three modes.

8.2.1 Problem Formulation

This section presents the studied positive switching discrete-time linear systems. Thus, throughout the section, the following systems are considered:

$$\begin{aligned} x(t+1) &= A_\alpha x(t) \\ x(0) &= x_0 \succeq 0, \end{aligned} \tag{8.1}$$

where $x \in \mathbb{R}^n$ is the state, with initial value x_0, t is a nonnegative integer that gives the time-sample, and α is a switching rule that indicates the active mode (subsystem), taking values in the finite set $\mathscr{I}$.

It is assumed that:

- The switching rule α is not known *a priori*, but its value is available at each sampling period.
- Matrices $A_i, i = 1,\ldots,N$ are positive and stable ($i.e., A_i \succeq 0; \rho(A_i) \prec 1$).

This assumption corresponds to practical implementations, where the switched system is supervised by a discrete-event system.

The following definitions, given in [78], will be used throughout the chapter:

Definition 8.1. *System (8.1) is said to be positive if, given any nonnegative initial state x_0, the corresponding trajectory remains in the positive orthant for all integer t, i.e., $x(t) \in \mathbb{R}^n_+$.*

In this work, we are interested in the stability analysis for this class of positive hybrid systems. Upon introducing the indicator function:

$$\xi(t) = [\xi_1(t), \ldots, \xi_N(t)]^T \tag{8.2}$$

and upon assuming that when the system is in mode i, then $\xi_i(t) = 1$ and $\xi_j(t) = 0$ if $j \neq i$, one can write the system (8.1) as follows:

$$\begin{aligned} x(t+1) &= \sum_{i=1}^{N} \xi_i(t) A_i x(t), \\ x(0) &= x_0 \succeq 0. \end{aligned} \tag{8.3}$$

Although the existence of a common quadratic Lyapunov function for the various subsystems guarantees the asymptotic stability of the switching system, finding such a function is difficult [146] and might lead to conservative results. Thus, this chapter studies the multiple Lyapunov functions introduced in [54]. However, the idea of using the indicator function ξ_i to model the switching system (8.3) allows one to use a multiple Lyapunov function which can be seen as a common non quadratic Lyapunov function defined by

$$V(t,x) = x^T(t)\left(\sum_{i=1}^{N} \xi_i(t) P_i\right)x(t). \tag{8.4}$$

8.2.2 Preliminary Results

This section recalls three useful results for the sequel while characterizing the property of positivity of a linear system to make easy its extension to switching systems.

Other useful results for positive systems are also recalled.

Definition 8.2. *A square matrix N is said to be Z-matrix if all its off-diagonal elements are negative.*

Theorem 8.1. *[5] For a positive matrix M, $\mathbb{I} - M$ is an M-matrix if and only if there exists a positive vector v such that $(\mathbb{I} - M)v \succ 0$.*

Theorem 8.2. *[6] For a positive matrix M, $\mathbb{I} - M$ is an M-matrix if and only if there exists a diagonal positive definite matrix W such that matrix $M^T W M - W$ is negative definite.*

In order to adequately characterize positive linear systems, we propose the following examples. Consider a linear system given by:

$$x(t+1) = Ax(t). \tag{8.5}$$

First, consider the case of a non positive system with

$$A = \begin{bmatrix} 0.2 & 0 \\ -1. & 0.6 \end{bmatrix}. \tag{8.6}$$

Although this system is not positive, a diagonal matrix P can be found satisfying

$$A^T P A - P < 0, \tag{8.7}$$

given by

$$P = \begin{bmatrix} 1 & 0 \\ 0 & 0.5 \end{bmatrix}. \tag{8.8}$$

Second, consider a positive system with

$$A = \begin{bmatrix} 0.1 & 1 \\ 0 & 0.2 \end{bmatrix}. \tag{8.9}$$

Although this system is positive, a non diagonal matrix P can also be found satisfying

$$A^T PA - P < 0,$$

given by

$$P = \begin{bmatrix} 0.6725 & -0.1293 \\ -0.1293 & 1.2997 \end{bmatrix}. \tag{8.10}$$

In fact, the positive system (8.5), with A given by (8.9), admits the following positive matrix P:

$$P = \begin{bmatrix} 1 & 0 \\ 0 & 2 \end{bmatrix}.$$

However, a non positive system (8.5), with A given by

$$A = \begin{bmatrix} 1 & 0.5 \\ -0.1 & 0.2 \end{bmatrix},$$

does not admit a diagonal positive matrix P satisfying $A^T PA - P < 0$, although it is asymptotically stable. In order to distinguish between these two cases, the following should be noted:

- For an asymptotically stable positive linear system, it is always possible to find a diagonal positive matrix P satisfying $A^T PA - P < 0$.
- It is not always possible to find a diagonal positive matrix P satisfying $A^T PA - P < 0$ for a non positive asymptotically stable linear system.

Recall now the following result based on a non quadratic multiple Lyapunov function.

Theorem 8.3. *[123] The following three statements are equivalent:*

1) There exists a Lyapunov function of the form

$$v(t,x) = x^T(t) \sum_{i=1}^{N} \xi_i(t) A_i \lambda_i, \lambda_i \succ 0 \tag{8.11}$$

whose difference is negative definite, proving the stability of (8.3).

2) There exist N positive vectors λ_i such that

$$A_i^T \lambda_j \prec \lambda_i, \forall (i,j) \in \mathscr{I}^2. \tag{8.12}$$

3) There exist N diagonal positive matrices P_i such that

$$a_{ik}^T p_j - p_{ik} \prec 0, \forall (i,j) \in \mathscr{I}^2, k \in [1,\ n], \tag{8.13}$$

with $p_j = [p_{j1} \dots p_{jn}]^T$, $P_i = diag\{p_{i1} \dots p_{in}\}$ and a_{ik} the kth column vector of matrix A_i.

The objective of this section is to present a necessary and sufficient condition for a positive discrete-time switching system to admit a multiple Lyapunov function of type (8.4).

8.2.3 Stability Analysis with Multiple Lyapunov Functions

In this section, a necessary and sufficient condition for a positive switching system to admit a multiple Lyapunov function is presented.

Theorem 8.4. *The positive switching system (8.3) admits a multiple Lyapunov function of type (8.4) whose difference ΔV is negative definite if and only if there exist diagonal positive matrices $W_1, \dots, W_N$ such that*

$$A_i^T W_j A_i - W_i < 0, \forall (i,j) \in \mathscr{I}^2. \tag{8.14}$$

Proof: (Sufficiency): Let condition (8.14) be satisfied for the positive switching system (8.3) with positive diagonal matrices W_i. According to Theorem 3.3, function $V(t,x) = x^T(t)(\sum_{i=1}^N \xi_i(t) W_i) x(t)$ is a Lyapunov function of the switching system (8.3) whose difference ΔV is negative definite.

(Necessity): Assume that function $V(t,x)$ is a multiple Lyapunov function whose difference ΔV is negative definite for the positive switching system (8.3). This function ensures asymptotic stability of the positive switching system for any switching sequence $\alpha(t)$. This means that this function ensures, at the same time, the asymptotic stability of each positive subsystem. Hence, applying Theorem 3.3, condition (3.23) follows with matrices P_i non necessarily diagonal. That is, each matrix A_i is positive and stable which is equivalent to have $\mathbb{I} - A_i$ M-matrix [5]. Further, by virtue of Theorem 8.1, there exist positive vectors v_i such that

$$A_i v_i \prec v_i, \forall i \in \mathscr{I}. \tag{8.15}$$

It is always possible to select from vectors v_i satisfying (8.15), vectors such that

$$A_i v_i \prec v_j, \forall (i,j) \in \mathscr{I}^2. \tag{8.16}$$

A first way to realize this is to select vectors v_i such that

$$A_i v_i \prec w, w(k) = min_i\{v_i(k)\}, k = 1, \ldots, n. \tag{8.17}$$

A second way to realize (8.16) is to solve a linear programming $MX \prec 0$ where M and X are defined by (8.30).

Let

$$\varphi_{ij} = v_j - A_i v_i \succ 0. \tag{8.18}$$

The following developments can be addressed:

$$\begin{aligned} \Sigma &= \varphi_{ij}^T P_j \varphi_{ij} \succ 0 \\ &= v_j^T P_j v_j - v_i^T A_i^T P_j v_j - v_j^T P_j A_i v_i + v_i^T A_i^T P_j A_i v_i \end{aligned} \tag{8.19}$$

Using condition (3.23), one can write

$$\begin{aligned} 0 \prec \Sigma &\prec v_j^T P_j v_j - v_i^T A_i^T P_j v_j - v_j^T P_j A_i v_i + v_i^T P_i v_i \\ &\prec v_i^T \left[P_i v_i - A_i^T P_j v_j\right] + \left[v_i^T P_i - v_j^T P_j A_i\right] v_i + \left[v_j^T P_j v_j - v_i^T P_i v_i\right] \\ &\prec 2v_i^T \left[P_i v_i - A_i^T P_j v_j\right] + \left[v_j^T P_j v_j - v_i^T P_i v_i\right] \\ &\prec 2v_i^T \left[g_i - A_i^T g_j\right] + \left[v_j^T P_j v_j - v_i^T P_i v_i\right], \end{aligned} \tag{8.20}$$

with $g_i = P_i v_i$. The goal now is to show that

$$g_i - A_i^T g_j \succ 0. \tag{8.21}$$

First, consider the case $i = j$. It follows from (8.20) that in this case, $y_i = (\mathbb{I} - A_i^T)g_i \succ 0$. Recall now that condition (8.16) for $i = j$ means that matrix $\mathbb{I} - A_i$ is an M-matrix. This implies that matrix $\mathbb{I} - A_i^T$ is also an M-matrix ensuring that $(\mathbb{I} - A_i^T)^{-1}$ exists and is nonnegative [5]. Hence,

$$g_i = P_i v_i = (\mathbb{I} - A_i^T)^{-1} y_i \succ 0. \tag{8.22}$$

For the case, $i \neq j$, it is obvious that the term

$$\Psi_{ij} = \left[v_j^T P_j v_j - v_i^T P_i v_i\right], \tag{8.23}$$

has no constant sign for any selected vectors v_i satisfying (8.16) and $\forall (i, j) \in \mathscr{I}^2$. Consider the worst case of $\Psi_{ij} \succ 0$. The following developments can be presented:

$$\begin{aligned} v_j^T P_j v_j &\succ v_i^T P_i v_i, \\ v_j^T P_j v_j - v_j^T A_j^T P_i v_i &\succ v_i^T P_i v_i - v_j^T A_j^T P_i v_i \\ v_j^T \left[P_j v_j - A_j^T P_i v_i\right] &\succ \left[v_i^T - v_j^T A_j^T\right] P_i v_i \\ v_j^T \left[g_j - A_j^T g_i\right] &\succ \left[v_i^T - v_j^T A_j^T\right] g_i \end{aligned} \tag{8.24}$$

According to (8.22) and (8.16), $g_i \succ 0$ and $\left[v_i^T - v_j^T A_j^T\right] \succ 0$. Hence, $v_j^T \left[g_j - A_j^T g_i\right] \succ 0$ for any selected vectors v_j satisfying (8.16) and $\forall (i,j) \in \mathscr{I}^2$. Consequently, $A_j^T g_i \prec g_j, \forall (i,j) \in \mathscr{I}^2$. Finally, condition (8.21) is fulfilled for any selected vectors v_i satisfying (8.16).

Now, let us define the following diagonal matrices:

$$W_i = diag\left\{w_i(1), w_i(2), \ldots, w_i(n)\right\}, w_i(l) = \frac{g_i(l)}{v_i(l)}, l = 1, \ldots, n. \tag{8.25}$$

By developing $\Gamma_{ij} = W_i - A_i^T W_j A_i$ as follows:

$$\Gamma_{ij} v_i = W_i v_i - A_i^T W_j A_i v_i.$$

Using (8.16) and (8.25) while noting that by construction of W_i, $g_i = W_i v_i$, one obtains

$$\begin{aligned} \Gamma_{ij} v_i &= g_i - A_i^T W_j (v_j - \varphi_{ij}) \\ &= (g_i - A_i^T g_j) + A_i^T W_j \varphi_{ij}. \end{aligned} \tag{8.26}$$

Using (8.21), and remembering that $\varphi_{ij} \succ 0$, one can deduce that $\Gamma_{ij} v_i \succ 0, \forall (i,j) \in \mathscr{I}^2$. Then, for matrix W_i diagonal and positive, Γ_{ij} is a Z-matrix satisfying $\Gamma_{ij} v_i \succ 0$, with v_i a positive vector, this implies that Γ_{ij} is an M-matrix $\forall (i,j) \in \mathscr{I}^2$. Recall that Γ_{ij} is also symmetric, then matrix Γ_{ij} is positive definite. Consequently, there exist positive diagonal matrices W_i satisfying $W_i - A_i^T W_j A_i > 0, \forall (i,j) \in \mathscr{I}^2$. This completes the proof. □

The construction of positive diagonal matrices W_i is based on vectors v_i and g_i satisfying inequalities (8.16) and (8.21), respectively. These two conditions are sufficient for the existence of a multiple Lyapunov function with diagonal matrices for the positive switching system.

Corollary 8.1. *If there exist positive vectors $v_i, d_i, i = 1, \ldots, N$ such that*

$$\begin{cases} A_i v_i \prec v_j \\ A_i^T d_j \prec d_i, \end{cases} \tag{8.27}$$
$$\forall (i,j) \in \mathscr{I}^2,$$

then the positive switching system (8.3) admits a multiple Lyapunov function of type (8.4) whose difference ΔV is negative definite with matrices W_i given by

$$W_i = diag\left\{w_i(1), w_i(2), \ldots, w_i(n)\right\}, w_i(l) = \frac{d_i(l)}{v_i(l)}, l = 1, \ldots, n. \tag{8.28}$$

Proof: *Follows readily from the proof of Theorem 8.4.* □

In fact, the two sets of vectors v_i and g_i can be obtained easily by solving a linear program written upon (8.27) as given by the following linear programming:

$$\begin{pmatrix} M & 0 \\ 0 & N \\ -\mathbb{I} & 0 \\ 0 & -\mathbb{I} \end{pmatrix} \begin{pmatrix} X \\ Y \end{pmatrix} \prec 0, \tag{8.29}$$

with

$$M = \begin{bmatrix} A_1-\mathbb{I} & 0 & 0 & \dots \\ A_1 & -\mathbb{I} & 0 & \dots \\ A_1 & 0 & -\mathbb{I} & \dots \\ \dots & \dots & \dots & \dots \\ A_1 & \dots & \dots & -\mathbb{I} \\ -\mathbb{I} & A_2 & 0 & \dots \\ 0 & A_2-\mathbb{I} & 0 & \dots \\ 0 & A_2 & -\mathbb{I} & \dots \\ \dots & \dots & \dots & \dots \\ 0 & A_2 & \dots & -\mathbb{I} \\ \dots & \dots & \dots & \dots \\ -\mathbb{I} & 0 & \dots & A_N \\ 0 & -\mathbb{I} & \dots & A_N \\ \dots & \dots & \dots & \dots \\ 0 & 0 & \dots & A_N-\mathbb{I} \end{bmatrix}, \; X = \begin{bmatrix} v_1 \\ v_2 \\ v_3 \\ \dots \\ v_N \end{bmatrix}, \tag{8.30}$$

$$N = \begin{bmatrix} A_1^T-\mathbb{I} & 0 & 0 & \dots \\ -\mathbb{I} & A_1^T & 0 & \dots \\ \dots & \dots & \dots & \dots \\ -\mathbb{I} & 0 & \dots & A_1^T \\ A_2^T & -\mathbb{I} & 0 & \dots \\ 0 & A_2^T-\mathbb{I} & 0 & \dots \\ \dots & \dots & \dots & \dots \\ 0 & -\mathbb{I} & \dots & A_2^T \\ \dots & \dots & \dots & \dots \\ A_N^T & 0 & \dots & -\mathbb{I} \\ 0 & A_N^T & \dots & -\mathbb{I} \\ \dots & \dots & \dots & \dots \\ 0 & 0 & \dots & A_N^T-\mathbb{I} \end{bmatrix}, \; Y = \begin{bmatrix} d_1 \\ d_2 \\ d_3 \\ \dots \\ d_N \end{bmatrix}. \tag{8.31}$$

If this linear program is feasible in X and Y, one can easily construct diagonal matrices W_i by using (8.28).

Remark 8.1. *It is worth noting that the construction of positive diagonal matrices W_i satisfying inequalities (8.14) can also be obtained by solving LMIs of type (3.25) or LMIs (3.36) for unforced systems by imposing that the solutions $X_i, i = 1, \dots, N$ are diagonal.*

Example 8.1. *Consider a system described by (8.1), with the following sets of system matrices:*

$$A_1 = \begin{bmatrix} 0.5 & 0.2 \\ 0.1 & 0.6 \end{bmatrix}, \quad A_2 = \begin{bmatrix} 0.8 & 0.3 \\ 0.2 & 0.4 \end{bmatrix}, \quad A_3 = \begin{bmatrix} 0.4 & 0 \\ 0.1 & 0.5 \end{bmatrix}.$$

In order to compute vectors v_i satisfying (8.16), one can solve a linear programming $MX \prec 0$. One feasible solution is given by

$$v_1 = \begin{bmatrix} 8.9072 \\ 7.0028 \end{bmatrix}, \quad v_2 = \begin{bmatrix} 9.1534 \\ 5.2468 \end{bmatrix}, \quad v_3 = \begin{bmatrix} 8.8973 \\ 5.0925 \end{bmatrix}.$$

It can be noted here that $A_i v_i \prec w$, with $w = [8.8973 \;\; 5.0925]^T = min_i\{v_i(k)\}, k = 1,\ldots,n; i = 1,\ldots,3$ is fulfilled. According to Theorem 3.3, there exist non diagonal positive matrices P_i satisfying (3.23):

$$P_1 = \begin{bmatrix} 0.1400 & -0.0282 \\ -0.0282 & 0.1590 \end{bmatrix}, \quad P_2 = \begin{bmatrix} 0.1894 & -0.0198 \\ -0.0198 & 0.1449 \end{bmatrix},$$

$$P_3 = \begin{bmatrix} 0.1310 & -0.0328 \\ -0.0328 & 0.1444 \end{bmatrix}.$$

One can notice that the set of vectors g_i can be obtained by $g_i = P_i v_i$, as used in the proof of Theorem 8.4. These vectors are positive. The corresponding diagonal matrices W_i, obtained by using (8.25), form a feasible solution and are given by:

$$W_1 = \begin{bmatrix} 0.1178 & 0 \\ 0 & 0.1231 \end{bmatrix}, \quad W_2 = \begin{bmatrix} 0.1781 & 0 \\ 0 & 0.1104 \end{bmatrix},$$

$$W_3 = \begin{bmatrix} 0.1122 & 0 \\ 0 & 0.0871 \end{bmatrix}.$$

According to Remark 8.1, another way to compute matrices W_i is to solve directly LMIs (3.25) by imposing X_i diagonal matrices which are equivalent to the conditions of Theorem 8.4. This leads to the following matrices W_i:

$$W_1 = \begin{bmatrix} 1.3755 & 0 \\ 0 & 1.3789 \end{bmatrix}, \quad W_2 = \begin{bmatrix} 1.2923 & 0 \\ 0 & 1.0471 \end{bmatrix},$$

$$W_3 = \begin{bmatrix} 0.9659 & 0 \\ 0 & 1.0669 \end{bmatrix}.$$

In order to compute vectors v_i and d_i satisfying (8.27), one can solve the linear programming (8.29). One feasible solution is given by:

$$v_1 = \begin{bmatrix} 8.9072 \\ 7.0028 \end{bmatrix}, \quad v_2 = \begin{bmatrix} 9.1534 \\ 5.2468 \end{bmatrix}, \quad v_3 = \begin{bmatrix} 8.8973 \\ 5.0925 \end{bmatrix},$$

$$d_1 = \begin{bmatrix} 3.3933 \\ 2.7131 \end{bmatrix}, \quad d_2 = \begin{bmatrix} 2.9859 \\ 3.5265 \end{bmatrix}, \quad d_3 = \begin{bmatrix} 2.7414 \\ 2.3065 \end{bmatrix}.$$

The corresponding matrices W_i, given by (8.25), are as follows:

$$W_1 = \begin{bmatrix} 3.8096 & 0 \\ 0 & 3.8743 \end{bmatrix}, \; W_2 = \begin{bmatrix} 3.2621 & 0 \\ 0 & 6.7211 \end{bmatrix},$$
$$W_3 = \begin{bmatrix} 3.0812 & 0 \\ 0 & 4.5293 \end{bmatrix}.$$

All the stability conditions $A_i^T W_j A_i - W_i < 0, i, j = 1,3$ are checked.

A second example is now studied to compare the obtained results with the ones of Theorem 8.3.

Example 8.2. *The following example taken from [123] is studied:*

$$A_1 = \begin{bmatrix} a & 0.8 \\ 0.3 & 0.3 \end{bmatrix}, \; A_2 = \begin{bmatrix} b & 0.2 \\ 0.6 & 0.8 \end{bmatrix}$$

By varying $a \in [0, \; 2], b \in [0, \; 2]$ with a step of 0.1 and testing conditions of Theorem 8.4 by using LMIs (3.23) with X_i diagonal matrices, conditions of Corollary 8.1 given by (8.29) and those of Theorem 8.3, the following figure is obtained.

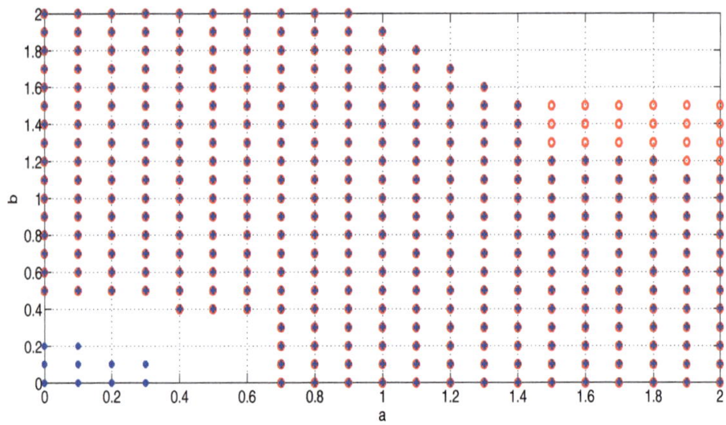

Fig. 8.1 Cases where conditions of Theorem 8.4 with LMIs (3.25) with X_i diagonal matrices (*), conditions of Corollary 8.1 given by (8.29) (+) and conditions of Theorem 8.3 (o) are satisfied

Figure 8.1 presents the cases where conditions of Theorem 8.4 by using LMIs (3.25) with X_i diagonal matrices, conditions of Corollary 8.1 given by (8.29) and those of Theorem 8.3 are feasible. This shows that the problem of existence of a multiple Lyapunov function of the form (8.4), with $P_i, i = 1, \ldots, N$ are positive diagonal matrices by using conditions of Corollary 8.1 under a linear program given by (8.29), has almost the same feasibility as the one of existence of a Lyaunov function *of type* $v(t,x) = x^T(t)\sum_{i=1}^{N}\xi_i(t)A_i\lambda_i, \lambda_i \succ 0$. *However, the problem of computing diagonal matrices* P_i *of Theorem 8.4 by using LMIs (3.25) with* X_i *diagonal matrices is more conservative. Finally, notice that the necessary and sufficient condition of Theorem 8.3 can be written under a linear programming of the form* $NY \prec 0, Y \succ 0$ *where* N, Y *are defined by (8.29). This constitutes a part of our global linear program (8.29), which explains the result obtained in Figure 8.1.*

A second example is studied for its particular stability in the following example.

Example 8.3. *Consider the switching system composed of two subsystems given by*

$$A_1 = \begin{bmatrix} 0 & 0.5 \\ 1 & 0 \end{bmatrix}, A_2 = \begin{bmatrix} 0 & 0.25 \\ \sqrt{5} & 0 \end{bmatrix}$$

Matrices A_1 *and* A_2 *are positive and stable. The simulation of two trajectories of this positive switching system, plotted in Figure 8.2, shows that the system is asymptotically stable.*

However, this system does not admit a multiple Lyapunov function of type (8.4) with diagonal matrices P_i*, even asymptotically stable. It seems to be a counterexample of Theorem 8.4. But, in fact, this system admits: neither a multiple Lyapunov function with non diagonal matrices* P_i *of type (8.4), computed with LMIs (3.23) nor with relaxed LMIs (3.36) for unforced systems, nor a common quadratic Lyapunov function. Besides, the multiple Lyapunov function proposed in Theorem 8.3 is not adequate. This particularity can be explained by the fact that these conditions of stability, with common or multiple Lyapunov functions, are only sufficient.*

This section has studied the problem of stability of positive switching linear discrete-time systems by using multiple Lyapunov functions. Necessary and sufficient conditions for positive switching discrete-time systems to admit multiple Lyapunov functions of type (8.4) are given. A linear program is presented to construct such multiple Lyapunov functions. The application of these results to an example composed of three subsystems is presented, showing the usefulness of the proposed conditions. A comparison study with the work of [123], upon a numerical example, is also presented. Finally, a particular example is presented to show the limitations of some sufficient conditions of stability of switching systems in general.

8.3 Stabilization of Positive Switching Linear Discrete-Time Systems

A necessary and sufficient condition for positive switching discrete-time linear systems to admit multiple Lyapunov functions was presented in the previous section. This condition is used in this section to synthesize the required controllers by using a set of LMIs for both state feedback and output feedback control. The extension to saturated switching systems is presented for unsaturating controllers.

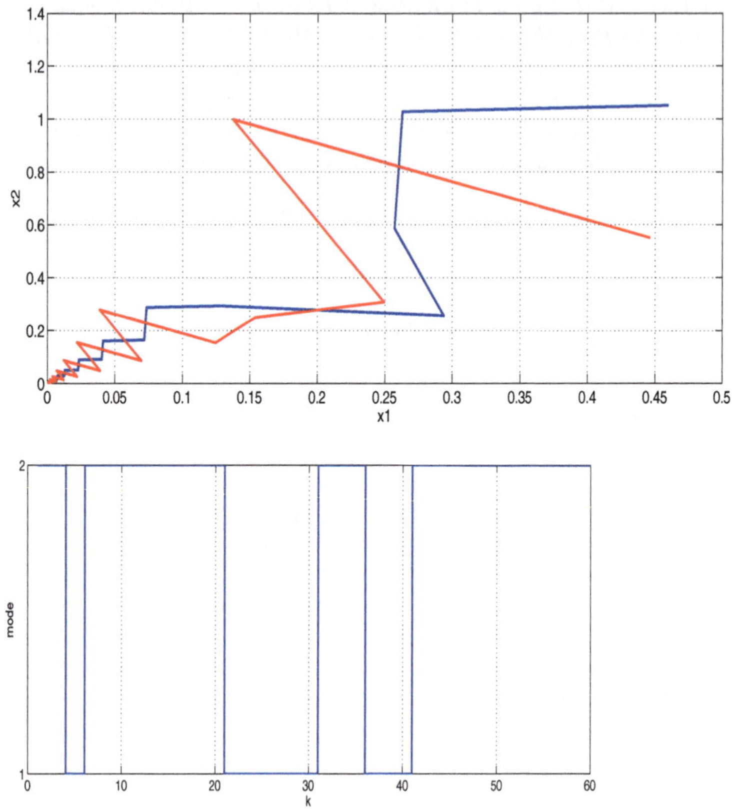

Fig. 8.2 Two Trajectories of the system in the positive orthant and the arbitrary sequence of switching

8.3.1 Problem Formulation

This section gives the description of the class of systems under consideration, namely, positive switching discrete-time linear systems. An equivalent description of such systems, based on the indicator function, is also presented.

Thus, throughout the section, the following switching discrete-time systems are considered:

$$\begin{aligned} x(t+1) &= A_\alpha x(t) + B_\alpha u(t) \\ y(t) &= C_\alpha x(t) \\ x(0) &= x_0 \succeq 0, \end{aligned} \tag{8.32}$$

where $x(t) \in \mathbb{R}^n$ is the state, with initial value x_0, $u(t) \in \mathbb{R}^m$ is the control, t is a non-negative integer that gives the sample index, and α is a switching rule that indicates the active mode (subsystem), taking values in the finite set $\mathscr{I} := \{1,\ldots,N\}$.

It is assumed that:

1) The switching system is stabilizable;
2) Matrices C_α are of full rank;
3) Only one subsystem is active at each time;
4) The switching rule α is not known *a priori* but its value is available at each sampling period.

These assumptions are frequent in the literature on switching systems, as they are valid for many systems found in practice: In particular, the last one corresponds to practical implementations, where the switched system is supervised by a discrete-event system (as is frequently the case in batch process, manufacturing, traffic control, etc.).

The following definitions given in [78] will be used throughout this section:

Definition 8.3. *System (8.32) is said to be controlled positive relative to an initial state $x_0 \succeq 0$ if there exists a control strategy such that the corresponding trajectory remains in the positive orthant for each integer t: i.e., $x(t) \in \mathbb{R}^n_+$.*

This work deals with the synthesis of stabilizing controllers for this class of hybrid systems, such that the resulting closed-loop system is positive and asymptotically stable. A feedback control law with gain that depends on the active mode is used. For the state feedback case, this is given by

$$u(t) = F_\alpha x(t), \tag{8.33}$$

so the closed-loop system can be written in this case as

$$x(t+1) = (A_\alpha + B_\alpha F_\alpha)x(t). \tag{8.34}$$

Upon introducing the indicator function:

$$\xi(t) = [\xi_1(t), \dots, \xi_N(t)]^T, \tag{8.35}$$

and assuming that when the system is in mode i, then $\xi_i(t) = 1$ and $\xi_j(t) = 0$ if $j \neq i$, one can write the closed-loop system (8.34) as follows:

$$x(t+1) = \sum_{i=1}^{N} \xi_i(t)\,(A_i + B_i F_i)\,x(t). \tag{8.36}$$

Alternatively, if output feedback is used:

$$u(t) = K_\alpha y(t), \tag{8.37}$$

$$\begin{aligned} x(t+1) &= (A_\alpha + B_\alpha K_\alpha C_\alpha)x(t) && (8.38)\\ &= \sum_{i=1}^{N} \xi_i(t)\,(A_i + B_i K_i C_i)\,x(t). && (8.39) \end{aligned}$$

8.3.2 Synthesis of Stabilizing Controllers by State Feedback Control

8.3.2.1 Unsaturated Case

In this subsection, only unsaturated switching systems are considered. The synthesis of the required controller by state feedback control for system (8.34) is given under LMI form. It is assumed here that the state is available.

First, Theorem 3.3 is used to provide a methodology to synthesize the controller according to the following result given in [34].

Theorem 8.5. *For the switching system (8.32), the following two statements are equivalent:*

i) For $1 \le i \le N$ there exist vectors $z_{i1}, \dots, z_{in} \in \mathbb{R}^m$ and diagonal matrices $X_i = diag\{\chi_{i1}, \dots, \chi_{in}\}$, with positive scalars χ_{ik}, such that:

$$\begin{bmatrix} X_i & \begin{bmatrix} \chi_{i1} a_{i1}^T + z_{i1}^T B_i^T \\ \cdots \\ \chi_{in} a_{in}^T + z_{in}^T B_i^T \end{bmatrix} \\ * & X_j \end{bmatrix} > 0, \ 1 \le j \le N, \tag{8.40}$$

$$\chi_{ik} a_{ik} + B_i z_{ik} \succeq 0, \ 1 \le k \le n. \tag{8.41}$$

where a_{ik} is the kth column vector of the matrix A_i (i.e., $A_i = [a_{i1} \ \dots \ a_{ik} \ \dots \ a_{in}]$).

ii) There exists a state feedback control $u(t) = F_i x(t)$ such that the closed-loop switching system (8.36) is positive, and admits a multiple Lyapunov function of type (8.4) whose difference ΔV is negative definite, proving asymptotic stability for any switching sequence $\alpha(t)$. The matrix F_i is then of the form

$$F_i = \begin{bmatrix} \frac{1}{\chi_{i1}} z_{i1} & \frac{1}{\chi_{i2}} z_{i2} & \cdots & \frac{1}{\chi_{in}} z_{in} \end{bmatrix}. \tag{8.42}$$

Proof:

Assume that *ii*) holds, so system (8.36) is positive and the functional $V(t,x)$ defined in (8.4) is a Lyapunov function whose difference ΔV is negative definite. Then, by virtue of Theorem 8.4, there exist diagonal positive definite matrices P_i satisfying $(A_i + B_i F_i)^T P_j (A_i + B_i F_i) - P_i < 0, \forall (i,j) \in \mathscr{I} \times \mathscr{I}$. By applying the Schur complement, the following equivalent inequality can be obtained:

$$\begin{bmatrix} P_i & (A_i + B_i F_i)^T P_i \\ * & P_j \end{bmatrix} > 0, \forall (i,j) \in \mathscr{I} \times \mathscr{I}, \tag{8.43}$$

which is also equivalent to

$$\begin{bmatrix} X_i & (A_i X_i + B_i Y_i)^T \\ * & X_j \end{bmatrix} > 0, \tag{8.44}$$

with $P_i = X_i^{-1}$ and $Y_i = F_iX_i$. Let $X_i = diag\{\chi_{i1}, \dots, \chi_{in}\} \succ 0$, then (8.42) can be written as $F_i = [z_{i1} \ z_{i2} \ \dots \ z_{in}]X_i^{-1}$, so $Y_i = F_iX_i = [z_{i1} \ z_{i2} \ \dots \ z_{in}]$. Since X_i is diagonal, $A_iX_i + B_iY_i$ can be expressed easily as follows: $A_iX_i + B_iY_i = [\chi_{i1}a_{i1} + B_iz_{i1} \ \dots \ \chi_{in}a_{in} + B_iz_{in}]$. By substituting in (8.44), one obtains inequalities (8.40). Furthermore, if the system (8.36) is positive, then $A_i + B_iF_i = A_i + B_iY_iX_i^{-1} \succeq 0$. Once post-multiplied with X_i, which is a positive matrix, this gives $A_iX_i + B_iY_i \succeq 0$. That is, $\chi_{is}a_{is} + B_iz_{is} \succeq 0$. Then, inequalities (8.41) are obtained.

Now, assuming that i) holds, let $P_i = X_i^{-1}$, $Y_i = [z_{i1}, \dots, z_{in}]$ and $F_i = Y_iX_i^{-1}$. The solution X_i of inequality (8.40) is positive definite. Since X_i is diagonal, then all its diagonal elements χ_{ik} are necessarily positive. It follows from this choice that $A_iX_i + B_iY_i = [\chi_{i1}a_{i1} + B_iz_{i1} \ \dots \ a_{in}\chi_{in} + B_iz_{in}] \succeq 0$. Thus, inequality (8.40) can be rewritten as (8.44) which is equivalent to (8.43). Using the Schur complement leads to $(A_i + B_iF_i)^T P_j(A_i + B_iF_i) - P_i < 0$ with P_i diagonal matrices. Besides, post-multiplying $A_iX_i + B_iY_i \succeq 0$ (using inequalities (8.44)) by $X_i^{-1} \succ 0$ leads to $A_i + B_iF_i \succeq 0$. Using Theorem 8.4, this implies that the switching positive system (8.36) admits a multiple Lyapunov function of type $V(t,x) = x^T(t)(\sum_{i=1}^N \xi_i(t)P_i)x(t)$, ensuring that the system is asymptotically stable at the origin for any arbitrary sequence of switching $\alpha(t)$. This completes the proof. □

Remark 8.2. *It is worth noting that Theorem 8.5 concerns the controlled positivity of the switching system: matrices A_i and B_i are not assumed to be positive. Moreover, matrices A_i are not supposed to be stable either, but nonetheless, the designed controllers give asymptotic stability and positivity for the closed-loop switching system. Thus, according to Definition 8.3, System (8.32) is said to be controlled positive.*

Remark 8.3. *The formulation in Theorem 8.5 concerns just an LMI plus an additional linear condition, so additional constraints can be easily added to consider particular problems. For example, it is possible to impose the nonnegativity of the control law by just adding constraints on the variables $z_1, \dots, z_n$ to make them nonnegative. In this case, when some of the matrices A_i or B_i are not positive, we are once again dealing with controlled positivity. However, when all matrices A_i and B_i are positive, we are dealing with the positivity of the switching system (8.32), according to Definition 8.3.*

Remark 8.4. *The conditions in Theorem 8.5 can be easily solved using off-the-shelf software, where a suitable cost function can be used to select a specific solution from the feasible ones. For example, to select a controller from those that give positivity and stability to the closed-loop system, the following maximization of the trace of matrices X_i can be used:*

$$(Pb.1) : \begin{cases} \sup_{X_i, z_{ij}} Trace(X_i) \\ s.t. \ (8.40), (8.41). \end{cases}$$

Example 8.4. *Consider a system described by (8.32), with the following sets of system matrices:*

$$A_1 = \begin{bmatrix} 0.3 & -0.1 \\ 0.4 & -0.2 \end{bmatrix}, A_2 = \begin{bmatrix} 0.1 & 0.2 \\ 0.3 & -0.4 \end{bmatrix}, A_3 = \begin{bmatrix} -0.2 & -0.1 \\ -0.6 & -0.3 \end{bmatrix},$$

$$B_1 = \begin{bmatrix} 2 & -1 \\ -0.6 & 1 \end{bmatrix}, B_2 = \begin{bmatrix} -1 & 4 \\ -3 & -1 \end{bmatrix}, B_3 = \begin{bmatrix} 2 & -0.9 \\ -0.7 & 2 \end{bmatrix}.$$

One feasible solution of the conditions of Theorem 8.5 gives the following matrices P_i:

$$P_1 = \begin{bmatrix} 0.0047 & 0 \\ 0 & 0.0042 \end{bmatrix}, P_2 = \begin{bmatrix} 0.0040 & 0 \\ 0 & 0.0045 \end{bmatrix},$$
$$P_3 = \begin{bmatrix} 0.0047 & 0 \\ 0 & 0.0042 \end{bmatrix},$$

with the following controller gains F_i:

$$F_1 = \begin{bmatrix} 0.0131 & 0.7027 \\ -0.0365 & 0.9749 \end{bmatrix}, F_2 = \begin{bmatrix} -0.0287 & -0.2726 \\ 0.0861 & -0.0769 \end{bmatrix},$$
$$F_3 = \begin{bmatrix} 0.5810 & 0.4255 \\ 0.6864 & 0.4668 \end{bmatrix}.$$

With these gains, the corresponding closed-loop matrices $A_{ci} = A_i + B_i F_i$ are as follows:

$$A_{c1} = \begin{bmatrix} 0.3627 & 0.3306 \\ 0.3556 & 0.3533 \end{bmatrix}, A_{c2} = \begin{bmatrix} 0.4729 & 0.1648 \\ 0.3000 & 0.4948 \end{bmatrix},$$
$$A_{c3} = \begin{bmatrix} 0.3442 & 0.3309 \\ 0.3661 & 0.3357 \end{bmatrix},$$

which can be seen to be positive and stable, with eigenvalues:

$$\sigma(A_{c1}) = \{0.7009; 0.0151\},$$
$$\sigma(A_{c2}) = \{0.7065; 0.2612\},$$
$$\sigma(A_{c3}) = \{0.6881; -0.0081\}.$$

It can easily be seen that all the main inequalities $A_{ci}^T P_j A_{ci} - P_i < 0, (i, j = 1, 2, 3)$, which ensure the asymptotic stability of the closed-loop switching system, are satisfied.

Simulation of this system in closed-loop starting from four different initial states under different arbitrary switching sequences give the state-space trajectories depicted in Figure 8.3, whereas Figure 8.4 presents one of the used arbitrary switching sequences.

8.3.2.2 Saturated Case

The obtained results in the previous section can be extended easily to saturated switching positive systems by using the results of Chapter VI. We assume in this

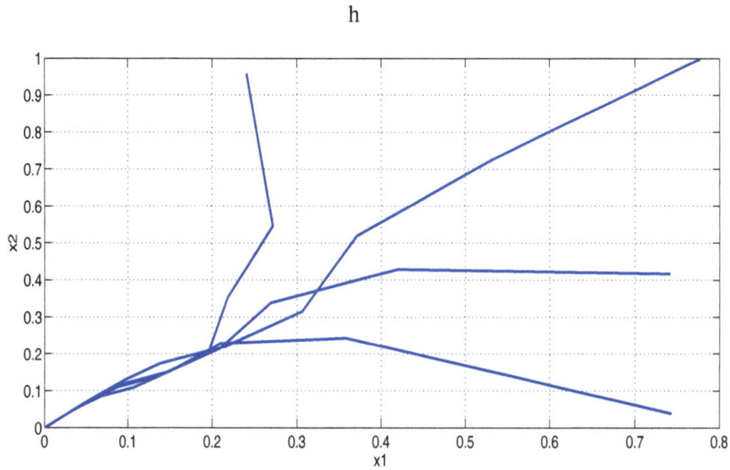

Fig. 8.3 Four trajectories in the positive orthant of the state space of the closed-loop switching system (8.45)

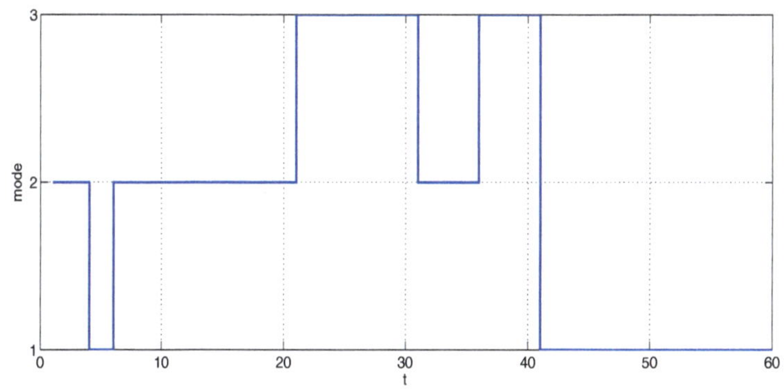

Fig. 8.4 An arbitrary switching sequence corresponding to one of the trajectories in Figure 8.3

subsection that the control is constrained between -1 and 1. We restrict ourselves to design unsaturating controllers.

Corollary 8.2. *If there exist vectors $z_{i1}, \dots, z_{in} \in \mathbb{R}^m$ and diagonal matrices $X_i = diag\{\chi_{i1}, \dots, \chi_{in}\}$, with positive scalars χ_{ik}, such that:*

$$\begin{bmatrix} X_i & \begin{bmatrix} \chi_{i1} a_{i1}^T + z_{i1}^T B_i^T \\ \cdots \\ \chi_{in} a_{in}^T + z_{in}^T B_i^T \end{bmatrix} \\ * & X_j \end{bmatrix} > 0, \tag{8.45}$$

$$\chi_{ik}a_{ik} + B_i z_{ik} \succeq 0, \tag{8.46}$$

$$\begin{bmatrix} 1 & Y_{li} \\ * & X_i \end{bmatrix} > 0, \tag{8.47}$$

$$\forall (i,j) \in \mathscr{I} \times \mathscr{I}, \forall l \in [1,m], \forall k \in [1,n],$$

then, the saturated switching system is asymptotically stable for any arbitrary switching sequence while guaranteeing positivity of the state with matrices F_i of the form

$$F_i = \begin{bmatrix} \frac{1}{\chi_{i1}} z_{i1} & \frac{1}{\chi_{i2}} z_{i2} & \cdots & \frac{1}{\chi_{in}} z_{in} \end{bmatrix}. \tag{8.48}$$

Proof: Follows obviously by considering Theorems 8.5 and 6.2 of Chapter VI. □

Remark 8.5. *In order to design saturating controllers, one has to use the equivalent form of the saturated switching system ($A_iX_i + B_iD_{is}Y_i + B_iD_{is}^{-}Z_i$) instead of ($A_iX_i + B_iY_i$) as presented in Chapter VI while replacing the LMI (8.47) by (6.28).*

8.3.3 Synthesis of Stabilizing Controllers by Output Feedback

Output feedback control plays a very important role in many control applications for switched systems, as knowledge of the state values is usually difficult, so an observer that takes into account the switching character, is needed [58]. The purpose of this section is then to extend the approach proposed in the previous section, for the design of controllers such that the resulting closed-loop system is positive and asymptotically stable, without using any reconstruction of the unavailable states. More precisely, in this section, we are interested in the stabilization of positive switching discrete-time linear systems by using static output feedback control [65], which is the most adequate for practical implementations, but known to be additionally complex for design, due to the necessity of including the output equation in the formulation.

8.3.3.1 Unsaturated Case

In this subsection, only unsaturated switching systems are considered. For this, the result of Theorem 8.4 can be used to synthesize the controller by output feedback according to the following result of [29]:

Theorem 8.6. *If there exist vectors $z_{i1}, \ldots, z_{im}$ in $\mathbb{R}^p$ and positive scalars $\chi_{i1}, \ldots, \chi_{in}$, $v_{i1}, \ldots, v_{ip}$ such that:*

$$\begin{bmatrix} X_i & [\zeta_{i1}, \ldots, \zeta_{in}]^T \\ * & X_j \end{bmatrix} > 0, \tag{8.49}$$

$$C_iX_i = V_iC_i \tag{8.50}$$

$$\zeta_{is} = \chi_{is}a_{is} + (\sum_{k=1}^{m} b_{ik}z_{ik}^T)c_{is} \succeq 0, \tag{8.51}$$

$$\forall (i,j) \in \mathscr{I}^2, s \in [1 \ n],$$

with $X_i = diag\{\zeta_{i1},\ldots,\zeta_{in}\}$, $A_i = [a_{i1} \ \ldots \ a_{in}]$, $B_i = [b_{i1} \ \ldots \ b_{in}]$, $C_i = [c_{i1} \ \ldots \ c_{in}]$, *then the closed-loop switching system (8.39) is positive and asymptotically stable at the origin for any switching sequence* $\alpha(t)$, *where* K_i *is given by*

$$K_i = [z_{i1} \ \ldots \ z_{im}]^T diag\left\{\frac{1}{v_{i1}},\ldots,\frac{1}{v_{ip}}\right\}. \tag{8.52}$$

Proof: The proof is parallel to that of Theorem 8.5: it is just necessary to use Theorem 8.4 with the closed-loop system (8.39) and an inequality similar to (3.33), obtained by using Equality (8.50). Then, using the proposed decompositions of system matrices, the matrix $A_i + B_iK_iC_i$ can be developed easily to obtain $(A_i + B_iK_iC_i)X_i = (A_iX_i + B_iK_iV_iC_i) = (A_iX_i + B_iY_iC_i) = [\zeta_{i1},\ldots,\zeta_{in}]$, with $K_iV_i = Y_i$. □

Remark 8.6. *It should be pointed out that the result of Theorem 8.6 does not require all the system matrices* A_i, B_i *and* C_i *to be positive (in fact matrices* A_i *might not be stable). However, the designed controllers allow asymptotic stability and positivity of the switching closed-loop system. Thus, Theorem 8.6 concerns the controlled positivity of the switching system.*

Remark 8.7. *The nonnegativity of the output feedback control law can also be taken into account by adding constraints on the variables:* $z_i \succeq 0, i = 1,\ldots,m$ *(provided that matrices* C_i *are positive). This corresponds to the concept of positivity in Definition 8.3.*

The output feedback result can be given under another formulation:

Corollary 8.3. *If there exist vectors* $z_{i1},\ldots,z_{im}$ *in* $\mathbb{R}^p$ *and positive scalars* d_i *such that:*

$$\begin{bmatrix} d_i\mathbb{I}_n & [\zeta_{i1},\ldots,\zeta_{in}]^T \\ * & d_j\mathbb{I}_n \end{bmatrix} > 0, \tag{8.53}$$

$$\zeta_{is} = d_ia_{is} + (\sum_{k=1}^{m} b_{ik}z_{ik}^T)c_{is} \succeq 0, \tag{8.54}$$

$$\forall (i,j) \in \mathscr{I}^2, s \in [1 \ n],$$

where $A_i = [a_{i1} \ \ldots \ a_{in}]$, $B_i = [b_{i1} \ \ldots \ b_{in}]$, $C_i = [c_{i1} \ \ldots \ c_{in}]$, *then the closed-loop switching system (8.39) is positive and asymptotically stable at the origin for any switching sequence* $\alpha(t)$, *where* K_i *is given by*

$$K_i = \left[\frac{1}{d_i}z_{i1} \ \cdots \ \frac{1}{d_i}z_{im}\right]^T. \tag{8.55}$$

Proof: Let $X_i = diag\{\chi_{i1},\ldots,\chi_{in}\}$ be given: then equality $C_iX_i = V_iC_i, i \in \mathscr{I}$ has a solution in V_i of the form $V_i = diag\{v_{i1},\ldots,v_{ip}\}$ if and only if $\chi_{i1} = \chi_{i2} = \cdots = \chi_{in} = v_{i1} = v_{i2} = \cdots = v_{ip}$. In fact, developing this equality with X_i and V_i diagonal matrices leads to $c^i_{lr}\chi^i_r = v^i_l c^i_{lr}, r = 1,\ldots,n; l = 1,\ldots,p; i \in \mathscr{I}$. That is, $\chi^i_r = v^i_l = d_i, r = 1,\ldots,n; l = 1,\ldots,p$. The reciprocal is obvious. Using the obtained solutions in Theorem 8.6, expressions (8.53), (8.54) and (8.55) follow. □

Example 8.5. *Consider the same switching system as in Example 8.4, with the following output matrices:*

$$C_1 = \begin{bmatrix} 1 & 0.5 \end{bmatrix}, C_2 = \begin{bmatrix} 1 & 1 \end{bmatrix}, C_3 = \begin{bmatrix} 0.5 & 1 \end{bmatrix}.$$

Applying the results of Theorem 8.6, one feasible solution is $v_{11} = 3063.52$, $v_{12} = 3912.9, v_{13} = 3049.1$, *which leads to the following controller gains:*

$$K_1 = \begin{bmatrix} 0.4776 \\ 0.7230 \end{bmatrix}, K_2 = \begin{bmatrix} -0.1412 \\ -0.0167 \end{bmatrix}, K_3 = \begin{bmatrix} 0.5936 \\ 0.8240 \end{bmatrix}.$$

Then, the closed-loop system matrices $A_{ci} = A_i + B_i K_i C_i$ *are as follows:*

$$A_{c1} = \begin{bmatrix} 0.5322 & 0.0161 \\ 0.8365 & 0.0182 \end{bmatrix}, A_{c2} = \begin{bmatrix} 0.1742 & 0.2742 \\ 0.7404 & 0.0404 \end{bmatrix},$$
$$A_{c3} = \begin{bmatrix} 0.0228 & 0.3457 \\ 0.0163 & 0.9325 \end{bmatrix},$$

which are positive and stable with eigenvalues: $\sigma(A_{c1}) = \{0.56; -0.0067\}$, $\sigma(A_{c2}) = \{0.56; -0.35\}$, $\sigma(A_{c3}) = \{0.94; 0.02\}$.

The corresponding Lyapunov matrices in Theorem 8.6 are $P_1 = 0.0003264\mathbb{I}_2$, $P_2 = 0.0002556\mathbb{I}_2$ *and* $P_3 = 0.0003280\mathbb{I}_2$*: It can be easily seen that the claim of Corollary 8.3 is checked. Besides, all the inequalities* $A_{ci}^T P_j A_{ci} - P_i < 0, (i, j = 1, 2, 3)$ *are fulfilled, ensuring the asymptotic stability of the switching discrete-time system. Four trajectories of the switching system are depicted in Figure 8.5. Figure 8.6 presents the switching sequence corresponding to one of these trajectories.*

8.3.3.2 Saturated Case

The obtained results in the previous section can be extended easily to saturated switching positive systems by using the results of Chapter VI. We assume again in this subsection that the control is constrained between -1 and 1. We restrict ourselves to designing unsaturating controllers by output feedback.

Corollary 8.4. *If there exist vectors* $z_{i1}, \ldots, z_{im}$ *in* $\mathbb{R}^p$ *and positive scalars* d_i *such that:*

$$\begin{bmatrix} d_i \mathbb{I}_n & [\zeta_{i1}, \ldots, \zeta_{in}]^T \\ * & d_j \mathbb{I}_n \end{bmatrix} > 0, \tag{8.56}$$

$$\zeta_{is} = d_i a_{is} + (\sum_{k=1}^{m} b_{ik} z_{ik}^T) c_{is} \succeq 0, \tag{8.57}$$

$$\begin{bmatrix} 1 & (Y_i C_i)_l \\ * & X_i \end{bmatrix} > 0, \tag{8.58}$$

$$\forall (i, j) \in \mathcal{I}^2, s \in [1, n], l \in [1, m],$$

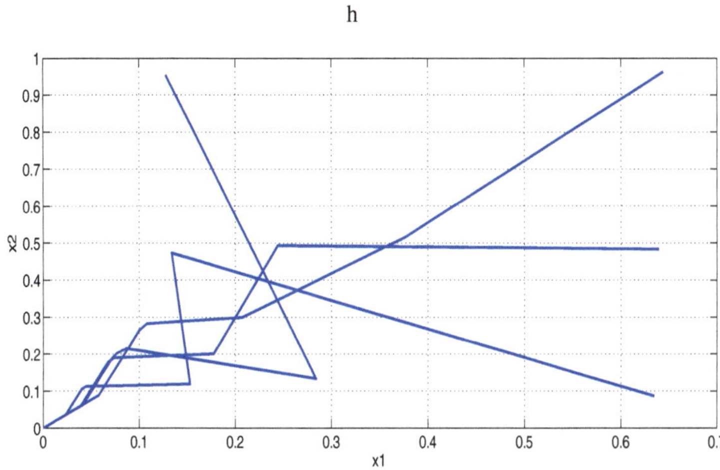

Fig. 8.5 Four trajectories in the positive orthant of the state space of the closed-loop switching system (8.56)

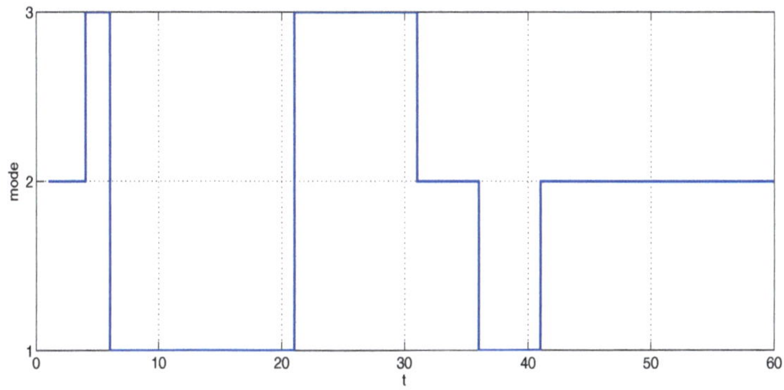

Fig. 8.6 An arbitrary switching sequence corresponding to one of the trajectories in 8.5

then the closed-loop switching system (8.39) is positive and asymptotically stable at the origin for any switching sequence $\alpha(t)$, where K_i is given by

$$K_i = \left[\frac{1}{d_i} z_{i1} \ \cdots \ \frac{1}{d_i} z_{im} \right]^T .$$

Proof: Follows obviously by considering Theorem 8.3 and Corollary 6.3 of Chapter VI. □

Remark 8.8. *In order to design saturating controllers, one has to use the equivalent form of the saturated switching system $(A_iX_i + B_iD_{is}Y_iC_i + B_iD_{is}^{-}Z_i)$ instead of $(A_iX_i + B_iY_iC_i)$ as presented in Chapter VI while replacing the LMI (8.58) by (6.34).*

8.4 Conclusion

This chapter has studied the problem of stability and stabilizability of positive switching linear discrete-time systems by using multiple Lyapunov functions. Necessary and sufficient conditions for positive switching discrete-time systems to admit multiple Lyapunov functions are given. These conditions are proposed in LMI form and can be used directly to synthesize the required controllers by state and output feedback control. The application of these results to examples composed by three subsystems is presented showing the usefulness of the proposed conditions. Further, these conditions can also be used to impose positiveness on the closed-loop switching system even if none of the open-loop subsystems is positive. It must be pointed out that the proposed methodology is constructive, in the sense that it can be easily extended to other switching systems, such as switching systems with delays [152]. Finally, the extension to saturated switching systems is presented for unsaturating controllers in both the cases of state feedback and output feedback controls.

Chapter 9
Stabilization of Discrete 2 D Switching Systems

9.1 Introduction

In many modeling problems of physical processes, a 2D switching representation is needed. One can cite 2D physically based model for advanced power bipolar devices [107] and heat flux switching and modulating in thermal transistor [124]. This class of systems can correspond to 2D state pace or 2D time space switching systems.

In the last two decades, the two-dimensional (2D) system theory has been paid considerable attention by many researchers. The 2D linear models were introduced in the seventies [87, 88, 93] and have found many applications, such as in digital data filtering, image processing [141], modeling of partial differential equations [127], etc. It is well known that 2D systems can be represented by different models as Roesser model, Fornasini-Marchesini model and Attasi model. For a complete description of these models and methods to transform the system from a model type to another, one can consult [109].

In connection with Roesser and Fornasini-Marchesini models, some important problems such as realization, controllability, minimum energy control, have been extensively investigated (see for example [88, 110]). On the other hand, the stabilization problem is not fully investigated and still not completely solved.

The stability of 2D discrete linear systems can be reduced to checking the stability of 2D characteristic polynomial [111]. This appears to be a difficult task for the control synthesis problem. In the literature, various types of easily checkable but only sufficient conditions for asymptotic stability and stabilization problems for 2D discrete linear systems have been proposed [91, 101, 125, 164].

This chapter is interested to discrete 2D time switching systems described with model Roesser. To the best of our knowledge, no works have directly considered 2D switching systems with arbitrary switching sequence before the work of [33]. One can only cite the work of [158] where the process of switching is considered as a Markovian jumping one. First of all, this work focuses on the understanding of the switching occurring for the 2D systems. To this end, a new time basis definition is proposed. As a consequence of this, an adequate indicator function is derived to model the discrete 2D time switching systems. This background enables one to

A. Benzaouia: Saturated Switching Systems, LNCIS 426, pp. 217–245.
springerlink.com

study the stabilizability problem of switching systems when the switching sequence is arbitrary. Hence, common Lyapunov quadratic and multiple Lyapunov functions are used for this class of systems for the first time in this work. In this context, sufficient conditions of stabilizability are presented. Furthermore, these conditions are presented in the form of a set of LMIs for state feedback and output feedback controls. The extension of the obtained results to the saturated 2D discrete switching systems is also presented for both the cases.

9.2 State Feedback Control

9.2.1 Problem Statement

In this section, the system under study is presented. A particular attention is given to understanding the switching of 2D systems. Hence, to model discrete switching systems, a new time basis definition together with the adequate indicator function is developed.

This section deals with switching systems described by the following Roesser model:

$$\begin{aligned} x^{+}(k,l) &= A_{\alpha}x(k,l)+B_{\alpha}u(k,l) \\ x^{h}(0,l) &= f(l), x^{v}(k,0)=g(k) \end{aligned} \tag{9.1}$$

with

$$x(k,l)=\begin{bmatrix} x^{h}(k,l) \\ x^{v}(k,l) \end{bmatrix}, x^{+}(k,l)=\begin{bmatrix} x^{h}(k+1,l) \\ x^{v}(k,l+1) \end{bmatrix}, \tag{9.2}$$

$$A_{\alpha}=\begin{bmatrix} A_{11}^{\alpha} & A_{12}^{\alpha} \\ A_{21}^{\alpha} & A_{22}^{\alpha} \end{bmatrix}, B_{\alpha}=\begin{bmatrix} B_{1}^{\alpha} \\ B_{2}^{\alpha} \end{bmatrix}, \tag{9.3}$$

where $x^{h}(k,l)$ is the horizontal state in $\mathbb{R}^{n_1}$, $x^{v}(k,l)$ is the vertical state in $\mathbb{R}^{n_2}$, $x(k,l)$ is the whole state in $\mathbb{R}^{n}$ with $n=n_1+n_2$, $u(k,l)$ is the control vector in $\mathbb{R}^{m}$. α is a switching rule which takes its values in the finite set $\mathscr{I} := \{1,\dots,N\}$, N is the number of subsystems (modes), k and l are integers in $\mathbb{Z}_{+}$. Matrices $A_{11}^{\alpha} \in \mathbb{R}^{n_1\times n_1}, A_{12}^{\alpha} \in \mathbb{R}^{n_1\times n_2}, A_{21}^{\alpha} \in \mathbb{R}^{n_2\times n_1}, A_{22}^{\alpha} \in \mathbb{R}^{n_2\times n_2}, B_{1}^{\alpha} \in \mathbb{R}^{n_1\times m}, B_{2}^{\alpha} \in \mathbb{R}^{n_2\times m}$ are constant.

It is assumed that:

- The switching system is stabilizable;
- At each time only one subsystem is active;
- The switching rule α is not known *a priori* but its value is available at each sampling period.

As reported by the cited references, the third assumption corresponds to practical implementations where the switching system is supervised by a discrete-event system, or operator, allowing for the value of α to be known at each sampling period in real time.

The evolution of the 2D system (9.1) shows that (k,l) is varying according to one direction at each time: to compute the horizontal state (resp. the vertical state) at coordinate $(k+1,l)$(resp. $(k,l+1)$) one needs to know both the horizontal and vertical states at coordinate (k,l). For example, to compute the states at $k+l=2$ (coordinates of the points $(2,0)$, $(1,1)$ and $(0,2)$ situated on the same line in Figure 9.1), one needs to use the states for all the possible values of (k,l) with $k+l=2$ (coordinates of the points $(1,0)$ and $(0,1)$ situated on the same below line in Figure 9.1). That is, to compute the states $x^h(k+1,l), x^v(k,l+1), k+l=1$, one has to compute the states $x^h(1,1), x^v(1,1), x^h(2,0), x^v(0,2)$ by using $x^h(0,1), x^v(0,1), x^h(1,0), x^v(1,0)$. This fact can be shown by Figure 9.1 for $k,l=0,\ldots,3$. Thus, the 2D system causality imposes an incrementing depending only on $\kappa=k+l$. For this, the switch can be assumed to occur only at each sampling of k or l, that is, each κ. The following definition extends the one given for 1D in Chapter *III* based on κ.

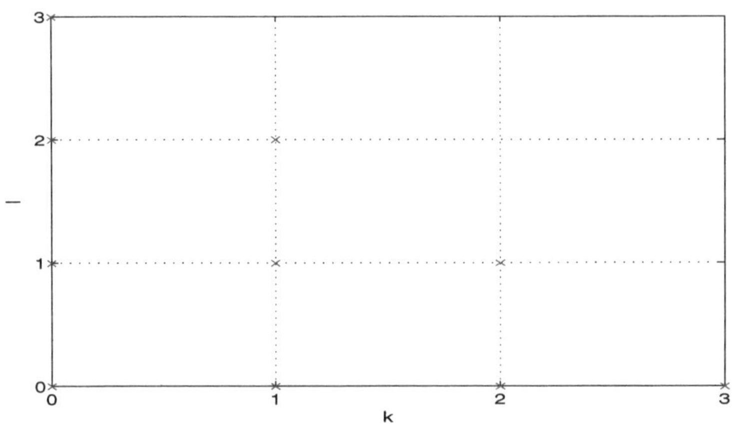

Fig. 9.1 How to compute the states at coordinates $(k+1,l)$ and $(k,l+1)$ situated on the same line with respect to the previous ones at coordinates (k,l) for $k,l=0,\ldots,3$ situated on the same previous line

Definition 9.1. *A hybrid time basis τ is a finite or infinite sequence of sets $I_i=\{\kappa\in\mathbb{Z}_+ : \kappa_i\leq\kappa\leq\bar{\kappa}_i\}$, with $\bar{\kappa}_i=\kappa_{i+1}$ for $i\in\mathscr{L}=\{0,\ldots,L\}$, $\bigcup_{i\in\mathscr{L}} I_i\subseteq\mathbb{Z}_+$, where the system (9.1) evolves with $\alpha=i$. If the sequence is finite, one may take $card(\tau)=L+1<\infty$ and $\bar{\kappa}_L=\infty$, then, I_L may be of the form $I_L=\{\kappa\in\mathbb{Z}_+ : \kappa_L\leq\kappa<\infty\}$.*

In this work, we are interested in the synthesis of a stabilizing controller for this class of hybrid systems. The following state feedback control law is used:

$$u(k,l)=F_\alpha x(k,l) \tag{9.4}$$

with

$$F_\alpha = [F_\alpha^1 \;\; F_\alpha^2] \tag{9.5}$$

where matrices $F_\alpha^1 \in \mathbb{R}^{m \times n_1}, F_\alpha^2 \in \mathbb{R}^{m \times n_2}$ which writes the closed-loop system as

$$x^+(k,l) = [A_\alpha + B_\alpha F_\alpha]x(k,l). \tag{9.6}$$

Since the switching is assumed to occur only at each κ, the following indicator function is proposed:

$$\xi(\kappa) = [\xi_1(\kappa), \ldots, \xi_N(\kappa)]^T, \tag{9.7}$$

if the switching system is in mode i, then $\xi_i(\kappa) = 1$ and $\xi_j(\kappa) = 0$ for $j \neq i$, so one can write the closed-loop system (9.6) as follows:

$$x^+(k,l) = \sum_{i=1}^{N} \xi_i(\kappa)[A_i + B_i F_i]x(k,l) = A^c(\xi)x(k,l), \tag{9.8}$$

with

$$A^c(\xi) = \sum_{i=1}^{N} \xi_i(\kappa)(A_i + B_i F_i). \tag{9.9}$$

Note that $\xi_i(\kappa) \geq 0; \sum_{i=1}^{N} \xi_i(\kappa) = 1$. Hence, the problem we are dealing with consists of designing the gains $F(\alpha)$ that stabilize the 2D closed-loop switching system (9.8). Our goal is to propose an algorithm that uses the LMI framework, to facilitate the computation of the feedback controller gains.

9.2.2 Preliminary Results

A useful stability result for 2D linear systems, which is based on Lyapunov function is recalled in this section. For this, consider the following 2D discrete autonomous system:

$$x^+(k,l) = \bar{A}x(k,l). \tag{9.10}$$

Theorem 9.1. *[4] The 2D system (9.10) is asymptotically stable if there exists a positive definite matrix of the form*

$$P = \begin{bmatrix} P_1 & 0 \\ 0 & P_2 \end{bmatrix}, \tag{9.11}$$

with $P_1 \in \mathbb{R}^{n_1 \times n_1}$ and $P_2 \in \mathbb{R}^{n_2 \times n_2}$, such that,

$$\bar{A}^T P \bar{A} - P < 0 \tag{9.12}$$

and the following is a Lyapunov function of the system:

$$V(x(k,l)) = x^T(k,l)Px(k,l). \tag{9.13}$$

Note that this result can be obtained with the following rate of increase

$$\triangle V(x(k,l)) = x^{+T}(k,l)Px^{+}(k,l) - x^T(k,l)Px(k,l)\ .$$

9.2.3 Stabilizability Analysis and Controller Design

This section presents sufficient conditions of asymptotic stability using either a common quadratic or multiple Lyapunov functions for 2D discrete switching systems given under Roesser model. The design using LMIs of the corresponding stabilizing controllers is also established. The main results of this section for the unsaturated systems were published first in [33].

9.2.3.1 Common Quadratic Lyapunov Function

The following result proposes a necessary and sufficient condition for the 2D closed-loop switching system (9.8) to admit a common quadratic Lyapunov function.

Theorem 9.2. *The following two statements are equivalent:*

i) There exists a Lyapunov function of the form (9.13) whose difference is negative definite, proving global asymptotic stability of the 2D switching system (9.8) for any arbitrary sequence of switching.
ii) There exist positive definite matrices $P_1 \in \mathbb{R}^{n_1 \times n_1}$ and $P_2 \in \mathbb{R}^{n_2 \times n_2}$ such that,

$$\begin{bmatrix} P_1 & 0 & (\Phi_{11}^i)^T & (\Phi_{21}^i)^T \\ * & P_2 & (\Phi_{12}^i)^T & (\Phi_{22}^i)^T \\ * & * & P_1^{-1} & 0 \\ * & * & * & P_2^{-1} \end{bmatrix} > 0, \forall i \in \mathscr{I} \tag{9.14}$$

with

$$\begin{array}{l} \Phi_{11}^i = A_{11}^i + B_1^i F_1^i, \Phi_{12}^i = A_{12}^i + B_1^i F_2^i \\ \Phi_{21}^i = A_{21}^i + B_2^i F_1^i, \Phi_{22}^i = A_{22}^i + B_2^i F_2^i \end{array} \tag{9.15}$$

and $*$ denoting the transpose of the corresponding off-diagonal term.

Proof: $(ii) \rightarrow (i)$: Consider the Lyapunov function candidate $V(x(k,l))$ given by (9.13) with the form (9.11) and compute its rate of increase:

$$\begin{aligned} \triangle V(x(k,l)) &= x^{+T}(k,l)Px^{+}(k,l) - x^T(k,l)Px(k,l) \\ &= x^T(k,l)\left[A^c(\xi)^T PA^c(\xi) - P\right]x(k,l)\ . \end{aligned}$$

This rate of increase is definite negative if

$$A^c(\xi)^T PA^c(\xi) - P < 0. \tag{9.16}$$

Using Schur complement, one obtains

$$\begin{bmatrix} P & A^c(\xi)^T P \\ * & P \end{bmatrix} > 0. \tag{9.17}$$

Substituting (9.9) into (9.17) leads to

$$\sum_{i=1}^{N} \xi_i(\kappa) \begin{bmatrix} P & [A_i + B_i F_i]^T P \\ * & P \end{bmatrix} > 0. \tag{9.18}$$

A sufficient condition to have (9.18) is given by

$$\begin{bmatrix} P & [A_i + B_i F_i]^T P \\ * & P \end{bmatrix} > 0, \forall i \in \mathscr{I}. \tag{9.19}$$

Further, pre-and post-multiplying (9.19) by $diag\{P^{-1},\ P^{-1}\}$ while substituting matrices A_i,B_i, F_i and P according to (9.3), (9.5) and (9.11), respectively, inequalities (9.14) are directly obtained. That is, the 2D switching system (9.8) is globally asymptotically stable for any sequence of switching.

$(i) \to (ii)$: Assume that function $V(x(k,l))$ given by (9.13) with the form (9.11) is a Lyapunov function for system (9.8). Then, its rate of increase $\triangle V(x(k,l)) < 0$. This implies that inequality (9.18) holds true. If the switching system is in mode i, one has $\xi_j(\kappa) = 1$, for $j = i$ and $\xi_j(\kappa) = 0$ for $j \neq i$. That is, (9.19) is satisfied. Following the same reasoning as before, condition (9.14) is obtained. □

This result can be used to synthesize the stabilizing controller under LMIs form.

Theorem 9.3. *If there exist positive definite matrices $X_1 \in \mathbb{R}^{n_1 \times n_1}$, $X_2 \in \mathbb{R}^{n_2 \times n_2}$ and matrices $Y_1^i \in \mathbb{R}^{m \times n_1}$ and $Y_2^i \in \mathbb{R}^{m \times n_2}$ such that*

$$\begin{bmatrix} X_1 & 0 & (\Psi_{11}^i)^T & (\Psi_{21}^i)^T \\ * & X_2 & (\Psi_{12}^i)^T & (\Psi_{22}^i)^T \\ * & * & X_1 & 0 \\ * & * & * & X_2 \end{bmatrix} > 0, \forall i \in \mathscr{I} \tag{9.20}$$

with

$$\Psi_{11}^i = A_{11}^i X_1 + B_1^i Y_1^i, \Psi_{12}^i = A_{12}^i X_2 + B_1^i Y_2^i$$
$$\Psi_{21}^i = A_{21}^i X_1 + B_2^i Y_1^i, \Psi_{22}^i = A_{22}^i X_2 + B_2^i Y_2^i$$

then, the 2D switching system (9.8) is globally asymptotically stable for any arbitrary sequence of switching; the corresponding controller gains are given by

$$F_i = [Y_1^i X_1^{-1} \quad Y_2^i X_2^{-1}] \tag{9.21}$$

with $P_1 = X_1^{-1}$ *and* $P_2 = X_2^{-1}$.

Proof: The sufficient condition of stability of the switching system is given by (9.16). Pre-and post-multiplying by $X = P^{-1}$, leads to

$$X - XA^c(\xi)^T X^{-1} A^c(\xi) X > 0.$$

Applying Schur complement gives

$$\begin{bmatrix} X & [A^c(\xi)X]^T \\ * & X \end{bmatrix} > 0. \tag{9.22}$$

Substituting (9.9) into (9.22), leads to

$$\sum_{i=1}^{N} \xi_i(\kappa) \begin{bmatrix} X & [A_i X + B_i F_i X]^T \\ * & X \end{bmatrix} > 0. \tag{9.23}$$

Let $Y_i = F_i X$. A sufficient condition to have (9.23) is,

$$\begin{bmatrix} X & [A_i X + B_i Y_i]^T \\ * & X \end{bmatrix} > 0, \forall i \in \mathscr{I}. \tag{9.24}$$

By substituting matrices A_i, B_i according to (9.3), $X = diag\{X_1, \ X_2\}$ with $X_1 = P_1^{-1}, X_2 = P_2^{-1}$ and $Y_i = [Y_1^i \ Y_2^i]$ the LMIs (9.20) are directly obtained. Finally, the stabilizing controller gains of the 2D switching system (9.8) are given by $F_i = Y_i X^{-1}$. Expression (9.21) follows. □

Example 9.1. *In a real world, some dynamical processes in gas absorption, water stream heating and air drying can be described by the Darboux equation. Consider the case where this practical system is composed of two modes as given in [158].*

$$A_1 = \begin{bmatrix} -2.2 & 0.5 \\ -0.2 & -1.8 \end{bmatrix}; B_1 = \begin{bmatrix} 10.3 \\ 0.5 \end{bmatrix}; A_2 = \begin{bmatrix} -1.8 & 0.6 \\ -0.3 & -1.2 \end{bmatrix}; B_2 = \begin{bmatrix} 0.3 \\ 0.4 \end{bmatrix}.$$

Assume in this case that the switching from one mode to another is arbitrary. The system is then discretized with a sampling time and space taken as $T = 0.1$.

The LMIs (9.20) are feasible and the common Lyapunov function is given by:

$$P = \begin{bmatrix} 1.0786 & 0 \\ 0 & 0.8783 \end{bmatrix}; F_1 = \begin{bmatrix} -6.5882 & -12.5000 \end{bmatrix}; F_2 = \begin{bmatrix} -9.3600 & -14.8000 \end{bmatrix}.$$

Example 9.2. *Consider now the example of two long transmission lines; each one can be represented by Figure 9.2.*

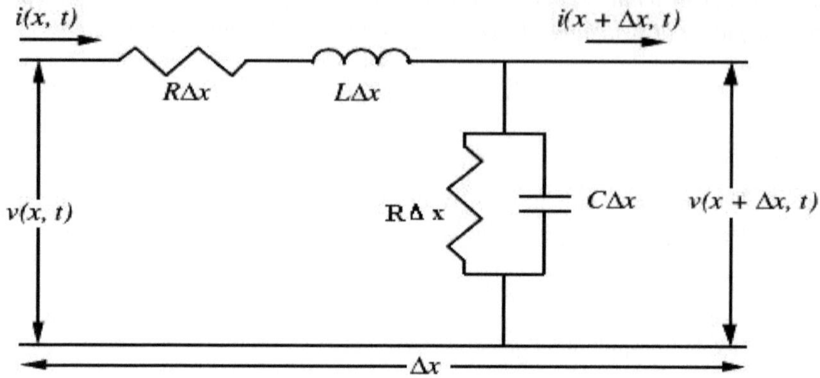

Fig. 9.2 Transmission line

We assume that one can switch from a line to another arbitrarily. The system equation as presented in [109] is given by:

$$\begin{aligned}\frac{\partial u(x,t)}{\partial x} &= L(\alpha)\frac{\partial i(x,t)}{\partial t} + R(\alpha)i(x,t) + v(x,t) \\ \frac{\partial i(x,t)}{\partial x} &= C(\alpha)\frac{\partial u(x,t)}{\partial t},\end{aligned} \tag{9.25}$$

where $C(\alpha), L(\alpha), R(\alpha)$ *are respectively the capacity, the inductance and the resistor of the portion of the line of length* $\triangle x$, $v(x,t)$ *is the control assumed here to be the voltage per meter in upside of the line. Using the approximations:*

$$\begin{aligned}\frac{\partial u(x,t)}{\partial x} &\simeq \frac{u(k,l) - u(k-1,l)}{\triangle x} \\ \frac{\partial u(x,t)}{\partial t} &\simeq \frac{u(k,l+1) - u(k,l)}{\triangle t}\end{aligned} \tag{9.26}$$

one can use the following change of coordinates for the current and the voltage: $u^h(k,l) = u(k-1,l)$ *and* $u^v(k,l) = u(k,l)$, $a(\alpha) = \frac{\triangle t}{L(\alpha)\triangle x}$, $b(\alpha) = 1 - \frac{R(\alpha)\triangle t}{L(\alpha)}$, $d(\alpha) = \frac{\triangle t}{C(\alpha)\triangle x}$ *and* $c(\alpha) = \frac{\triangle t}{L(\alpha)}$. *Define:*

$$x^h(k,l) = \begin{bmatrix} u^h(k,l) \\ i^h(k,l) \end{bmatrix}, x^v(k,l) = \begin{bmatrix} u^v(k,l) \\ i^v(k,l) \end{bmatrix}.$$

The system (9.25) can then be transformed into a 2D switching Roesser model given by (9.1) with:

$$A_1 = \begin{bmatrix} 0 & 0 & 1 & 0 \\ 0 & 0 & 0 & 1 \\ 0 & -a_1 & b_1 & a_1 \\ -d_1 & 0 & d_1 & 1 \end{bmatrix}; B_1^T = \begin{bmatrix} 0 \; 0 \; 0 \; -c_1 \end{bmatrix};$$

$$A_2 = \begin{bmatrix} 0 & 0 & 1 & 0 \\ 0 & 0 & 0 & 1 \\ 0 & -a_2 & b_2 & a_2 \\ -d_2 & 0 & d_2 & 1 \end{bmatrix}; B_2^T = \begin{bmatrix} 0 \; 0 \; 0 \; -c_2 \end{bmatrix}.$$

In this example, $n_1 = 2, n_2 = 2$ *and* $m = 1$. *For* $a_1 = 0.5, b_1 = -0.5, c_1 = 0.1, d_1 = 0.8, a_2 = 0.8, b_2 = -0.1, c_2 = 0.4, d_2 = 0.6$, *the LMI solver of Matlab is used to solve LMIs (9.20) leading to feasible solutions in* X_1, X_2, Y_1 *and* Y_2. *The matrix* P *and the gain controllers* F_1, F_2 *are then derived by using relations (9.21) as follows:*

$$P_1 = \begin{bmatrix} 0.0087 & -0.0009 \\ -0.0009 & 0.0200 \end{bmatrix}, P_2 = \begin{bmatrix} 0.0171 & -0.0073 \\ -0.0073 & 0.0561 \end{bmatrix},$$
$$F_1 = \begin{bmatrix} 1.4932 \; -0.2713 \; -1.5372 \; -2.2233 \end{bmatrix},$$
$$F_2 = \begin{bmatrix} 7.9918 \; -0.6640 \; -8.6579 \; -9.3156 \end{bmatrix}$$

The trajectories of the states, which are asymptotically stable, are not plotted.

Consider now the following numerical example:

$$A_1 = \begin{bmatrix} 3 & 0 & 1 \\ 1 & -4 & 1 \\ -1 & 5 & 1 \end{bmatrix}; B_1 = \begin{bmatrix} 1 & 0 \\ 0 & 1 \\ 1 & 1 \end{bmatrix},$$

$$A_2 = \begin{bmatrix} 1 & 0 & -1 \\ 0 & 2 & 1 \\ 0.4 & 1 & -1 \end{bmatrix}, B_2 = \begin{bmatrix} 1 & 1 \\ 0 & 1 \\ 1 & 0 \end{bmatrix}.$$

The LMIs (9.20) are not feasible.

Assume now that the discrete switching system given by (9.1) is saturated as follows:

$$\begin{aligned} x^+(k,l) &= A_\alpha x(k,l) + B_\alpha sat(u(k,l)) \\ x^h(0,l) &= f(l), x^v(k,0) = g(k), \end{aligned} \tag{9.27}$$

where $sat(.)$ is a normalized saturation function. We only restrict ourselves to design unsaturating controller by using state feedback control (9.4) with matrix F decomposed according to (9.5). Applying the same tools developed during the previous chapters, we can present the following result.

Corollary 9.1. *For a positive scalar* ρ, *if there exist positive definite matrices* $X_1 \in \mathbb{R}^{n_1\times n_1}$, $X_2 \in \mathbb{R}^{n_2\times n_2}$ *and matrices* $Y_1^i \in \mathbb{R}^{m\times n_1}$ *and* $Y_2^i \in \mathbb{R}^{m\times n_2}$ *such that*

$$\begin{bmatrix} X_1 & 0 & (\Psi_{11}^i)^T & (\Psi_{21}^i)^T \\ * & X_2 & (\Psi_{12}^i)^T & (\Psi_{22}^i)^T \\ * & * & X_1 & 0 \\ * & * & * & X_2 \end{bmatrix} > 0, \forall i \in \mathscr{I} \tag{9.28}$$

$$\begin{bmatrix} 1/\rho & Y_{1l}^i & Y_{2l}^i \\ * & X_1 & 0 \\ * & 0 & X_2 \end{bmatrix} > 0, l \in [1,\ m] \tag{9.29}$$

with

$$\Psi_{11}^i = A_{11}^i X_1 + B_1^i Y_1^i, \Psi_{12}^i = A_{12}^i X_2 + B_1^i Y_2^i$$
$$\Psi_{21}^i = A_{21}^i X_1 + B_2^i Y_1^i, \Psi_{22}^i = A_{22}^i X_2 + B_2^i Y_2^i$$

then, the 2D saturated switching system (9.8) is asymptotically stable for any arbitrary sequence of switching and $\forall x_o \in \varepsilon(X^{-1},\ \rho)$; *the corresponding controller gains are given by*

$$F_i = [Y_1^i X_1^{-1} \quad Y_2^i X_2^{-1}] \tag{9.30}$$

with $P_1 = X_1^{-1}$ *and* $P_2 = X_2^{-1}$.

Proof: The proof follows readily by writing the inclusion relation $\varepsilon(P,\rho) \subset \mathscr{L}(F_i)$ under LMI form, with

$$\mathscr{L}(F_i) = \{x \in \mathbb{R}^n / |F_l^i x| \leq 1 \ \} \tag{9.31}$$

where F_l^i denotes the lth row of matrix F_i and ρ is a positive scalar. □

9.2.3.2 Multiple Lyapunov Function

Although the existence of a common quadratic Lyapunov function for the various subsystems guarantees the asymptotic stability of the switching system, such a function is not always possible and might lead to conservative results. Thus, this subsection studies the problem of stabilizability by using multiple Lyapunov functions. These multiple Lyapunov functions are considered to be a strong tool in the analysis of the stability of switching systems. Consider the following multiple Lyapunov function candidate:

$$\vartheta(x(k,l)) = x^T(k,l)P(\xi)x(k,l) = x^T(k,l)(\sum_{i=1}^{N} \xi_i(\kappa)P_i)x(k,l). \tag{9.32}$$

Define

$$P_i = \begin{bmatrix} P_1^i & 0 \\ 0 & P_2^i \end{bmatrix}. \tag{9.33}$$

The following result proposes a necessary and sufficient condition for the 2D closed-loop switching system (9.8) to admit a multiple Lyapunov function.

Theorem 9.4. *The following two statements are equivalent:*

i) There exists a Lyapunov function of the form (9.32) whose difference is negative definite, proving global asymptotic stability of the 2D switching system (9.8) for any arbitrary sequence of switching.
ii) There exist positive definite matrices $P_1^i \in \mathbb{R}^{n_1 \times n_1}$, $P_2^i \in \mathbb{R}^{n_2 \times n_2}, i = 1, \ldots, N$ such that

$$\begin{bmatrix} P_1^i & 0 & (\Phi_{11}^i)^T & (\Phi_{21}^i)^T \\ * & P_2^i & (\Phi_{12}^i)^T & (\Phi_{22}^i)^T \\ * & * & (P_1^j)^{-1} & 0 \\ * & * & * & (P_2^j)^{-1} \end{bmatrix} > 0, \forall (i,j) \in \mathscr{I}^2 \tag{9.34}$$

where Φ_{ij} are given by (9.15).

Proof: $(ii) \rightarrow (i)$: Consider the Lyapunov function candidate $\vartheta(x(k,l))$ given by (9.32) with the form (9.33) and compute its rate of increase:

$$\begin{aligned} \triangle \vartheta(x(k,l)) &= x^{+T}(k,l) P_+(\xi) x^+(k,l) - x^T(k,l) P(\xi) x(k,l) \\ &= x^T(k,l) \left[A^c(\xi)^T P_+(\xi) A^c(\xi) - P(\xi) \right] x(k,l), \end{aligned}$$

with

$$P_+(\xi) = \sum_{j=1}^{N} \xi_j(\kappa+1) \begin{bmatrix} P_1^j & 0 \\ 0 & P_2^j \end{bmatrix}.$$

Note that,

$$P_+(\xi) = \sum_{j=1}^{N} \xi_j(\kappa+1) P_j \tag{9.35}$$

This rate of increase is definite negative if

$$A^c(\xi)^T P_+(\xi) A^c(\xi) - P(\xi) < 0. \tag{9.36}$$

Using Schur complement, one obtains

$$\begin{bmatrix} P(\xi) & A^c(\xi)^T P_+(\xi) \\ * & P_+(\xi) \end{bmatrix} > 0. \tag{9.37}$$

Substituting (9.9) into (9.37), leads to

$$\sum_{i=1}^{N}\sum_{j=1}^{N}\xi_i(\kappa)\xi_j(\kappa+1)\begin{bmatrix} P_i & [A_i+B_iF_i]^T P_j \\ * & P_j \end{bmatrix} > 0.$$

It follows that the set of inequalities

$$\begin{bmatrix} P_i & [A_i+B_iF_i]^T P_j \\ * & P_j \end{bmatrix} > 0, \forall (i,j) \in \mathscr{I}^2 \tag{9.38}$$

forms a sufficient condition to have $\triangle\vartheta(x(k,l)) < 0$. Further, pre- and post-multiplying (9.38) by $diag\{P_i^{-1},\ P_j^{-1}\}$ while substituting matrices A_i,B_i, F_i and P_j according to (9.3), (9.5) and (9.33) respectively, inequalities (9.34) are directly obtained. That is the 2D switching system (9.8) is globally asymptotically stable for any sequence of switching.

$(i) \rightarrow (ii)$: Assume that function $\vartheta(x(k,l))$ given by (9.32) with the form (9.33) is a Lyapunov function for system (9.8). Then, its rate of increase $\triangle\vartheta(x(k,l)) < 0$. This implies that inequality (9.38) holds true. If the switching system is in mode i at κ and mode j at $\kappa+1$, one has $\xi_s(\kappa) = 1, \xi_r(\kappa+1) = 1$, for $s = i, r = j$ and $\xi_s(\kappa) = 0, \xi_r(\kappa) = 0$ for $s \neq i, r \neq j$. That is, (9.38) is satisfied. Following the same reasoning as before, condition (9.34) is obtained. □

Remark 9.1. • *It is worth noting that inequalities (9.34) can be given under the following compact form:*

$$\begin{bmatrix} P_i & (A_i+B_iF_i)^T \\ * & P_j^{-1} \end{bmatrix} > 0, \forall (i,j) \in \mathscr{I}^2. \tag{9.39}$$

- *The results of [33] can be obtained as a particular case of Theorem 9.4.*

The synthesis of the controller can then be derived.

Theorem 9.5. *If there exist positive definite matrices $X_1^i \in \mathbb{R}^{n_1 \times n_1}$, $X_2^i \in \mathbb{R}^{n_2 \times n_2}$ and matrices $Y_1^i \in \mathbb{R}^{m \times n_1}$ and $Y_2^i \in \mathbb{R}^{m \times n_2}$ such that*

$$\begin{bmatrix} X_1^i & 0 & (\Pi_{11}^i)^T & (\Pi_{21}^i)^T \\ * & X_2^i & (\Pi_{12}^i)^T & (\Pi_{22}^i)^T \\ * & * & X_1^j & 0 \\ * & * & * & X_2^j \end{bmatrix} > 0, \forall (i,j) \in \mathscr{I}^2 \tag{9.40}$$

with

$$\Pi_{11}^i = A_{11}^i X_1^i + B_1^i Y_1^i, \Pi_{12}^i = A_{12}^i X_2^i + B_1^i Y_2^i$$
$$\Pi_{21}^i = A_{21}^i X_1^i + B_2^i Y_1^i, \Pi_{22}^i = A_{22}^i X_2^i + B_2^i Y_2^i$$

then, the switching system (9.8) is globally asymptotically stable for any arbitrary sequence of switching; the controller gains are given by

$$F_i = [Y_1^i (X_1^i)^{-1} \quad Y_2^i (X_2^i)^{-1}] \tag{9.41}$$

with $P_1^i = (X_1^i)^{-1}$ *and* $P_2^i = (X_2^i)^{-1}$.

Proof: By pre-and post-multiplying inequality (9.39) by $diag\{P_i^{-1}, \mathbb{I}\}$ while letting $X_i = P_i^{-1}$ with $P_1^i = (X_1^i)^{-1}$ and $P_2^i = (X_2^i)^{-1}$, $Y_i = F_i X_i$, it follows that

$$\begin{bmatrix} X_i & (A_i X_i + B_i Y_i)^T \\ * & X_j \end{bmatrix} > 0, \forall (i,j) \in \mathscr{I}^2 . \tag{9.42}$$

By substituting matrices A_i,B_i and P_j according to (9.3) and (9.33) respectively, and using $X_i = diag\{X_1^i \quad X_2^i\}$, $Y_i = [Y_1^i \; Y_2^i]$ the LMIs (9.40) are directly obtained. Finally, the stabilizing controller gains of the 2D switching system (9.8) are given by $F_i = Y_i X_i^{-1}$. Expression (9.41) follows. □

Example 9.3. *Considering the same example as in Example 9.2, the LMI solver of Matlab is used to solve LMIs (9.40) leading to feasible solutions in* $X_1^1, X_1^2, X_2^1, X_2^2, Y_1$ *and* Y_2. *The matrices* P_1, P_2 *and the gain controllers* F_1, F_2 *are then derived by using relations (9.41) as follows:*

$$P_1^1 = \begin{bmatrix} 0.4009 & 0.0989 \\ 0.0989 & 0.7837 \end{bmatrix}; P_1^2 = \begin{bmatrix} 3.9334 & 2.9613 \\ 2.9613 & 3.2969 \end{bmatrix};$$
$$P_2^1 = \begin{bmatrix} 0.4394 & 0.0790 \\ 0.0790 & 0.4562 \end{bmatrix}; P_2^2 = \begin{bmatrix} 2.0094 & 0.9370 \\ 0.9370 & 1.7379 \end{bmatrix};$$
$$F_1 = \begin{bmatrix} -0.9879 & -0.5563 & 2.7369 & 2.9319 \end{bmatrix};$$
$$F_2 = \begin{bmatrix} -0.8221 & -1.1374 & 13.0216 & 13.9148 \end{bmatrix}.$$

Example 9.4. *The second example of Example 9.2 is studied. For the same data, the LMIs (9.40) are feasible. The obtained solutions are given by:*

$$P_1^1 = \begin{bmatrix} 0.0269 & -0.0531 \\ -0.0531 & 0.1086 \end{bmatrix}, P_2^1 = 0.0011,$$
$$P_1^2 = \begin{bmatrix} 0.0049 & -0.0142 \\ -0.0142 & 0.0476 \end{bmatrix}, P_2^2 = 0.0016$$
$$F_1 = \begin{bmatrix} -0.6407 & -4.2421 & -0.5241 \\ -0.0040 & 2.2068 & -0.8029 \end{bmatrix},$$
$$F_2 = \begin{bmatrix} -0.8375 & 1.1870 & 1.7299 \\ 0.0985 & -2.4938 & -1.1626 \end{bmatrix}$$

Figure 9.3 plots the evolution of the three states $x_1^h(k,l)$, $x_2^h(k,l)$, $x^v(k,l)$ and the corresponding arbitrary sequence of switching represented by the last figure. One can notice that the 2D closed-loop switching system is asymptotically stable.

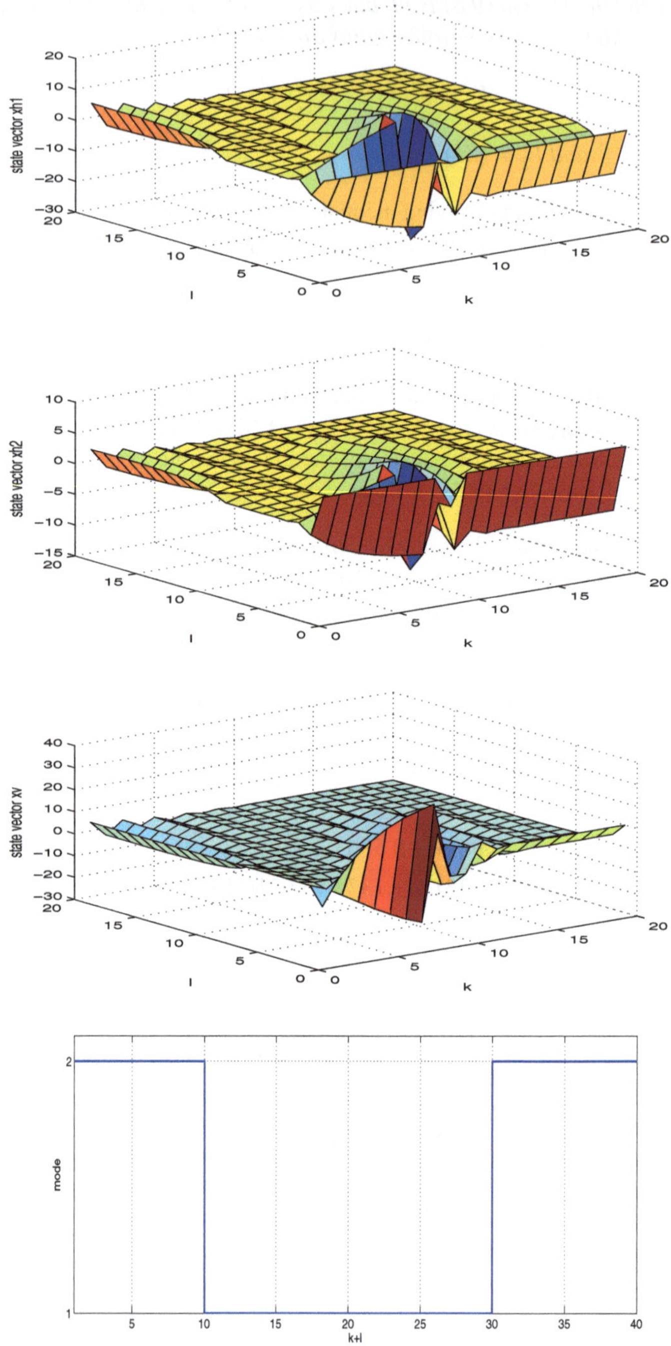

Fig. 9.3 The trajectory of the states $x_1^h(k,l)$, $x_2^h(k,l)$, $x^v(k,l)$ and the corresponding sequence of switching obtained with Theorem 9.5

Comment 9.1. *The second example of Example 9.4 shows that the studied 2D switching system is asymptotically stable for any sequence of switching without admitting a common quadratic Lyapunov function. This confirms the claim proposed by many authors about this fact. Thus, the results of Theorem 9.5 constitute a strong tool to study stabilizability of 2D discrete switching systems with arbitrary sequence of switching.*

Assume now that the 2D switching system is normally saturated. The result of Theorem 9.5 can be extended to the saturated 2D switching systems by adding an inclusion LMI based on the multiple Lyapunov function by the following result.

Corollary 9.2. *For a positive scalar ρ, if there exist positive definite matrices $X_1^i \in \mathbb{R}^{n_1 \times n_1}$, $X_2^i \in \mathbb{R}^{n_2 \times n_2}$ and matrices $Y_1^i \in \mathbb{R}^{m \times n_1}$ and $Y_2^i \in \mathbb{R}^{m \times n_2}$ such that*

$$\begin{bmatrix} X_1^i & 0 & (\Pi_{11}^i)^T & (\Pi_{21}^i)^T \\ * & X_2^i & (\Pi_{12}^i)^T & (\Pi_{22}^i)^T \\ * & * & X_1^j & 0 \\ * & * & * & X_2^j \end{bmatrix} > 0, \forall (i,j) \in \mathscr{I}^2 \tag{9.43}$$

$$\begin{bmatrix} 1/\rho & Y_{1l}^i & Y_{2l}^i \\ * & X_1^i & 0 \\ * & 0 & X_2^i \end{bmatrix} > 0, l \in [1,\ m] \tag{9.44}$$

with

$$\Pi_{11}^i = A_{11}^i X_1^i + B_1^i Y_1^i, \Pi_{12}^i = A_{12}^i X_2^i + B_1^i Y_2^i$$
$$\Pi_{21}^i = A_{21}^i X_1^i + B_2^i Y_1^i, \Pi_{22}^i = A_{22}^i X_2^i + B_2^i Y_2^i$$

then, the saturated 2D switching system (9.8) is asymptotically stable for any arbitrary sequence of switching and $\forall x_o \in \bigcup_i^N \varepsilon(X_i^{-1},\ \rho)$; the controller gains are given by

$$F_i = [Y_1^i (X_1^i)^{-1} \quad Y_2^i (X_2^i)^{-1}] \tag{9.45}$$

with $P_1^i = (X_1^i)^{-1}$ and $P_2^i = (X_2^i)^{-1}$.

Proof: The proof follows readily by writing the inclusion relation $\varepsilon(P_i, \rho) \subset \mathscr{L}(F_i)$ under LMI form, with

$$\mathscr{L}(F_i) = \{x \in \mathbb{R}^n / |F_l^i x| \leq 1\ \} \tag{9.46}$$

where F_l^i denotes the lth row of matrix F_i and ρ is a positive scalar. □

Example 9.5. *Consider the same example as Example 9.3. The obtained results $P_i^1, Y_i^2, i = 1,2$ are used to check inequalities (9.43) which are definite positive. As a consequence, the ellipsoid sets $\varepsilon(P_i,\ 1)$ are well contained inside the polyhedral sets $\mathscr{L}(F_i)$ ensuring that the saturated switching 2D system is asymptotically stable.*

The obtained results of Example 9.4 are also used to check inequalities (9.43) which are definite positive. The same consequence as for the previous example is deduced.

Remark 9.2. *It must be pointed out that the problem of saturated 2D continuous systems was studied in [102]. Besides, the inclusion inequality (9.43), which is equivalent to* $F_l^i P_i^{-1}(F_l^i)^T \leq 1/\rho$*, can be developed as*

$$(F_1^i)_l (P_1^i)^{-1} (F_1^i)_l^T + (F_2^i)_l (P_2^i)^{-1} (F_2^i)_l^T \leq 1/\rho.$$

It is obvious that this inequality does not imply $(F_1^i)_l (P_1^i)^{-1} (F_1^i)_l^T \leq 1/\rho_1$ *and* $(F_2^i)_l (P_2^i)^{-1} (F_2^i)_l^T \leq 1/\rho_2$*. However, the reciprocal is true provided that* $\rho_1 + \rho_2 = 1$*. This means that the inclusion condition* $\varepsilon(P_i, \rho) \subset \mathscr{L}(F_i)$ *must be globally satisfied. Nevertheless, the two local inclusions* $\varepsilon(P_i^1, \rho_1) \subset \mathscr{L}(F_i^1)$ *and* $\varepsilon(P_i^2, \rho_2) \subset \mathscr{L}(F_i^2)$ *with* $\rho_1 + \rho_2 = 1$ *are sufficient to realize the global inclusion.*

This result can be easily extended to positive switching 2D systems by using the obtained results in Chapter *VIII* as follows:

Corollary 9.3. *For a positive scalar* ρ*, if there exist diagonal positive matrices* $X_1^i \in \mathbb{R}^{n_1 \times n_1}$*,* $X_2^i \in \mathbb{R}^{n_2 \times n_2}$ *and matrices* $Y_1^i \in \mathbb{R}^{m \times n_1}$ *and* $Y_2^i \in \mathbb{R}^{m \times n_2}$ *such that*

$$\begin{bmatrix} X_1^i & 0 & (\Pi_{11}^i)^T & (\Pi_{21}^i)^T \\ * & X_2^i & (\Pi_{12}^i)^T & (\Pi_{22}^i)^T \\ * & * & X_1^j & 0 \\ * & * & * & X_2^j \end{bmatrix} > 0, \forall (i,j) \in \mathscr{I}^2 \tag{9.47}$$

$$\begin{bmatrix} 1/\rho & Y_{1l}^i & Y_{2l}^i \\ * & X_1^i & 0 \\ * & 0 & X_2^i \end{bmatrix} > 0, l \in [1,\ m] \tag{9.48}$$

$$\Pi_{11}^i = A_{11}^i X_1^i + B_1^i Y_1^i \succ 0, \tag{9.49}$$

$$\Pi_{12}^i = A_{12}^i X_2^i + B_1^i Y_2^i \succ 0, \tag{9.50}$$

$$\Pi_{21}^i = A_{21}^i X_1^i + B_2^i Y_1^i \succ 0, \tag{9.51}$$

$$\Pi_{22}^i = A_{22}^i X_2^i + B_2^i Y_2^i \succ 0, \tag{9.52}$$

then, the saturated 2D switching system (9.8), with $x_0 \succ 0$*, is asymptotically stable while the state is positive for any arbitrary sequence of switching and* $\forall x_o \in \bigcup_i^N \varepsilon(X_i^{-1},\ \rho)$*; the controller gains are given by:*

$$F_i = [Y_1^i (X_1^i)^{-1} \quad Y_2^i (X_2^i)^{-1}] \tag{9.53}$$

with $P_1^i = (X_1^i)^{-1}$ *and* $P_2^i = (X_2^i)^{-1}$*.*

Proof: Follows readily by using the proof of Theorem 8.5. □

In this section, 2D discrete switching systems represented by Roesser model with state feedback control are studied. The analysis of the switching phenomena associated to 2D discrete systems is developed. A new indicator function is then adopted.

These tools enable one to analyze the stabilizability of this class of systems. Hence, common quadratic and multiple Lyapunov functions are used to obtain sufficient conditions of asymptotic stabilization. These conditions are then used to synthesize the required controllers under LMI formulations. Three numerical examples illustrate the results.

9.3 Output Feedback Control

In this section, the problem of static output feedback control (SOFC) for 2D discrete switching systems is addressed. General sufficient conditions of asymptotic stability are developed with constraining equalities. In order to relax the conservatism, a classical method already used for 1D systems and recalled in Chapter *I*, based on the decomposition of matrix C, is proposed. This method enables one to obtain sufficient conditions of asymptotic stability for 2D systems under LMI form and without any restrictive equality involving matrix C. However, this technique can be applied only for a class of 2D systems having a particular decomposition of matrix C, due to the asymptotic stability constraint of 2D systems to have a Lyapunov function of diagonal block form.

9.3.1 Problem Formulation

In this section, the system under study is presented. Hence, to model a discrete switching systems, a new time basis definition together with the adequate indicator function is developed in the previous section.

This section deals with switching systems described by the following Roesser model:

$$\begin{aligned} x^+(k,l) &= A_\alpha x(k,l) + B_\alpha u(k,l) \\ y(k,l) &= C_\alpha x(k,l) \\ x^h(0,l) &= f(l) \\ x^v(k,0) &= g(k) \end{aligned} \tag{9.54}$$

with

$$\begin{aligned} x(k,l) &= \begin{bmatrix} x^h(k,l) \\ x^v(k,l) \end{bmatrix}, \\ x^+(k,l) &= \begin{bmatrix} x^h(k+1,l) \\ x^v(k,l+1) \end{bmatrix}, \\ A_\alpha &= \begin{bmatrix} A_{11}^\alpha & A_{12}^\alpha \\ A_{21}^\alpha & A_{22}^\alpha \end{bmatrix}, B_\alpha = \begin{bmatrix} B_1^\alpha \\ B_2^\alpha \end{bmatrix}, \end{aligned} \tag{9.55}$$

$$C_\alpha = \begin{bmatrix} C_1^\alpha & C_2^\alpha \end{bmatrix} \tag{9.56}$$

where $x^h(k,l)$ is the horizontal state in $\mathbb{R}^{n_1}$, $x^v(k,l)$ is the vertical state in $\mathbb{R}^{n_2}$, $x(k,l)$ is the whole state in $\mathbb{R}^n$ with $n = n_1 + n_2$, $u(k,l)$ is the control vector in $\mathbb{R}^m$, and

$y(k,l)$ is the output vector in $\mathbb{R}^p$. α is a switching rule which takes its values in the finite set $\mathscr{I} := \{1,\ldots,N\}$, N is the number of subsystems (modes), k and l are integers in $\mathbb{Z}_+$. It is assumed that:

- The switching system is stabilizable;
- At each time only one subsystem is active;
- Matrices C_α are of full rank row;
- The switching rule α is not known *a priori* but its value is available at each sampling period.

As reported by the cited references, the fourth assumption corresponds to practical implementations where the switching system is supervised by a discrete-event system, or operator, allowing for the value of α to be known at each sampling period in real time.

In this work, we are interested in the synthesis of a stabilizing controller for this class of hybrid systems. The following output feedback control law is used:

$$u(k,l) = K_\alpha y(k,l), \tag{9.57}$$

and write the closed-loop system as

$$x^+(k,l) = [A_\alpha + B_\alpha K_\alpha C_\alpha]x(k,l). \tag{9.58}$$

Since the switching is assumed to occur only at each κ, the following indicator function is proposed:

$$\xi(\kappa) = [\xi_1(\kappa),\ldots,\xi_N(\kappa)]^T, \tag{9.59}$$

if the switching system is in mode i, then $\xi_i(\kappa) = 1$ and $\xi_j(\kappa) = 0$ for $j \neq i$, so one can write the closed-loop system (9.58) as follows:

$$\begin{aligned} x^+(k,l) &= \sum_{i=1}^{N} \xi_i(\kappa)[A_i + B_i K_i C_i]x(k,l) \\ &= A^c(\xi)x(k,l), \end{aligned} \tag{9.60}$$

with

$$A^c(\xi) = \sum_{i=1}^{N} \xi_i(\kappa)(A_i + B_i K_i C_i). \tag{9.61}$$

Note that $\xi_i(\kappa) \geq 0; \sum_{i=1}^{N} \xi_i(\kappa) = 1$. Hence, the problem we are dealing with consists of designing the gains $K(\alpha)$ that stabilize the 2D closed-loop switching system (9.60). Our goal is to propose an algorithm that uses the LMI framework, to facilitate the computation of the feedback controller gains.

9.3.2 Preliminaries

This section is devoted to some preliminary lemmas useful to the developments in the sequel. It consists in a lemma which makes possible to use a singular value decomposition (SVD) of matrix C.

Since matrix C is assumed to be of full rank row p, one can always decompose C as follows:

$$C = C_l[\mathbb{I}_p \ \ 0]C_r, \tag{9.62}$$

where matrices $C_l \in \mathbb{R}^{p\times p}$ and $C_r \in \mathbb{R}^{n\times n}$ are nonsingular. In fact, these matrices can be obtained with SVD decomposition leading to have C_r as a unitary matrix satisfying $C_r^T = C_r^{-1}$. This decomposition was already used in Chapter *II* for 1D systems.

Lemma 9.1. *The equality $CX = WC$ with X a symmetric matrix, admits a solution in W if*

$$X = C_r^T \begin{bmatrix} X_1 & 0 \\ 0 & X_2 \end{bmatrix} C_r \tag{9.63}$$

where the solution is of the form $W = C_lX_1C_l^{-1}$, with $X_1 \in \mathbb{R}^{p\times p}$, $X_2 \in \mathbb{R}^{(n-p)\times(n-p)}$ an arbitrary matrix and C_r is given by the decomposition of matrix C.

Proof: Let matrix X be given by (9.63). By substituting (9.63) and (9.62) into $CX = WC$, the following developments follow:

$$\begin{aligned} CX &= C_l[\mathbb{I}_p \ \ 0]C_rC_r^T \begin{bmatrix} X_1 & 0 \\ 0 & X_2 \end{bmatrix} C_r \\ &= C_lX_1[\mathbb{I}_p \ \ 0]C_r \\ &= C_lX_1C_l^{-1}C_l[\mathbb{I}_p \ \ 0]C_r \\ &= WC \end{aligned}$$

with $W = C_lX_1C_l^{-1}$.

9.3.3 Stabilizability Analysis and Controller Design

9.3.3.1 Common Quadratic Lyapunov Function

Consider now the 2D discrete system (9.54). The result of Theorem 9.2 can be used to synthesize the stabilizing controller under LMIs form by using a common quadratic Lyapunov function $V(x(k,l) = x(k,l)^T Px(k,l)$ with $P = diag\{P_1, P_2\}$ as a particular case.

Theorem 9.6. *If there exist matrices $Y_i \in \mathbb{R}^{m \times p}$, non singular matrix $W_i \in \mathbb{R}^{p \times p}$, positive definite matrices $X_1 \in \mathbb{R}^{n_1 \times n_1}$, $X_2 \in \mathbb{R}^{n_2 \times n_2}$ such that the following LMI holds true :*

$$\begin{bmatrix} X_1 & 0 & (\Sigma_{11}^i)^T & (\Sigma_{21}^i)^T \\ 0 & X_2 & (\Sigma_{12}^i)^T & (\Sigma_{22}^i)^T \\ * & * & X_1 & 0 \\ * & * & 0 & X_2 \end{bmatrix} > 0, \tag{9.64}$$

$$C_1^i X_1 = W_i C_1^i \tag{9.65}$$

$$C_2^i X_2 = W_i C_2^i \tag{9.66}$$

$$\forall i \in \mathscr{I}$$

where

$$\begin{aligned} \Sigma_{11}^i &= A_{11}^i X_1 + B_1^i Y_i C_1^i, \\ \Sigma_{12}^i &= A_{12}^i X_2 + B_1^i Y_i C_2^i, \\ \Sigma_{21}^i &= A_{21}^i X_1 + B_2^i Y_i C_1^i, \\ \Sigma_{22}^i &= A_{22}^i X_2 + B_2^i Y_i C_2^i, \end{aligned} \tag{9.67}$$

then, the 2D discrete switching system (9.60) is asymptotically stable for any arbitrary sequence of switching, the gain controllers are given by

$$K_i = Y_i W_i^{-1} . \tag{9.68}$$

Proof: By virtue of Theorem 9.2, a sufficient condition of asymptotic stability of the 2D discrete closed-loop system (9.60) is given by inequalities (9.14) while replacing F_i by $K_i C_i = K_i [C_1^i \;\; C_2^i]$ and using common Lyapunov function. In this case, matrices Φ_{ij} become:

$$\begin{aligned} \Phi_{11}^i &= A_{11}^i + B_1^i K_i C_1^i \\ \Phi_{12}^i &= A_{12}^i + B_1^i K_i C_2^i \\ \Phi_{21}^i &= A_{21}^i + B_2^i K_i C_1^i \\ \Phi_{22}^i &= A_{22}^i + B_2^i K_i C_2^i \end{aligned} \tag{9.69}$$

By pre-and post-multiplying (9.14) by $diag\{X_1, X_2, \mathbb{I}, \mathbb{I}\}$ and letting $X_i = P_i^{-1}$, one obtains

$$\begin{bmatrix} X_1 & 0 & (\Phi_{11}^i X_1)^T & (\Phi_{21}^i X_1)^T \\ 0 & X_2 & (\Phi_{12}^i X_2)^T & (\Phi_{22}^i X_2)^T \\ * & * & X_1 & 0 \\ * & * & 0 & X_2 \end{bmatrix} > 0 . \tag{9.70}$$

Using equalities (9.65), (9.66) and $Y_i = K_i W_i$, (9.64) is obtained. □

Example 9.6. *Consider the 2D discrete switching system given by (9.54) with* $n_1 = 2, n_2 = 1, m = 2, p = 2$, *where*

$$A_1 = \begin{bmatrix} 0.5 & 0 & 0.1 \\ 0 & 1 & 0 \\ 0 & 0 & 0.1 \end{bmatrix}, \ B_1 = \begin{bmatrix} 1 & 0 \\ 0 & 1 \\ -0.1 & 0.1 \end{bmatrix}, C_1 = \begin{bmatrix} 1 & 1 & 0 \\ 0 & 1 & 0 \end{bmatrix},$$
$$A_2 = \begin{bmatrix} 0.2 & 0 & -0.1 \\ 0 & 0.1 & 0.1 \\ 1 & 1 & 0.1 \end{bmatrix}, B_2 = \begin{bmatrix} 1 & 0.1 \\ 0 & 1 \\ 1 & 0 \end{bmatrix}, C_2 = \begin{bmatrix} 1 & 0 & 0 \\ 0 & 1 & 0 \end{bmatrix}.$$

The LMIs (9.64), (9.65) and (9.66) are feasible. A solution is given by:

$$X_1 = 1E+3 \begin{bmatrix} 3.0475 & 0.0000 \\ 0.0000 & 2.1984 \end{bmatrix}, X_2 = 3.0475E+3,$$
$$K_1 = \begin{bmatrix} -0.2878 & 0.2807 \\ -0.0052 & -0.9865 \end{bmatrix},$$
$$K_2 = \begin{bmatrix} -0.5998 & -0.3579 \\ 0.0221 & -0.0514 \end{bmatrix}$$

The trajectories of the 2D switching system obtained with Theorem 9.6 are not plotted.

The result of Theorem 9.6 is given with constraining equalities (9.65) and (9.66). In order to relax this result, the following corollary is presented for a class of 2D switching systems.

Corollary 9.4. *If a subsystem* $s \in \mathscr{I}$ *has matrix* C_s *decomposable as (9.62) with* $C_r^s = \mathbb{I}_n$ *and* $p = n_1$, *then the corresponding conditions (9.64)-(9.66) in Theorem 9.6 are replaced by*

$$\begin{bmatrix} X_1 & 0 & (\Psi_{11}^s)^T & (\Psi_{21}^s)^T \\ 0 & X_2 & (\Psi_{12}^s)^T & (\Psi_{22}^s)^T \\ * & * & X_1 & 0 \\ * & * & 0 & X_2 \end{bmatrix} > 0, \tag{9.71}$$

where

$$\begin{aligned} \Psi_{11}^s &= A_{11}^s X_1 + B_1^s Y_s C_l^s, \\ \Psi_{12}^s &= A_{12}^s X_2, \\ \Psi_{21}^s &= A_{21}^s X_1 + B_2^s Y_s C_l^s, \\ \Psi_{22}^s &= A_{22}^s X_2, \end{aligned} \tag{9.72}$$

with $X_1 \in \mathbb{R}^{p \times p}$, $X_2 \in \mathbb{R}^{(n-p)\times(n-p)}$; *the corresponding gain controller is given by*

$$K_s = Y_s C_l^s X_1^{-1} (C_l^s)^{-1}. \tag{9.73}$$

Proof: In this case one has, $C_s = [C_1^s \; C_2^s] = [C_l^s \; 0]$. It suffices to replace $C_1^s = C_l^s$ and $C_2^s = 0$ in (9.67) to obtain (9.71) with (9.72). One can notice that in this case, due to the asymptotic stability constraint of 2D systems to have a Lyapunov function of diagonal block form, matrix C_r is assumed to be equal to identity implying that p has to be equal to n_1. Thus, equality $C_sX = W_sC_s$ is reduced to $C_l^sX_1 = W_sC_l^s$. According to Lemma 9.1, the solution of equality $C_l^sX_1 = W_sC_l^s$ is given by $W_s = C_l^sX_1(C_l^s)^{-1}$. Then the expression of the output controller gain is derived. □

Example 9.7. *Let us consider the example studied in Example 9.6. The LMIs (9.64), (9.65) and (9.66) for mode 1 and (9.71) for mode 2 are feasible, due to the fact that matrix C_2 can be decomposed according to $C_2 = [C_l^2 \; 0]$. An obtained solution is given by:*

$$X_1 = \begin{bmatrix} 803.8942 & 0.0000 \\ 0.0000 & 363.8036 \end{bmatrix}, X_2 = 803.8942,$$
$$K_1 = \begin{bmatrix} -0.2911 & 0.2807 \\ -0.0035 & -0.9935 \end{bmatrix},$$
$$K_2 = \begin{bmatrix} -0.6001 & -0.5045 \\ 0.0106 & -0.1038 \end{bmatrix}.$$

Figure 9.4 plots the evolution of the states $x^h(k,l)$ and $x^v(k,l)$ obtained with Corollary 9.4. It is clear that the system is asymptotically stable and the states converge toward zero, as suggested by the approach.

Assume now that the discrete switching system given by (9.54) is saturated as follows:

$$\begin{aligned} x^+(k,l) &= A_\alpha x(k,l) + B_\alpha sat(u(k,l)) && (9.74) \\ x^h(0,l) &= f(l), x^v(k,0) = g(k), \\ y(k) &= C_\alpha x(k) && (9.75) \end{aligned}$$

where $sat(.)$ is a normalized saturation function. We only restrict ourselves to design unsaturating controller by using output feedback control (9.57). Applying the same tools developed during the previous chapters, we can present the following result.

Corollary 9.5. *For a positive scalar ρ, if there exist positive definite matrices $X_1 \in \mathbb{R}^{n_1 \times n_1}$, $X_2 \in \mathbb{R}^{n_2 \times n_2}$ and matrices $Y_1^i \in \mathbb{R}^{m \times n_1}$ and $Y_2^i \in \mathbb{R}^{m \times n_2}$ such that*

$$\begin{bmatrix} X_1 & 0 & (\Sigma_{11}^i)^T & (\Sigma_{21}^i)^T \\ 0 & X_2 & (\Sigma_{12}^i)^T & (\Sigma_{22}^i)^T \\ * & * & X_1 & 0 \\ * & * & 0 & X_2 \end{bmatrix} > 0, \qquad (9.76)$$
$$C_1^iX_1 = W_iC_1^i \qquad (9.77)$$
$$C_2^iX_2 = W_iC_2^i \qquad (9.78)$$

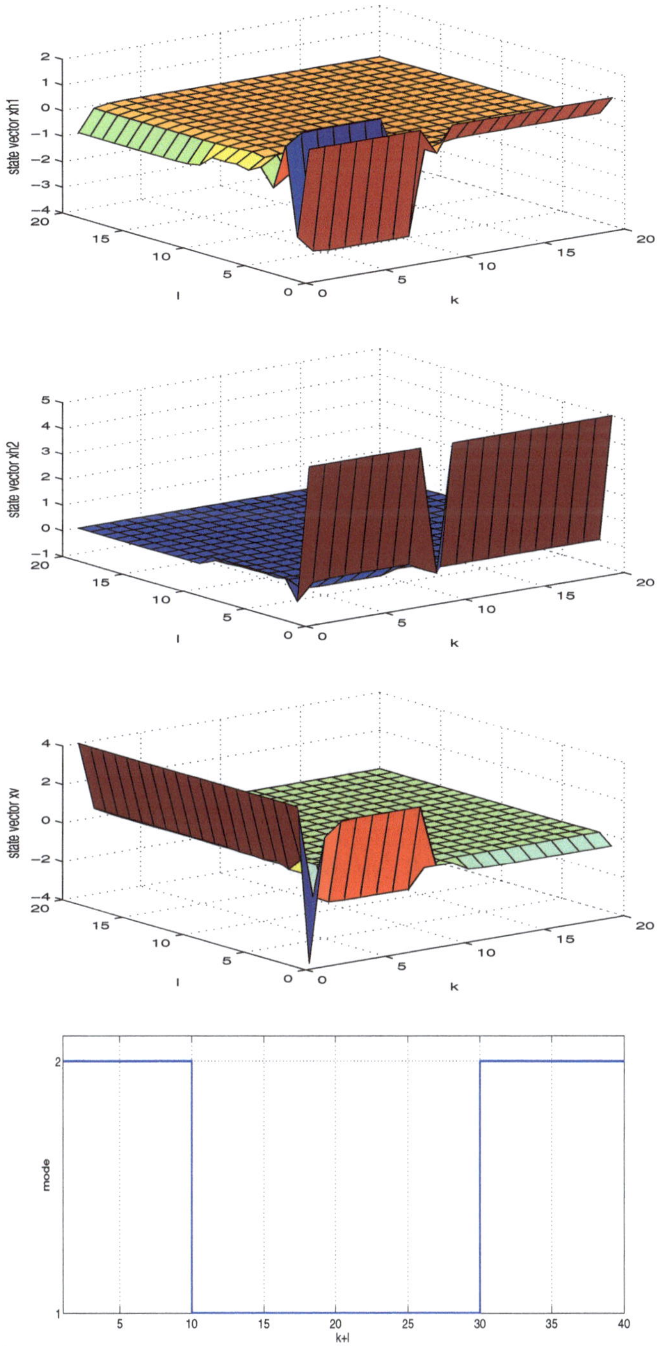

Fig. 9.4 The states evolution of $x_1^h(k,l)$, $x_2^h(k,l)$, $x^v(k,l)$ and the corresponding sequence of switching obtained with Corollary 9.4

$$\begin{bmatrix} 1/\rho & (Y_iC_1^i)_l & (Y_iC_2^i)_l \\ * & X_1 & 0 \\ * & 0 & X_2 \end{bmatrix} > 0, l \in [1,\ m] \tag{9.79}$$
$$\forall i \in \mathscr{I}$$

with

$$\Psi_{11}^i = A_{11}^i X_1 + B_1^i Y_1^i, \Psi_{12}^i = A_{12}^i X_2 + B_1^i Y_2^i \Psi_{21}^i = A_{21}^i X_1 + B_2^i Y_1^i \Psi_{22}^i = A_{22}^i X_2 + B_2^i Y_2^i$$

then, the 2D saturated switching system (9.60) is asymptotically stable for any arbitrary sequence of switching and $\forall x_o \in \varepsilon(X^{-1},\ \rho)$*; the corresponding controller gains are given by*

$$K_i = Y_i W_i^{-1} \tag{9.80}$$

with $P_1 = X_1^{-1}$ *and* $P_2 = X_2^{-1}$.

Proof: The proof follows readily by writing the inclusion relation $\varepsilon(P,\rho) \subset \mathscr{L}(K_iC_i)$ under LMI form, with

$$\mathscr{L}(K_iC_i) = \{x \in \mathbb{R}^n / |(K_iC_i)_l x| \leq 1\ \} \tag{9.81}$$

where $(K_iC_i)_l$ denotes the lth row of matrix K_iC_i and ρ is a positive scalar. According to [53], the inclusion relation $\varepsilon(P,\rho) \subset \mathscr{L}(K_iC_i)$ is satisfied if and only if

$$\rho(K_iC_i)_l P^{-1}(K_iC_i)_l^T < 1, l \in [1,\ m]. \tag{9.82}$$

Using (9.77)-(9.78), $C_i = [C_1^i\ \ C_2^i]$, (9.80), the inequality (9.82) can be successively developed as

$$\begin{aligned} \rho(K_iC_iP^{-1})_l P(K_iC_iP^{-1})_l^T &< 1 \\ \rho(K_iC_iX)_l X^{-1}(K_iC_iX)_l^T &< 1 \\ \rho(K_iC_1^iX_1)_l X^{-1}(K_iC_2^iX_2)_l^T &< 1 \\ \rho(K_iW_iC_1^i)_l X^{-1}(K_iW_iC_2^i)_l^T &< 1 \\ \rho(Y_iC_1^i)_l X^{-1}(Y_iC_2^i)_l^T &< 1. \end{aligned}$$

Applying Schur complement to the last inequality, LMI (9.79) follows. □

9.3.3.2 Multiple Lyapunov Function

Consider now the 2D discrete switching system (9.1). The result of Theorem 9.6 can be used to synthesize the stabilizing controller under LMIs form by using multiple Lyapunov functions.

Theorem 9.7. *If there exist matrices* $Y_i \in \mathbb{R}^{m\times p}$*, non singular matrix* $W_i \in \mathbb{R}^{p\times p}$*, positive definite matrices* $X_1^i \in \mathbb{R}^{n_1\times n_1}$*,* $X_2^i \in \mathbb{R}^{n_2\times n_2}$ *such that the following LMI holds true :*

$$\begin{bmatrix} X_1^i & 0 & (\Gamma_{11}^i)^T & (\Gamma_{21}^i)^T \\ 0 & X_2^i & (\Gamma_{12}^i)^T & (\Gamma_{22}^i)^T \\ * & * & X_1^j & 0 \\ * & * & 0 & X_2^j \end{bmatrix} > 0, \tag{9.83}$$

$$C_1^i X_1^i = W_i C_1^i \tag{9.84}$$

$$C_2^i X_2^i = W_i C_2^i \tag{9.85}$$

$$\forall (i,j) \in \mathscr{I}^2$$

where

$$\begin{aligned} \Gamma_{11}^i &= A_{11}^i X_1^i + B_1^i Y_i C_1^i, \\ \Gamma_{12}^i &= A_{12}^i X_2^i + B_1^i Y_i C_2^i, \\ \Gamma_{21}^i &= A_{21}^i X_1^i + B_2^i Y_i C_1^i, \\ \Gamma_{22}^i &= A_{22}^i X_2^i + B_2^i Y_i C_2^i, \end{aligned} \tag{9.86}$$

then, the 2D discrete switching system (9.60) is asymptotically stable for any arbitrary sequence of switching ; the gain controllers are given by

$$K_i = Y_i W_i^{-1} \, . \tag{9.87}$$

Proof: The proof follows a similar reasoning as the proof of Theorem 9.4 based on Theorem 9.6. □

Example 9.8. *Considering the 2D discrete switching system given by (9.1),*

$$A_1 = \begin{bmatrix} 0.4 & 0 & 0.1 \\ 1 & -0.4 & 1 \\ -0.1 & 0.5 & 0.1 \end{bmatrix}, B_1 = \begin{bmatrix} 1 & 0 \\ 0 & 1 \\ 1 & 1 \end{bmatrix}, C_1 = \begin{bmatrix} 1 & 1 & 1 \\ 0 & 1 & 0 \end{bmatrix},$$

$$A_2 = \begin{bmatrix} 0.1 & 0 & -0.1 \\ 0 & 0.2 & 1 \\ 0.4 & 0.1 & -0.1 \end{bmatrix}, B_2 = \begin{bmatrix} 1 & 1 \\ 0 & 1 \\ 1 & 0 \end{bmatrix}, C_2 = \begin{bmatrix} 1 & 0 & 0 \\ 0 & 1 & 0 \end{bmatrix}.$$

The LMIs (9.64), (9.65) and (9.66) are not feasible while the LMIs (9.83), (9.84) and (9.85) are feasible. A solution is given by:

$$X_1^1 = 1E+4 \begin{bmatrix} 0.1465 & 0.0000 \\ 0.0000 & 1.1405 \end{bmatrix}, X_2^1 = 0.1465E+4,$$

$$X_1^2 = 1E+4 \begin{bmatrix} 1.1092 & -0.0000 \\ -0.0000 & 1.1092 \end{bmatrix}, X_2^2 = 0.4349E+4,$$

$$K_1 = \begin{bmatrix} -0.1295 & 0.0338 \\ -0.0059 & -0.3047 \end{bmatrix}, K_2 = \begin{bmatrix} -0.3404 & -0.0488 \\ 0.1803 & -0.0250 \end{bmatrix}.$$

The trajectories of the 2D switching system obtained with Theorem 9.7 are not plotted.

The result of Theorem 9.7 is given with constraining equalities (9.84) and (9.85). In order to relax this result, the following corollary is presented for a class of 2D systems.

Corollary 9.6. *If a subsystem $s \in \mathscr{I}$ has matrix C_s decomposable as (9.62) with $C_r^s = \mathbb{I}_n$ and $p = n_1$ then the corresponding conditions (9.83)-(9.85) in Theorem 9.7 are replaced by*

$$\begin{bmatrix} X_1^s & 0 & (\Theta_{11}^s)^T & (\Theta_{21}^s)^T \\ 0 & X_2^s & (\Theta_{12}^s)^T & (\Theta_{22}^s)^T \\ * & * & X_1^j & 0 \\ * & * & 0 & X_2^j \end{bmatrix} > 0, \tag{9.88}$$

where

$$\begin{aligned} \Theta_{11}^s &= A_{11}^s X_1^s + B_1^s Y_s C_l^s, \\ \Theta_{12}^s &= A_{12}^s X_2^s, \\ \Theta_{21}^s &= A_{21}^s X_1^s + B_2^s Y_s C_l^s, \\ \Theta_{22}^s &= A_{22}^s X_2^s, \end{aligned} \tag{9.89}$$

with $X_1^s \in \mathbb{R}^{p \times p}$, $X_2^s \in \mathbb{R}^{(n-p)\times(n-p)}$; the corresponding gain controller is given by

$$K_s = Y_s C_l^s (X_1^s)^{-1} (C_l^s)^{-1}. \tag{9.90}$$

Proof: Obvious. □

Example 9.9. *Let us consider the example studied in Example 9.8. The LMIs (9.83), (9.84) and (9.85) for mode 1 and (9.88) for mode 2 are feasible. An obtained solution is given by:*

$$\begin{aligned} X_1^1 &= 1E+4 \begin{bmatrix} 0.3535 & -0.0000 \\ -0.0000 & 2.6444 \end{bmatrix}, X_2^1 = 0.3535E+4, \\ X_1^2 &= 1E+4 \begin{bmatrix} 2.0912 & -0.0061 \\ -0.0061 & 3.3561 \end{bmatrix}, X_2^2 = 1.2176E+4, \\ K_1 &= \begin{bmatrix} -0.1258 & 0.0357 \\ -0.0016 & -0.3167 \end{bmatrix}, K_2 = \begin{bmatrix} -0.3294 & -0.0309 \\ 0.1853 & -0.0159 \end{bmatrix}. \end{aligned}$$

Figure 9.5 plots the evolution of the states $x^h(k,l)$, $x^v(k,l)$ and the corresponding sequence of switching obtained with Corollary 9.6. It is clear that the switching system is asymptotically stable and the states converge toward zero, as suggested by the approach.

Assume now that the discrete switching system is given by (9.74). The following result can be easily obtained.

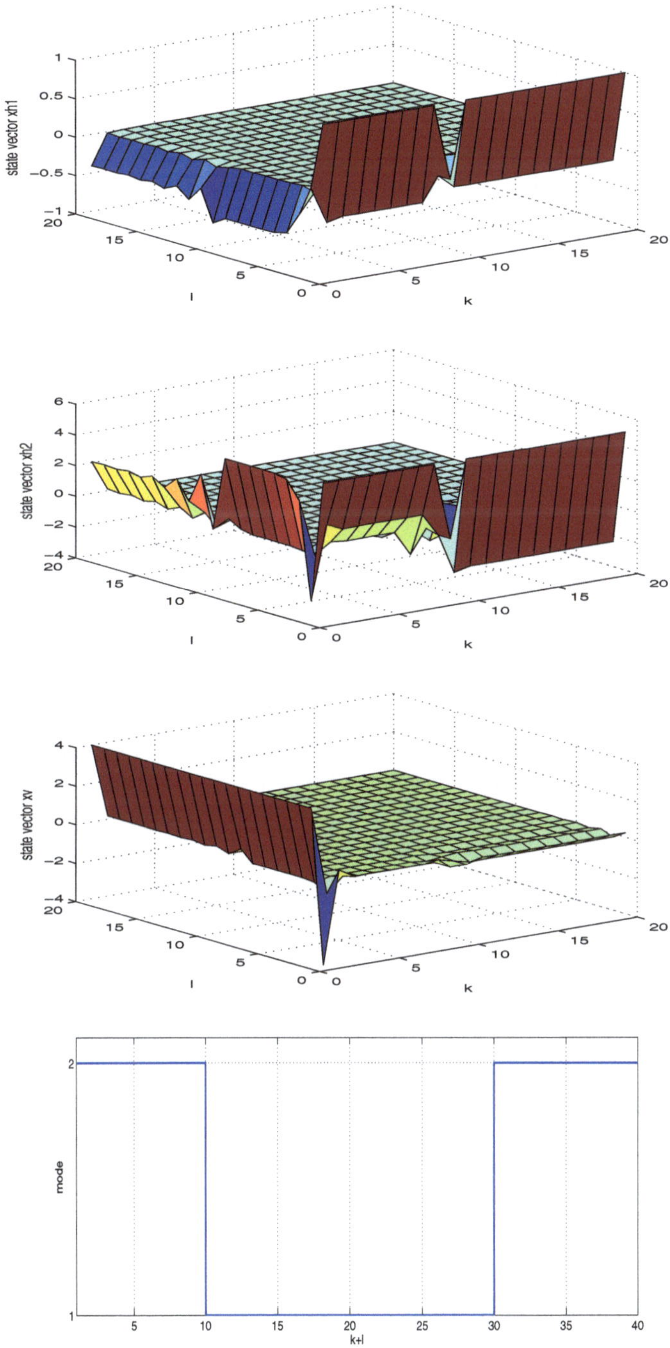

Fig. 9.5 The states evolution of $x_1^h(k,l)$, $x_2^h(k,l)$, $x^v(k,l)$ and the corresponding sequence of switching with Corollary 9.6

Corollary 9.7. *For a positive scalar ρ, if there exist matrices $Y_i \in \mathbb{R}^{m\times p}$, non singular matrix $W_i \in \mathbb{R}^{p\times p}$, positive definite matrices $X_1^i \in \mathbb{R}^{n_1\times n_1}$, $X_2^i \in \mathbb{R}^{n_2\times n_2}$ such that the following LMIs hold true :*

$$\begin{bmatrix} X_1^i & 0 & (\Gamma_{11}^i)^T & (\Gamma_{21}^i)^T \\ 0 & X_2^i & (\Gamma_{12}^i)^T & (\Gamma_{22}^i)^T \\ * & * & X_1^j & 0 \\ * & * & 0 & X_2^j \end{bmatrix} > 0, \tag{9.91}$$

$$C_1^i X_1^i = W_i C_1^i \tag{9.92}$$

$$C_2^i X_2^i = W_i C_2^i \tag{9.93}$$

$$\forall (i,j) \in \mathscr{I}^2$$

$$\begin{bmatrix} 1/\rho & (Y_iC_1^i)_l & (Y_iC_2^i)_l \\ * & X_1^i & 0 \\ * & 0 & X_2^i \end{bmatrix} > 0, l \in [1,\ m] \tag{9.94}$$

where

$$\begin{aligned} \Gamma_{11}^i &= A_{11}^i X_1^i + B_1^i Y_i C_1^i, \\ \Gamma_{12}^i &= A_{12}^i X_2^i + B_1^i Y_i C_2^i, \\ \Gamma_{21}^i &= A_{21}^i X_1^i + B_2^i Y_i C_1^i, \\ \Gamma_{22}^i &= A_{22}^i X_2^i + B_2^i Y_i C_2^i, \end{aligned} \tag{9.95}$$

then, the 2D saturated switching system (9.60), with $x_0 \succ 0$, is asymptotically stable for any arbitrary sequence of switching and $\forall x_o \in \bigcup_i^N \varepsilon(X_i^{-1},\ \rho)$; the corresponding controller gains are given by

$$K_i = Y_i W_i^{-1} \tag{9.96}$$

with $P_1 = X_1^{-1}$ and $P_2 = X_2^{-1}$.

Proof: The proof follows readily by writing the inclusion relation $\varepsilon(P_i,\rho) \subset \mathscr{L}(K_iC_i)$ under LMI form, with

$$\mathscr{L}(K_iC_i) = \{x \in \mathbb{R}^n / |(K_iC_i)_l x| \leq 1\ \} \tag{9.97}$$

where $(K_iC_i)_l$ denotes the lth row of matrix K_iC_i and ρ is a positive scalar. The same reasoning as in the case of common Lyapunov function is followed to obtain LMI (9.94). Recall that the set $\bigcup_i^N \varepsilon(X_i^{-1},\ \rho)$ is a set of stability as shown in the previous chapters for saturated switching discrete-time systems. □

This result can be easily extended to positive switching 2D systems by using the obtained results in Chapter VIII as follows:

Corollary 9.8. *For a positive scalar ρ, if there exist matrices $Y_i \in \mathbb{R}^{m\times p}$, positive scalars d_i such that the following LMIs hold true :*

$$\begin{bmatrix} d_i\mathbb{I} & 0 & (\Gamma_{11}^i)^T & (\Gamma_{21}^i)^T \\ 0 & d_i\mathbb{I} & (\Gamma_{12}^i)^T & (\Gamma_{22}^i)^T \\ * & * & d_j\mathbb{I} & 0 \\ * & * & 0 & d_j\mathbb{I} \end{bmatrix} > 0, \tag{9.98}$$

$$\begin{bmatrix} 1/\rho & (Y_iC_1^i)_l & (Y_iC_2^i)_l \\ * & d_i\mathbb{I} & 0 \\ * & 0 & d_i\mathbb{I} \end{bmatrix} > 0, \tag{9.99}$$

$$\Gamma_{11}^i = d_iA_{11}^i + B_1^iY_iC_1^i \succ 0, \tag{9.100}$$

$$\Gamma_{12}^i = d_iA_{12}^i + B_1^iY_iC_2^i, \succ 0 \tag{9.101}$$

$$\Gamma_{21}^i = d_iA_{21}^i + B_2^iY_iC_1^i \succ 0, \tag{9.102}$$

$$\Gamma_{22}^i = d_iA_{22}^i + B_2^iY_iC_2^i \succ 0, \tag{9.103}$$

$$\forall (i,j) \in \mathscr{I}^2, l \in [1,\ m], \tag{9.104}$$

then, the 2D saturated switching system (9.60) is asymptotically stable while the state is positive for any arbitrary sequence of switching and $\forall x_o \in \bigcup_i^N \varepsilon(X_i^{-1}, \rho)$*; the corresponding controller gains are given by*

$$K_i = d_i^{-1}Y_i \tag{9.105}$$

with $P_1 = P_2 = d_i^{-1}\mathbb{I}$.

Proof: Follows readily by using the proof of Corollary 8.3. □

9.4 Conclusion

In this chapter, the problem of stabilizability of the 2D discrete switching systems has been studied using static output feedback control. General conditions of asymptotic stability under LMI form involving constraining equalities are first presented. A classical method already used for 1D systems [18] and recalled in Chapter I, based on the decomposition of matrix C, is proposed. This method enables one to obtain sufficient conditions of asymptotic stability for 2D switching systems under LMI form and without any restrictive equality involving matrix C. However, this technique can be applied only for a class of 2D switching systems having a particular decomposition of matrix C, due to the asymptotic stability constraint of 2D systems to have a Lyapunov function of diagonal form. Numerical examples are provided to illustrate the results. The extensions of the obtained results to the saturated 2D discrete switching systems and the positive saturated 2D discrete switching systems are also presented for the state feedback case and the output feedback case respectively.

Chapter 10
Switching Takagi-Sugeno Systems

10.1 Introduction

The objective of this chapter is to study discrete-time switching nonlinear systems . Each subsystem is written as an equivalent T-S fuzzy models. Since the introduction of T-S fuzzy models by Takagi and Sugeno [147] in 1985, fuzzy model control has been extensively studied because T-S fuzzy models provide an effective representation of complex nonlinear systems [28, 31, 38, 39, 41, 81, 83, 134, 135]. In the literature, few works were interested on this kind of systems as pointed out in the general introduction. One can consult [148, 166] where a detailed study of the problem of stability and controller switching for switching fuzzy systems is presented. In all these previous works, only a common Lyapuniv function is used for all the T-S fuzzy subsystems. Even for switching linear systems, it was proven that the use of such a common Lyapunov function leads to conservative results [54, 146]. In this chapter, sufficient conditions of stabilizability, by state feedback control, are obtained by using two types of multiple Lyapunov function. The first one called switching Lyapunov function is multiple for the switching system while common for the T-S fuzzy subsystems. The second, called switched Lyapunov function is multiple for both the T-S fuzzy subsystems and the switching modes. This choice allows a general approach which may be less conservative than the one obtained with switching Lyapunov function. These conditions are then worked out to be presented under LMIs form. The case of saturated switching T-S fuzzy systems is also investigated by using only unsaturating controllers.

10.2 Stabilization of Switching T-S Systems by Multiple Switching Lyapunov Function

10.2.1 Problem Formulation

In this section, we give a precise problem statement for the class of systems under consideration, namely, discrete-time switching T-S fuzzy linear systems. An

A. Benzaouia: Saturated Switching Systems, LNCIS 426, pp. 247–274.
springerlink.com

equivalent description of such systems, based on the indicator function is provided in this work.

Thus, we consider nonlinear systems described by:

$$
\begin{aligned}
x(t+1) &= f_\sigma x(t) + g_\sigma u(t) \\
x(0) &= x_0
\end{aligned}
\tag{10.1}
$$

where $x \in \mathbb{R}^n$ is the state, $u \in \mathbb{R}^m$ is the control and σ a switching rule, which takes its values in the finite set $\mathscr{I} := \{1,\ldots,N\}$, $t \in \mathbb{Z}_+$, f and g are some Lipschitz functions.

It is assumed that:

- The switching system is stabilizable;
- Only one subsystem is activated at each switch;
- The switching rule is not known *a priory*, but $\alpha(t)$ is available at each t.

The third assumption corresponds to practical implementations where the switched system is supervised by a discrete-event system, or operator, allowing for $\sigma(t)$ to be known in real time.

In this work, we are interested by the synthesis of stabilizing controllers for this class of nonlinear hybrid systems.

Each nonlinear subsystem i $(i = 1;\ldots,N)$ could be represented by a T-S fuzzy model described by
Plant Rule $s(s = 1,2,\cdots,r_i)$: IF M^i_{s1} is z_{i1} and $\cdots$ and M^i_{sL} is z_{iL} THEN

$$
\begin{aligned}
x(t+1) &= A_{is}x(t) + B_{is}u(t) \\
x(0) &= x_0,
\end{aligned}
\tag{10.2}
$$

where r_i is the number of IF-THEN rules of the subsystem i. M^i_{sl} is the fuzzy set, $z_{sl}(t)$, for $l = 1, \ldots, L$ are the premise variables. Matrices A_{is} and B_{is} are constant, of appropriate size, and satisfy the following assumption:

H) $(A_{is}, B_{is}), i = 1,\ldots,N; s = 1,\ldots,r_i$ are stabilizable.

Define the following variables:

$$
\theta_i = \left[\theta_{i1} \ldots \theta_{ir_i}\right]^T,
$$

where

$$
\theta_{is}(z(t)) = \prod_{l=1}^{L} M^i_{sl}(z(t)); h_{is}(z(t)) = \frac{\theta_{is}(z(t))}{\sum_{s=1}^{r_i}\theta_{is}(z(t))}
\tag{10.3}
$$

$$
\begin{aligned}
&h_{is}(z(t)) \geq 0, \sum_{s=1}^{r_i} h_{is}(z(t)) = 1 \\
&s = 1,\ldots,r_i.
\end{aligned}
\tag{10.4}
$$

Scalars $h_{is}(z(t))$ are called membership functions. By using the common used center-average defuzzifier, product interference and singleton fuzzifier, the T-S fuzzy systems can be inferred as

$$x(t+1) = \sum_{s=1}^{r_i} h_{is}(\theta(z(t)))[A_{is}x(t) + B_{is}u(t)] \tag{10.5}$$

$$= A_i(\theta)x(t) + B_i(\theta)u(t), \tag{10.6}$$

with

$$A_i(\theta) = \sum_{s=1}^{r_i} h_{is}(\theta(z(t)))A_{is}, \tag{10.7}$$

$$B_i(\theta) = \sum_{s=1}^{r_i} h_{is}(\theta(z(t)))B_{is}.$$

The design of state feedback stabilizing fuzzy controllers for the fuzzy system (10.2) is based on the parallel distributed compensation (PDC). The controller is then expressed by using the same membership function as the system to be controlled.

Controller Rule $s(s = 1,2,..,r_i)$: IF M^i_{s1} is z_{i1} and $\cdots$ and M^i_{sL} is z_{iL} THEN

$$u(t) = K_{is}x(t) \tag{10.8}$$

The overall state feedback control law is represented by:

$$u(t) = K_i(\theta)x(t) = \sum_{s=1}^{r_i} h_{is}(\theta(z(t)))K_{is}x(t). \tag{10.9}$$

In the sequel, for brevity we use $h_{is}(t)$ to denote $h_{is}(\theta(z(t)))$. The whole subsystem in closed-loop corresponding to the nonlinear mode i is then given by:

$$x(t+1) = [A_i(\theta)) + B_i(\theta)K_i(\theta)]\,x(t), \tag{10.10}$$

where $A_i(\theta)$ and $B_i(\theta)$ are given by (10.7).

Upon introducing the indicator function

$$\xi(t) = [\xi_1(t), \ldots, \xi_N(t)]^T, \tag{10.11}$$

where $\xi_i(t) = 1$ if the switching system is in mode i and $\xi_i(t) = 0$ if it is in a different mode, one can write the whole closed-loop system corresponding to (10.1) as follows:

$$x(t+1) = \sum_{i=1}^{N} \xi_i(t)[A_i(\theta) + B_i(\theta)K_i(\theta)]x(t), \tag{10.12}$$

where $A_i(\theta)$ and $B_i(\theta)$ are given by (10.7). Hence, the problem we are dealing with consist of designing the gains $K_i(\theta)$ that stabilize the closed-loop system (10.12). Our goal is to propose an algorithm that uses the LMI framework to facilitate the computation of the feedback control gains K_{is}.

The system in closed-loop (10.12) can be equivalently written as

$$x(t+1) = \sum_{i=1}^{N} \xi_i(t) A_{c_i}(\theta) x(t). \tag{10.13}$$

with $A_{c_i}(\theta) = A_i(\theta) + B_i(\theta)K_i(\theta)$.

The existence of a common Lyapunov function to the various subsystems guarantees the asymptotic stability of the switching system. In general, the search for such function is not always obvious [146]. Multiple Lyapunov functions were introduced in [54] and developed in Chapter *III*. They constitute a strong tool in the analysis of the stability of the hybrid systems, and in particular the switching systems. So, they will be used in this chapter.

A corresponding Lyapunov function for the system is then given by:

$$V(t,x) = x^T(t)(\sum_{i=1}^{N} \xi_i(t)P_i)x(t). \tag{10.14}$$

10.2.2 Conditions of Stabilization

In this section, the design of a stabilizing controller for the class of switching T-S fuzzy system is presented. We assume that the state is available, so a state feedback control can be performed.

Theorem 10.1. *If there exist N symmetric matrices $P_1, \dots, P_N$ such that:*

$$\begin{bmatrix} P_i & (A_{is} + B_{is}K_{il})^T P_j \\ * & P_j \end{bmatrix} > 0, \tag{10.15}$$
$$\forall (i,j) \in \mathscr{I} \times \mathscr{I}, l, s = 1, \dots, r_i,$$

then the switching T-S fuzzy system in closed-loop (10.12), is asymptotically stable at the origin for any sequence of switching $\alpha(t)$.

Proof: The rate of increase of the Lyapunov function candidate (10.14) along the trajectories of (10.13) is given by:

$$\Delta V(t,x(t)) = x^T(t+1)(\sum_{j=1}^{N} \xi_j(t+1)P_j)x(t+1) - x^T(t)(\sum_{i=1}^{N} \xi_i(t)P_i)x(t)$$

$$= x^T(t)\left[\left(\sum_{i=1}^{N}\xi_i(t)Ac_i(\theta)\right)^T\left(\sum_{j=1}^{N}\xi_j(t+1)P_j\right)\left(\sum_{i=1}^{N}\xi_i(t)Ac_i(\theta)\right)\right.$$
$$\left.-\sum_{i=1}^{N}\xi_i(t)P_i\right]x(t). \quad (10.16)$$

Let condition (10.15) be satisfied. At this level, for each i, multiply the inequalities (10.15) by $\xi_j(t+1)$ and sum $j = 1,\ldots,N$. Multiply the resulting inequalities by $\xi_i(t)$ and sum $i = 1,\ldots,N$. Multiply again the resulting inequalities by $h_{is}(t)$ and sum $s = 1,\ldots,r_i$. Multiply finally the resulting inequalities by $h_{il}(t)$ and sum $l = 1,\ldots,r_i$. As $\sum_{i=1}^{N}\xi_i(t) = \sum_{j=1}^{N}\xi_j(t+1) = \sum_{s=1}^{r_i}h_{is}(t) = \sum_{l=1}^{r_i}h_{il}(t) = 1$, one gets:

$$\Psi(i,j,s,l) = \begin{bmatrix} \sum_{i=1}^{N}\xi_i(t)P_i & \Gamma^T(\sum_{j=1}^{N}\xi_j(t+1)P_j) \\ * & \sum_{j=1}^{N}\xi_j(t+1)P_j \end{bmatrix} > 0, \quad (10.17)$$

with

$$\Gamma = \sum_{i=1}^{N}\sum_{s=1}^{r_i}\sum_{l=1}^{r_i}\xi_i(t)h_{is}(t)h_{il}(t)Ac_{isl}, \quad (10.18)$$
$$Ac_{isl} = A_{is} + B_{is}K_{il}.$$

The use of Schur complement allows us to write condition (10.17) under the equivalent form

$$\Gamma^T\left(\sum_{j=1}^{N}\xi_j(t+1)P_j\right)\Gamma - \sum_{i=1}^{N}\xi_i(t)P_i < 0, \quad (10.19)$$
$$\forall(i,j) \in \mathscr{I}\times\mathscr{I}, s,l = 1,\ldots,r_i.$$

Letting λ be the largest eigenvalue among all the above matrices, we obtain that

$$\Delta V(t,x(t)) \le \lambda x^T(t)x(t) < 0, \quad (10.20)$$

which ensures the desired result. Noting that condition (10.15) is also satisfied for $i = j$ and $s = l$, this implies that each subsystem is asymptotically stable at the origin. □

This stability result can be used for control synthesis as follows.

Theorem 10.2. *If there exists N symmetric matrices $X_1,\ldots,X_N$ and matrices $Y_{11},\ldots,$ Y_{Nr_N} solutions of the following LMIs:*

$$\begin{bmatrix} X_i & (A_{is}X_i + B_{is}Y_{il})^T \\ * & X_j \end{bmatrix} > 0, \quad (10.21)$$
$$\forall(i,j) \in \mathscr{I}\times\mathscr{I}, \forall(s,l) \in [1,r_i]^2,$$

then the switching T-S fuzzy system in closed-loop (10.12), with

$$K_{il} = Y_{il}X_i^{-1}, P_i = X_i^{-1} \tag{10.22}$$

is asymptotically stable at the origin for any sequence of switching $\alpha(t)$.

Proof: The inequalities (10.15) can be transformed equivalently by Schur complement to the following:

$$P_j \left[A_{is} + B_{is}K_{il}\right] P_i^{-1} \left[A_{is} + B_{is}K_{il})\right]^T P_j - P_j < 0. \tag{10.23}$$

By pre and post-multiplying (10.23) by P_j^{-1}, it follows that:

$$\left[A_{is} + B_{is}K_{il}\right] P_i^{-1} \left[A_{is} + B_{is}K_{il}\right]^T - P_j^{-1} < 0. \tag{10.24}$$

Letting $X_i = P_i^{-1}$, inequality (10.24) can then be rewritten as

$$\left[A_{is} + B_{is}K_{il}\right] X_i \left[A_{is} + B_{is}K_{il}\right]^T - X_j < 0,$$

That is, using $Y_{il} = K_{il}X_i$ gives

$$\left[A_{is}X_i + B_{is}Y_{il}\right] X_i^{-1} \left[A_{is}X_i + B_{is}Y_{il})\right]^T - X_j < 0.$$

The use of the Schur complement a second time leads to:

$$\begin{bmatrix} X_i & (A_{is}X_i + B_{is}Y_{il})^T \\ * & X_j \end{bmatrix} > 0. \tag{10.25}$$

Then, condition (10.15) is equivalent to the LMI (10.21). □

Remark 10.1. *The global system is considered as a switching system composed of N set of r_i T-S fuzzy subsystems each one, $i = 1, \dots, N$. The Lyapunov function used in this work is the multiple one. This function can be seen as a multiple one for the switching nonlinear system while for each set of T-S fuzzy subsystems, P_i is a common Lyapunov function.*

Remark 10.2. *In the case that matrix B is the same for all the T-S Fuzzy systems and $r_i = r$ as it is frequently the case, the LMIs (10.21) are transformed to:*

$$\begin{bmatrix} X_i & (A_{is}X_i + BY_{is})^T \\ * & X_j \end{bmatrix} > 0,$$
$$(i,j) \in \mathscr{I} \times \mathscr{I}, s \in [1,r].$$

In order to reduce the number of LMIs, one can use the technique of PDC control. This reduction is presented by the following result.

Theorem 10.3. *If there exist N symmetric matrices $X_1,\ldots,X_N$ and matrices $Y_{11},\ldots, Y_{Nr_N}$ solution of the following LMIs:*

$$\begin{bmatrix} X_i & (A_{is}X_i+B_{is}Y_{is})^T \\ * & X_j \end{bmatrix} > 0, \qquad (10.26)$$
$$\forall (i,j)\in \mathscr{I}\times\mathscr{I}, \forall s\in[1,r_i],$$

and

$$\begin{bmatrix} X_i & \frac{1}{2}(A_{is}X_i+B_{is}Y_{il}+A_{il}X_i+B_{il}Y_{is})^T \\ * & X_j \end{bmatrix} > 0, \qquad (10.27)$$
$$\forall (i,j)\in \mathscr{I}\times\mathscr{I}, \forall s<l\in[1,r_i],$$

then the switching T-S fuzzy system in closed-loop (10.12), with

$$K_{il} = Y_{il}X_i^{-1}, P_i = X_i^{-1}, \qquad (10.28)$$

is asymptotically stable at the origin for any sequence of switching $\alpha(t)$.

Proof: Matrix Γ given by (10.18) can be equivalently written by using the following well known developments:

$$\begin{aligned}\Gamma &= \sum_{i=1}^{N}\sum_{s=1}^{r_i}\sum_{l=1}^{r_i}\xi_i(t)h_{is}(t)h_{il}(t)Ac_{isl} \\ &= \sum_{i=1}^{N}\sum_{s=1}^{r_i}\xi_i(t)h_{is}^2(t)Ac_{iss}+\sum_{i=1}^{N}\sum_{s=1}^{r_i}\sum_{l>s}^{r_i}\xi_i(t)h_{is}(t)h_{il}(t)(Ac_{isl}+Ac_{ils}). \end{aligned} \qquad (10.29)$$

Recall that one can obviously have

$$\begin{aligned} 1 &= (\sum_{s=1}^{r_i}h_{is}(t))(\sum_{l=1}^{r_i}h_{il}(t)) \\ &= \sum_{s=1}^{r_i}\sum_{l=1}^{r_i}h_{is}(t)h_{il}(t) \\ &= \sum_{s=1}^{r_i}h_{is}^2(t)+2\sum_{s=1}^{r_i}\sum_{l>s}^{r_i}h_{is}(t)h_{il}(t) \end{aligned}$$

By replacing in (10.17), one can equivalently obtain

$$\Psi(i,j,s,l) =$$
$$\begin{bmatrix} \Sigma_{i=1}^{N}\Sigma_{s=1}^{r_i}h_{is}^2(t)\xi_i(t)P_i & (\Sigma_{i=1}^{N}\Sigma_{s=1}^{r_i}\xi_i(t)h_{is}^2(t)Ac_{iss})^T\,(\Sigma_{j=1}^{N}\xi_j(t+1)P_j) \\ * & \Sigma_{j=1}^{N}\Sigma_{s=1}^{r_i}h_{is}^2(t)\xi_j(t+1)P_j \end{bmatrix}$$

$$+\begin{bmatrix} 2\sum_{i=1}^{N}\sum_{s=1}^{r_i}\sum_{l>s}^{r_i} h_{is}(t)h_{il}(t)\xi_i(t)P_i & \left(\sum_{i=1}^{N}\sum_{s=1}^{r_i}\sum_{l>s}^{r_i}\xi_i(t)h_{is}(t)h_{il}(t)(Ac_{isl}+Ac_{ils})\right)^T\left(\sum_{j=1}^{N}\xi_j(t+1)P_j\right) \\ * & 2\sum_{j=1}^{N}\sum_{s=1}^{r_i}\sum_{l>s}^{r_i} h_{is}(t)h_{il}(t)\xi_j(t+1)P_j \end{bmatrix} > 0,$$

which can be written as

$$\Psi(i,j,s,l) = \sum_{i=1}^{N}\sum_{s=1}^{r_i}\sum_{j=1}^{N}\xi_i(t)\xi_j(t+1)h_{is}^2(t)\begin{bmatrix} P_i & (Ac_{iss})^T P_j \\ * & P_j \end{bmatrix} +$$

$$2\sum_{i=1}^{N}\sum_{s=1}^{r_i}\sum_{l>s}^{r_i}\sum_{j=1}^{N}\xi_i(t)\xi_j(t+1)h_{is}(t)h_{il}(t)\begin{bmatrix} P_i & \frac{(Ac_{isl}+Ac_{ils})^T}{2}P_j \\ * & P_j \end{bmatrix} > 0.$$

Finally, following the same developments as in Theorem 10.2, conditions (10.26) and (10.27) can be obtained as sufficient conditions for asymptotic stability of the swithing T-S fuzzy system (10.12) for any sequence of switching $\alpha(t)$. □

Example 10.1. *Consider the following numerical system composed of four modes and two T-S subsystems:*

Mode 1*:*

$$A_{11} = \begin{bmatrix} 0.5 & 0.5 \\ -0.2 & 0.4 \end{bmatrix}; A_{12} = \begin{bmatrix} 0.2 & -0.2 \\ 0.5 & 1 \end{bmatrix}; B_1 = \begin{bmatrix} 2 \\ -1 \end{bmatrix};$$

Mode 2*:*

$$A_{21} = \begin{bmatrix} 1 & 0.2 \\ -0.5 & 0.5 \end{bmatrix}; A_{22} = \begin{bmatrix} -0.2 & -1 \\ 0.2 & -0.5 \end{bmatrix}; B_2 = B_1;$$

Mode 3*:*

$$A_{31} = \begin{bmatrix} 1 & -0.2 \\ 0.5 & 0.4 \end{bmatrix}; A_{32} = \begin{bmatrix} -0.2 & -1 \\ -0.5 & -0.5 \end{bmatrix}; B_3 = B_1;$$

Mode 4*:*

$$A_{41} = \begin{bmatrix} -0.2 & -1 \\ 0.5 & 0.1 \end{bmatrix}; A_{41} = \begin{bmatrix} 1 & -0.5 \\ -0.1 & -0.2 \end{bmatrix}; B_4 = B_1.$$

the set of LMIs derived using common quadratic Lyapunov function, as in [163], is not feasible, whereas the LMIs (10.21) are feasible. A solution is the following:

$$P_1 = \begin{bmatrix} 0.1079 & 0.0285 \\ 0.0285 & 0.0733 \end{bmatrix};$$

$$P_2 = \begin{bmatrix} 0.2589 & -0.0084 \\ -0.0084 & 0.0839 \end{bmatrix};$$

$$P_3 = \begin{bmatrix} 0.1943 & 0.0503 \\ 0.0503 & 0.0839 \end{bmatrix};$$

$$P_4 = \begin{bmatrix} 0.1517 & 0.0219 \\ 0.0219 & 0.0771 \end{bmatrix}.$$

In this section, a sufficient condition of stabilizability of switching discrete-time T-S fuzzy systems is obtained by using a state feedback control. The main result of this section is given under LMI formulation. A numerical example is studied to show that the results obtained with common Lyapunov function are more restrictive than the results of this work.

10.3 Stabilization of Switching T-S Systems by Switched Lyapunov Function

In this section, a sufficient condition of stabilizability is obtained by using a switched Lyapunov function and state feedback control. The new proposed Lyapunov function is multiple for both the T-S fuzzy subsystems and the switching modes. This choice allows a general approach which may be less conservative than the one considered in the previous section. The obtained conditions are then worked out to be presented in LMI form.

10.3.1 Conditions of Stabilization

In this section, the design of stabilizing controllers for the class of switching T-S fuzzy systems is now presented. We assume that the state is available, so a state feedback control can be performed. Consider the following Lyapunov function candidate for the system given by:

$$\begin{aligned} V(t,x) &= x^T(t)(\sum_{i=1}^{N}\sum_{s=1}^{r_i}\xi_i(t)h_{is}(t)P_{is})x(t) \\ &= x^T(t)P(\xi,\theta)x(t) = x^T(t)(\sum_{i=1}^{N}\xi_i(t)P_i(\theta)x(t) \end{aligned} \tag{10.30}$$

This Lyapunov function candidate is multiple for both the T-S fuzzy subsystems and for the switching modes. Here it is called the switched Lyapunov function. The first result of this section can then be stated. This result is taken from [40].

Theorem 10.4. *If there exist $r_N N$ symmetric matrices $P_{11},\ldots,P_{Nr_N}$ such that:*

$$\begin{bmatrix} P_{il} & (A_{is}+B_{is}K_{il})^T P_{jk} \\ * & P_{jk} \end{bmatrix} > 0, \tag{10.31}$$
$$\forall (i,j) \in \mathscr{I}^2, l,s = 1,\ldots,r_i, k = 1,\ldots,r_j$$

then the switching T-S fuzzy system in closed-loop (10.12) is asymptotically stable at the origin for any sequence of switching $\alpha(t)$.

Proof: The rate of increase of the Lyapunov function candidate (10.30) along the trajectories of (10.13) is given by:

$$\Delta V(t,x(t)) = x^T(t+1)P_+(\xi,\theta)x(t+1) - x^T(t)P(\xi,\theta)x(t), \tag{10.32}$$

where $P_+(\xi,\theta) = \sum_{j=1}^{N}\sum_{s=1}^{r_j}\xi_j(t+1)h_{js}(t+1)P_{js}$.

$$\Delta V(t,x(t)) = x^T(t)\left[\left(\sum_{i=1}^{N}\xi_i(t)Ac_i(\theta)\right)^T P_+(\xi,\theta)\left(\sum_{i=1}^{N}\xi_i(t)Ac_i(\theta)\right) - P(\xi,\theta)\right]x(t).$$

Let condition (10.31) be satisfied. At this level, for each i, multiply the inequalities by $\xi_i(t)$ and sum for $i = 1,\ldots,N$. Multiply the resulting inequalities by $\xi_j(t+1)$ and sum for $j = 1,\ldots,N$. Multiply the resulting inequalities by $h_{jk}(t+1)$ and sum for $k = 1,\ldots,r_j$. Multiply the resulting inequalities by $h_{is}(t)$ again and sum for $s = 1,\ldots,r_i$. Finally, multiply the resulting inequalities by $h_{il}(t)$ and sum for $l = 1,\ldots,r_i$. As $\sum_{i=1}^{N}\xi_i(t) = \sum_{j=1}^{N}\xi_j(t+1) = \sum_{k=1}^{r_j}h_{jk}(t+1) = \sum_{s=1}^{r_i}h_{is}(t) = \sum_{l=1}^{r_i}h_{il}(t) = 1$, and one gets:

$$\Psi(i,j,l,s,k) = \begin{bmatrix} \sum_{i=1}^{N}\sum_{l=1}^{r_i}\xi_i(t)h_{il}(t)P_{il} & \Gamma^T\left(\sum_{j=1}^{N}\sum_{k=1}^{r_j}\xi_j(t+1)h_{jk}(t+1)P_{jk}\right) \\ * & \sum_{j=1}^{N}\sum_{k=1}^{r_j}\xi_j(t+1)h_{jk}(t+1)P_{jk} \end{bmatrix} > 0, \tag{10.33}$$

with

$$\begin{aligned} \Gamma &= \sum_{i=1}^{N}\sum_{s=1}^{r_i}\sum_{l=1}^{r_i}\xi_i(t)h_{is}(t)h_{il}(t)Ac_{isl}, \\ Ac_{isl} &= A_{is}+B_{is}K_{il}. \end{aligned} \tag{10.34}$$

The use of the Schur complement allows us to write condition (10.33) under the equivalent form

$$\Gamma^T\left(\sum_{j=1}^{N}\sum_{k=1}^{r_j}\xi_j(t+1)h_{jk}(t+1)P_{jk}\right)\Gamma - \sum_{i=1}^{N}\sum_{l=1}^{r_i}\xi_i(t)h_{il}(t)P_{il} < 0, \tag{10.35}$$
$$\forall (i,j) \in \mathscr{I}^2, s,l = 1,\ldots,r_i; k = 1,\ldots,r_j.$$

Letting λ be the largest eigenvalue among all the above matrices, we obtain that

$$\Delta V(t,x(t)) \le \lambda x^T(t)x(t) < 0, \tag{10.36}$$

which ensures the desired result. Noting that condition (10.31) is also satisfied for $s = l = k$, this implies that each T-S fuzzy subsystem is asymptotically stable at the origin. □

This stability result can be used for control synthesis as follows.

Theorem 10.5. *If there exist $r_N N$ symmetric matrices $X_{11},\dots,X_{Nr_N}$ and matrices $Y_{11},\dots,Y_{Nr_N}$ solutions of the following LMIs:*

$$\begin{bmatrix} X_{il} & (A_{is}X_{il}+B_{is}Y_{il})^T \\ * & X_{jk} \end{bmatrix} > 0, \tag{10.37}$$

$$\forall (i,j) \in \mathscr{I}^2, s,l = 1,\dots,r_i; k = 1,\dots,r_j,$$

then the switching T-S fuzzy system in closed-loop (10.12), with

$$K_{il} = Y_{il}X_{il}^{-1}, P_{il} = X_{il}^{-1}, \tag{10.38}$$

is asymptotically stable at the origin for any sequence of switching $\alpha(t)$.

Proof: The inequalities (10.31) can be transformed equivalently by the Schur complement to the following:

$$P_{jk}\left[A_{is}+B_{is}K_{il}\right]P_{il}^{-1}\left[A_{is}+B_{is}K_{il})\right]^T P_{jk} - P_{jk} < 0. \tag{10.39}$$

By pre and post-multiplying (10.39) by P_{jk}^{-1}, it follows that:

$$\left[A_{is}+B_{is}K_{il}\right]P_{il}^{-1}\left[A_{is}+B_{is}K_{il}\right]^T - P_{jk}^{-1} < 0. \tag{10.40}$$

Letting $X_{ik} = P_{ik}^{-1}$, inequality (10.40) can then be rewritten as

$$\left[A_{is}+B_{is}K_{il}\right]X_{il}\left[A_{is}+B_{is}K_{il}\right]^T - X_{jk} < 0.$$

That is, using $Y_{il} = K_{il}X_{il}$ gives

$$\left[A_{is}X_{il}+B_{is}Y_{il}\right]X_{il}^{-1}\left[A_{is}X_{il}+B_{is}Y_{il})\right]^T - X_{jk} < 0.$$

The use of the Schur complement a second time leads to:

$$\begin{bmatrix} X_{il} & (A_{is}X_{il}+B_{is}Y_{il})^T \\ * & X_{jk} \end{bmatrix} > 0. \tag{10.41}$$

Then, condition (10.31) is equivalent to LMI (10.37). □

Remark 10.3. *The global system is considered as a switching system composed of N sets of r_i T-S fuzzy subsystems each, $i = 1,\ldots,N$. The Lyapunov function used in this work is the switched one. This function can be seen as a multiple one for both the switching nonlinear system and the T-S fuzzy subsystems.*

Remark 10.4. *In the case that matrix B is the same for all the T-S fuzzy systems, LMIs (10.37) are transformed to:*

$$\begin{bmatrix} X_{is} & (A_{is}X_{is} + BY_{is})^T \\ * & X_{jk} \end{bmatrix} > 0$$
$$(i,j) \in \mathscr{I}^2, s = 1,\ldots,r_i; k = 1,\ldots,r_j.$$

In order to reduce the number of LMIs, one can use, the technique of non PDC control used in [95]. For this consider the following Lyapunov function candidate:

$$\begin{aligned} V(t,x) &= x^T(t)\left(\sum_{i=1}^{N}\sum_{s=1}^{r_i}\xi_i(t)h_{is}(t)P_{is}\right)^{-1} x(t) \\ &= x^T(t)P(\xi,\theta)^{-1}x(t) \end{aligned} \tag{10.42}$$

The non PDC control used in this case is given by:

$$\begin{aligned} u(t) &= K_i(\theta)P(\xi,\theta)^{-1}x(t) \\ &= \left(\sum_{s=1}^{r_i} h_{is}(\theta(z(t)))K_{is}\right)\left(\sum_{i=1}^{N}\sum_{s=1}^{r_i}\xi_i(t)h_{is}(t)P_{is}\right)^{-1} x(t), \\ &= F_i(\theta)x(t) \end{aligned} \tag{10.43}$$

The switching T-S fuzzy system in closed-loop becomes:

$$\begin{aligned} x(t+1) &= \sum_{i=1}^{N}\xi_i(t)(A_i(\theta) + B_i(\theta)K_i(\theta)P(\xi,\theta)^{-1})x(t), \\ &= \sum_{i=1}^{N}\xi_i(t)(A_i(\theta) + B_i(\theta)F_i(\theta))x(t) \\ &= \sum_{i=1}^{N}\xi_i(t)Ao_i(\theta)x(t), \\ &= Ao(\xi,\theta)x(t). \end{aligned} \tag{10.44}$$

This reduction is presented by the following result.

Theorem 10.6. *If there exist $r_N N$ symmetric matrices $P_{11},\ldots,\ P_{Nr_N}$ matrices $K_{11},\ldots,K_{Nr_N}$ solutions of the following LMIs:*

$$\Omega^k_{ijss} > 0, \tag{10.45}$$

$$\Omega^k_{ijsl} + \Omega^k_{ijls} > 0, \tag{10.46}$$

$$\forall (i,j) \in \mathscr{I}^2, s < l = 1,\ldots,r_i; k = 1,\ldots,r_j,$$

with

$$\Omega^k_{ijsl} = \begin{bmatrix} P_{is} & (A_{is}P_{il} + B_{is}K_{il})^T \\ * & P_{jk} \end{bmatrix}, \tag{10.47}$$

then the switching T-S fuzzy system in closed-loop (10.12) is asymptotically stable at the origin for any sequence of switching $\alpha(t)$.

Proof:
The rate of increase of the Lyapunov function candidate (10.42) is given by:

$$\Delta V(t,x(t)) = x^T(t)\left[Ao(\xi,\theta)^T P_+(\xi,\theta)^{-1} Ao(\xi,\theta) - P(\xi,\theta)^{-1}\right] x(t).$$

The rate of increase is definite negative if the following inequality holds:

$$\left[Ao(\xi,\theta)^T P_+(\xi,\theta)^{-1} Ao(\xi,\theta) - P(\xi,\theta)^{-1}\right] < 0. \tag{10.48}$$

Pre and post multiply inequality (10.48) by $P(\xi,\theta)$

$$\left[P(\xi,\theta)Ao(\xi,\theta)^T P_+(\xi,\theta)^{-1} Ao(\xi,\theta)P(\xi,\theta) - P(\xi,\theta)\right] < 0. \tag{10.49}$$

Inequality (10.49) can be equivalently written as

$$\begin{bmatrix} P(\xi,\theta) & P(\xi,\theta)Ao(\xi,\theta)^T \\ * & P_+(\xi,\theta) \end{bmatrix} > 0. \tag{10.50}$$

That is

$$\Psi(\xi,\theta) = \sum_{i=1}^{N}\sum_{j=1}^{N} \xi_i(t)\xi_j(t+1) \begin{bmatrix} P_i(\theta) & [Ao_i(\theta)P_i(\theta)]^T \\ * & P_{+j}(\theta) \end{bmatrix} > 0. \tag{10.51}$$

where $P_{+j}(\theta) = \sum_{s=1}^{r_j} h_{js}(t+1)P_{js}$. It is obvious that the following conditions are sufficient to have (10.51).

$$\begin{bmatrix} P_i(\theta) & [Ao_i(\theta)P_i(\theta)]^T \\ * & P_{+j}(\theta) \end{bmatrix} > 0, \tag{10.52}$$

By replacing Ao_i by its expression given by (10.44), it follows that

$$\Psi(\theta) = \begin{bmatrix} P_i(\theta) & [A_i(\theta)P_i(\theta) + B_i(\theta)F_i(\theta)P_i(\theta)]^T \\ * & P_{+j}(\theta) \end{bmatrix} > 0. \tag{10.53}$$

Taking account of (10.44), one can have

$$\Phi = [A_i(\theta)P_i(\theta) + B_i(\theta)K_i(\theta)]^T. \tag{10.54}$$

Matrix Φ can be equivalently written by using the following well known developments:

$$\begin{aligned} \Phi &= \sum_{s=1}^{r_i}\sum_{l=1}^{r_i} h_{is}(t)h_{il}(t)G_{isl} \\ &= \sum_{s=1}^{r_i} h_{is}^2(t)G_{iss} + \sum_{s=1}^{r_i}\sum_{l>s}^{r_i} h_{is}(t)h_{il}(t)(G_{isl} + G_{ils}), \end{aligned} \tag{10.55}$$

with $G_{isl} = A_{is}P_{il} + B_{is}K_{il}$. Recall that one can obviously have

$$\begin{aligned} 1 &= (\sum_{s=1}^{r_i} h_{is}(t))(\sum_{l=1}^{r_i} h_{il}(t)) \\ &= \sum_{s=1}^{r_i}\sum_{l=1}^{r_i} h_{is}(t)h_{il}(t) \\ &= \sum_{s=1}^{r_i} h_{is}^2(t) + 2\sum_{s=1}^{r_i}\sum_{l>s}^{r_i} h_{is}(t)h_{il}(t) \end{aligned}$$

Note also that one can develop as follows:

$$\begin{aligned} P_i(\theta) &= (\sum_{s=1}^{r_i} h_{is}(t))(\sum_{l=1}^{r_i} h_{il}(t)P_{il}) \\ &= \sum_{s=1}^{r_i} h_{is}^2(t)P_{is} + \sum_{s=1}^{r_i}\sum_{l>s}^{r_i} h_{is}(t)h_{il}(t)(P_{is} + P_{il}). \end{aligned}$$

By replacing in (10.53), one can equivalently obtain

$$\Psi(i,j,l,k,s) = \begin{bmatrix} \sum_{s=1}^{r_i} h_{is}^2(t)P_{is} & (\sum_{s=1}^{r_i} h_{is}^2(t)G_{iss})^T \\ * & \sum_{s=1}^{r_i} h_{is}^2(t)P_{+j}(\theta) \end{bmatrix} +$$

$$\begin{bmatrix} 2\sum_{s=1}^{r_i}\sum_{l>s}^{r_i} h_{is}(t)h_{il}(t)\frac{(P_{is}+P_{il})}{2} & 2\left(\sum_{s=1}^{r_i}\sum_{l>s}^{r_i} h_{is}(t)h_{il}(t)\frac{(G_{isl}+G_{ils})}{2}\right)^T \\ * & 2\sum_{s=1}^{r_i}\sum_{l>s}^{r_i} h_{is}(t)h_{il}P_{+j}(\theta) \end{bmatrix} > 0,$$

which can be written as

$$\Psi(i,j,l,k,s) = \sum_{s=1}^{r_i}\sum_{k=1}^{r_j} h_{is}^2(t)h_{jk}(t+1)\begin{bmatrix} P_{is} & G_{iss}^T \\ * & P_{jk} \end{bmatrix} +$$
$$2\sum_{k=1}^{r_j}\sum_{s=1}^{r_i}\sum_{l>s}^{r_i} h_{is}(t)h_{il}(t)h_{jk}(t+1)\begin{bmatrix} \frac{P_{is}+P_{il}}{2} & \frac{(G_{isl}+G_{ils})^T}{2} \\ * & P_{jk} \end{bmatrix} > 0. \qquad (10.56)$$

One can then deduce that the following conditions are sufficient for the asymptotic stability of the switching T-S fuzzy system

$$\begin{bmatrix} P_{is} & G_{iss}^T \\ * & P_{jk} \end{bmatrix} > 0 \qquad (10.57)$$
$$\begin{bmatrix} \frac{P_{is}+P_{il}}{2} & \frac{(G_{isl}+G_{ils})^T}{2} \\ * & P_{jk} \end{bmatrix} > 0. \qquad (10.58)$$

Finally, conditions (10.45) and (10.46) are sufficient conditions of asymptotic stability of the switching T-S fuzzy system (10.44) for any sequence of switching $\alpha(t)$. □

Remark 10.5. *In the case that matrix B is the same for all the T-S fuzzy systems, LMIs (10.45) and (10.46) are transformed to:*

$$\Omega_{ijss}^k > 0, \qquad (10.59)$$
$$\Omega_{ijsl}^k + \Omega_{ijls}^k > 0, \qquad (10.60)$$
$$\forall (i,j) \in \mathscr{I}^2, s < l = 1,\ldots,r_i; k = 1,\ldots,r_j,$$

with

$$\Omega_{ijsl}^k = \begin{bmatrix} P_{is} & (A_{is}P_{il} + BK_{il})^T \\ * & P_{jk} \end{bmatrix}, \qquad (10.61)$$

The surprising consequence is that the number of LMIs for this particular case obtained with non PDC control remains the same as for uncommon matrix B contrarily to the LMIs obtained for PDC control and noted with Remark 10.4.

Example 10.2. *Consider the following numerical system composed of two modes and four T-S subsystems:*

Mode 1*:*

$$A_{11} = \begin{bmatrix} 0.5 & -1 \\ a & 0.4 \end{bmatrix}; \; A_{12} = \begin{bmatrix} 0.2 & a \\ -1 & 1 \end{bmatrix};$$
$$A_{13} = \begin{bmatrix} 1 & -1 \\ b & 0.2 \end{bmatrix}; \; A_{14} = \begin{bmatrix} a & -1 \\ 1 & 0.2 \end{bmatrix};$$

Mode 2:

$$A_{21} = \begin{bmatrix} 1 & a \\ -b & 0.5 \end{bmatrix}; \; A_{22} = \begin{bmatrix} a & -1 \\ b & -0.6 \end{bmatrix};$$
$$A_{23} = \begin{bmatrix} a & b \\ -1 & 0.3 \end{bmatrix}; \; A_{24} = \begin{bmatrix} 0.4 & b \\ -1 & a \end{bmatrix};$$

with a common matrix $B = [1 \;\; 1]^T$. Scalars a, b are design parameters for testing the feasibility of the proposed LMIs (10.37). The feasibility of LMIs (10.37), LMIs with the semi-multiple Lyapunov function (10.14) of Section 10.2 and LMIs with the common Lyapunov function as used in [163] were checked varying scalars $a, b \in [-2, 2]$ with a step of 0.1. Figure 10.1 shows, respectively, the feasibility number of LMIs (10.37) marked with o, the second of the semi-Lyapunov function marked with $+$ and the third of common Lyapunov function marked with $\times$. Note that when LMIs with a common Lyapunov function or a semi-multiple Lyapunov function (10.14) are feasible, LMIs (10.37) are also feasible.

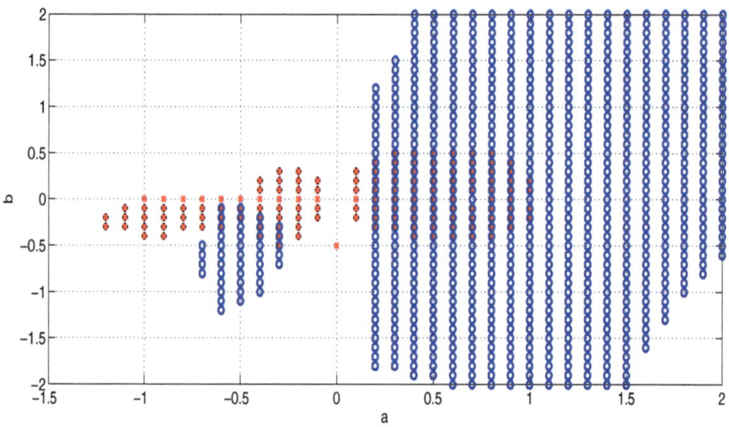

Fig. 10.1 Feasibility of LMIs (10.37) depending on scalars a, b

10.4 Saturated Discrete-Time Switching T-S Fuzzy Systems

In this section, the previous results obtained for the unsaturated systems are extended to saturated discrete-time switching T-S fuzzy systems . We restrict ourselves again to design only unsaturating controllers for its simple use for almost the same performances as the saturating controller method.

For this, system (10.5) becomes

$$x(t+1) = \sum_{i=1}^{N}\sum_{s=1}^{r_i} \xi_i h_{is}(\theta(z(t)))\left[A_{is}x(t) + B_{is}sat(u(t))\right]$$
$$= \sum_{i=1}^{N} \xi_i \left[A_i(\theta)x(t) + B_i(\theta)sat(u(t))\right]. \tag{10.62}$$

The saturation function is assumed here to be normalized, *i. e.*, $sat(u) = max(u,1)$. The following result presents sufficient conditions under LMI forms to compute un-saturating PDC state feedback controllers given by (10.9) for this class of systems.

Corollary 10.1. *For positive scalars ρ_i, if there exist $r_N N$ symmetric matrices $X_{11},\ldots, X_{Nr_N}$ and matrices $Y_{11},\ldots,Y_{Nr_N}$ solutions of the following LMIs :*

$$\begin{bmatrix} X_{il} & (A_{is}X_{il} + B_{is}Y_{il})^T \\ * & X_{jk} \end{bmatrix} > 0, \tag{10.63}$$
$$\forall (i,j) \in \mathscr{I}^2, s,l = 1,\ldots,r_i; k = 1,\ldots,r_j,$$
$$\begin{bmatrix} 1/\rho_i & (Y_{il})_\iota \\ * & X_{il} \end{bmatrix} > 0, \tag{10.64}$$
$$\forall i \in \mathscr{I}, l = 1,\ldots,r_i; \iota \in [1,\ m],$$

then the saturated switching T-S fuzzy system in closed-loop (10.62), with (10.9) and

$$K_{il} = Y_{il}X_{il}^{-1}, P_{il} = X_{il}^{-1}, \tag{10.65}$$

is asymptotically stable at the origin for any sequence of switching $\alpha(t)$ and $\forall x_0 \in \bigcup_{i=1}^{N}\left(\bigcap_{l=1}^{r_N} \varepsilon(P_{il},\rho_i)\right)$.

Proof: The proof follows readily by writing the inclusion relation $\varepsilon(P_{il},\rho) \subset \mathscr{L}(K_{il})$ under LMI form, with

$$\mathscr{L}(K_{il}) = \{x \in \mathbb{R}^n / |(K_{il})_\iota x| \leq 1\ \} \tag{10.66}$$

where $(K_{il})_\iota$ denotes the ιth row of matrix K_{il} and ρ_i is a positive scalar. For this class of systems, we can ensure that, for each mode i, all the trajectories emanating from $\left(\bigcap_{l=1}^{r_N} \varepsilon(P_{il},\rho_i)\right)$ are asymptotically stable for all the T-S fuzzy subsystems as showed in [81, 41]. Besides, for the switching system, all the trajectories emanating from $\bigcup_{i=1}^{N}\left(\bigcap_{l=1}^{r_N} \varepsilon(P_{il},\rho_i)\right)$ are asymptotically stable for the whole switching T-S fuzzy system. □

The result of Theorem 10.6 can also be extended to the saturated case.

Corollary 10.2. *If there exist $r_N N$ symmetric matrices $P_{11},\ldots, P_{Nr_N}$ matrices $K_{11},\ldots,K_{Nr_N}$ solutions of the following LMIs:*

$$\Omega^k_{ijss} > 0, \tag{10.67}$$

$$\Omega^k_{ijsl} + \Omega^k_{ijls} > 0, \tag{10.68}$$

$$\forall (i,j) \in \mathscr{I}^2, s < l = 1,\ldots,r_i; k = 1,\ldots,r_j,$$

$$\begin{bmatrix} 1/\rho_i & (K_{il})_\iota \\ * & P_{il} \end{bmatrix} > 0, \tag{10.69}$$

$$\forall i \in \mathscr{I}, l = 1,\ldots,r_i; \iota \in [1,\ m],$$

with

$$\Omega^k_{ijsl} = \begin{bmatrix} P_{is} & (A_{is}P_{il} + B_{is}K_{il})^T \\ * & P_{jk} \end{bmatrix}, \tag{10.70}$$

then the switching T-S fuzzy system in closed-loop (10.12) is asymptotically stable at the origin for any sequence of switching $\alpha(t)$.

Proof: To prove condition (10.69), one can use the result of [53] to write the inclusion condition $\varepsilon(P_i^{-1}(\theta),\rho_i) \subset \mathscr{L}(F_i(\theta))$ equivalently as:

$(F_i(\theta))_\iota P_i(\theta)(F_i^T(\theta))_\iota \le \rho_i^{-1}$. Using $F_i(\theta) = K_i(\theta)P_i^{-1}(\theta)$, one can obtain $(K_i(\theta))_\iota P_i^{-1}(\theta)(K_i^T(\theta))_\iota \le \rho_i^{-1}$, which can be transformed by Schur complement to

$$\begin{bmatrix} 1/\rho_i & (K_i(\theta))_\iota \\ * & P_i(\theta) \end{bmatrix} > 0,$$

$$\forall i \in \mathscr{I}, \iota \in [1,\ m].$$

That is,

$$\sum_{l=1}^{r_i} h_{il}(\theta(z(t))) \begin{bmatrix} 1/\rho_i & (K_{il})_\iota \\ * & P_{il} \end{bmatrix} > 0, \tag{10.71}$$

$$\forall i \in \mathscr{I}, \iota \in [1,\ m].$$

Finally, inequality (10.69) is sufficient to realize (10.71). □

10.5 Saturated Positive Switching T-S Fuzzy Discrete-Time Systems

This section is interested to a particular class of systems, switching T-S fuzzy discrete-time systems with saturated control and positive state. Remember that Chapter *VIII* studied positive switching discrete-time systems and saturated positive switching discrete-time systems. In the literature, positive T-S fuzzy systems were investigated for the first time in [37]. Using all the available results on positive switching systems and T-S fuzzy systems, the results of this section are then deduced.

For this, system (10.5) becomes

$$\begin{aligned} x(t+1) &= \sum_{i=1}^{N}\sum_{s=1}^{r_i} \xi_i h_{is}(\theta(z(t)))\left[A_{is}x(t)+B_{is}sat(u(t))\right] \\ &= \sum_{i=1}^{N} \xi_i\left[A_i(\theta)x(t)+B_i(\theta)sat(u(t))\right], \\ x_0 &\succ 0 \end{aligned} \tag{10.72}$$

The saturation function is assumed again to be normalized, *i. e.*, $sat(u) = max(u,1)$. The following result presents sufficient conditions under LMI forms to compute unsaturating PDC state feedback controllers given by (10.9) for this class of systems.

Corollary 10.3. *For positive scalars ρ_i, if there exist diagonal positive $r_N N$ matrices $X_{11},\ldots, X_{Nr_N}$ and matrices $Y_{11},\ldots,Y_{Nr_N}$ solutions of the following LMIs :*

$$\begin{bmatrix} X_{il} & (A_{is}X_{il}+B_{is}Y_{il})^T \\ * & X_{jk} \end{bmatrix} > 0, \tag{10.73}$$

$$\begin{bmatrix} 1/\rho_i & (Y_{il})_\iota \\ * & X_{il} \end{bmatrix} > 0, \tag{10.74}$$

$$A_{is}X_{il}+B_{is}Y_{il} \succ 0 \tag{10.75}$$

$$\forall (i,j) \in \mathscr{I}^2, s,l=1,\ldots,r_i; k=1,\ldots,r_j, \iota \in [1,\ m],$$

then the saturated switching T-S fuzzy system in closed-loop (10.72), with (10.9) and

$$K_{il} = Y_{il}X_{il}^{-1}, P_{il} = X_{il}^{-1}, \tag{10.76}$$

is asymptotically stable at the origin while the state is positive for any sequence of switching $\alpha(t)$ and $\forall x_0 \in \bigcup_{i=1}^{N}\left(\bigcap_{l=1}^{r_N} \varepsilon(P_{il},\rho_i)\right)$.

Proof: The proof follows readily from the previous results. □

10.6 Application to a Real Plant Model

In order to show the need for this study on positive systems, consider the process composed of two similar linked tanks of capacity 22 liters each given by Figure 10.2.

In Figure 10.2, $L_j, x_j, j=1,2$ hold for the levels in $meter(m)$ and $liter$, respectively, of tank j, u_j represents the flow in $liter/mn$ of pump j, q_{12} is the the flow between the two tanks and q_j the loss flow of each tank. We consider that q_{12} is an on/off valve. The flow of the two pump is limited to $5liter/mn$. The system is

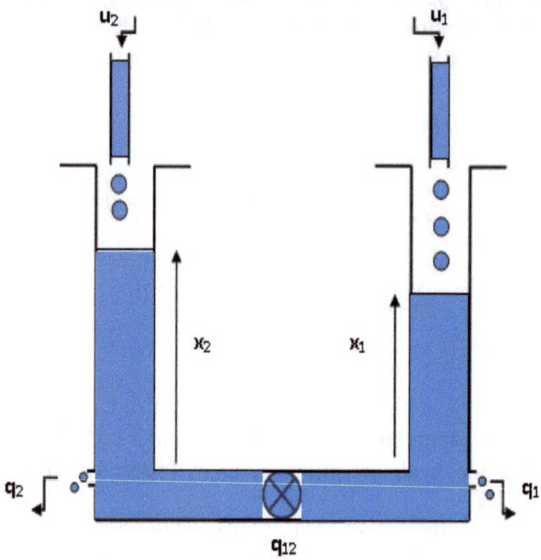

Fig. 10.2 Process composed of two linked tanks

equipped with two level sensors measuring liquid heights in the tanks and five flow-rates sensors measuring flows u_1, u_2, q_1, q_2 and q_{12}. The physical equations of the system, which can be described by two modes, are

Mode 1: Valve q_{12} is on:

$$\begin{aligned} S_1 \dot{L}_1(t) &= u_1(t) - q_{12}(t) - q_1(t) \\ S_2 \dot{L}_2(t) &= u_2(t) + q_{12}(t) - q_2(t) \end{aligned}$$

Mode 2: Valve q_{12} is off

$$\begin{aligned} S_1 \dot{L}_1(t) &= u_1(t) - q_1(t) \\ S_2 \dot{L}_2(t) &= u_2(t) - q_2(t), \end{aligned}$$

where L_j is the jth tank level in m. Applying the Torricelli law , one obtains:

$$\begin{aligned} q_1 &= \gamma_1 S_1 \sqrt{2gL_1} \\ q_2 &= \gamma_1 S_2 \sqrt{2gL_2} \\ q_{12} &= \gamma_{12} S_p \sqrt{2g|L_1 - L_2|} sign(x_1 - x_2) \end{aligned}$$

where γ_1, γ_2 and γ_{12} are physical constants, S_1, S_2 are the sections of the two tanks respectively, S_p is the section of the pipe linking the two tanks and g the gravity acceleration.

By using similar tanks fact $S_1 = S_2 = S$ and letting $x_j = SL_j$, which is the volume of tank j in *liter*, one obtains

$$\begin{aligned} q_1 &= \gamma_1 \sqrt{2gS}\sqrt{x_1} \\ q_2 &= \gamma_1 \sqrt{2gS}\sqrt{x_2} \\ q_{12} &= \gamma_{12} S_p \sqrt{\frac{2g}{S}} \sqrt{|x_1 - x_2|} sign(x_1 - x_2) \end{aligned}$$

The process model is then as follows:

$$\begin{aligned} \dot{x}_1(t) &= u_1 - R_1\sqrt{x_1} - \delta_i R_{12}\sqrt{|x_1 - x_2|} sign(x_1 - x_2) \\ \dot{x}_2(t) &= u_2 - R_2\sqrt{x_2} + \delta_i R_{12}\sqrt{|x_1 - x_2|} sign(x_1 - x_2), \end{aligned}$$

with $\delta_1 = 1$ for mode 1 and $\delta_2 = 0$ for mode 2. R_1, R_2 and R_{12} represents physical constants to be identified.

The obtained model is then nonlinear. Note that levels x_j and the feed flows u_j must always be positive.

To obtain a T-S fuzzy representation for this nonlinear system, the classical transformation $\sqrt{x_j} = \frac{x_j}{\sqrt{x_j}} = x_j z_j$ with $z_j = \frac{1}{\sqrt{x_j}}, j = 1,2$ is used. In this case, $\sqrt{|x_1 - x_2|} sign(x_1 - x_2) = |x_1 - x_2| sign(x_1 - x_2) \frac{1}{\sqrt{|x_1 - x_2|}} = (x_1 - x_2) \frac{z_1 z_2}{\sqrt{|z_2^2 - z_1^2|}}$ since $|x_1 - x_2| sign(x_1 - x_2) = x_1 - x_2$. The corresponding model is then given by:

$$\begin{aligned} \dot{x}(t) &= A_i(z_1, z_2)x(t) + Bu(t), \\ y(t) &= Cx(t) \end{aligned} \tag{10.77}$$

where matrix $A_i(z_1, z_2)$ has the general following form:

$$A_i(z_1, z_2) = \begin{bmatrix} -R_1 z_1 - \frac{\delta_i R_{12} z_1 z_2}{\sqrt{|z_1^2 - z_2^2|}} & \frac{\delta_i R_{12} z_1 z_2}{\sqrt{|z_1^2 - z_2^2|}} \\ \frac{\delta_i R_{12} z_1 z_2}{\sqrt{|z_1^2 - z_2^2|}} & -R_2 z_2 - \frac{\delta_i R_{12} z_1 z_2}{\sqrt{|z_1^2 - z_2^2|}} \end{bmatrix},$$

which is Metzler by construction , $B = \mathbb{I}_2$, $C = \mathbb{I}_2$ and $\delta_1 = 1, \delta_2 = 0$. Discretizing this system with a sampling time T by using Euler method, the following system is obtained:

$$\begin{aligned} x(t+1) &= Ad_i(z_1, z_2)x(t) + B_d u(t), \\ &= \sum_{l=1}^{4} h_{il} \left(Ad_{il} x(t) + B_d u(t) \right) \\ y(t) &= Cx(t), \end{aligned} \tag{10.78}$$

with $Ad_i(z_1, z_2) = \mathbb{I} + TA_i(z_1, z_2)$, $B_d = TB$ and $Ad_{il} = \mathbb{I} + TA_{il}$.

The objective is that the output y tracks a given reference y_r. Thus, keeping in mind the notation of (10.7), the following control is used: $u(t) = K_i(\theta)x(t) + F_i(\theta)y_r$, where controller gain $K(\theta)$ ensures the asymptotic stability together with the positivity in closed-loop while controller gain $F(\theta)$ achieves the tracking objective . In this case, one obtains $X(z) = (z\mathbb{I} - \hat{A}d_i(\theta))B_dF_i(\theta)Y_r(z)$, where matrix $\hat{A}d_i(\theta) = Ad_i(\theta) + B_dK_i(\theta)$. That is,

$$Y(z) = \frac{C\left[z\mathbb{I} - \hat{A}d_i(\theta)\right]^{-1} B_dF_i(\theta)y_r}{1 - z^{-1}},$$

where z stands for the Laplace variable in discrete-time. Using the final value theorem, one can deduce $y(\infty) = C\left[\mathbb{I} - \hat{A}d_i(\theta)\right]^{-1} B_dF_i(\theta)y_r$. If one chooses $F_i(\theta) = B_d^{-1}(\mathbb{I} - \hat{A}d_i(\theta))$, matrices B_d and C being, in our case, diagonal and identity respectively, the tracking objective will be reached with $y(\infty) = y_r$. Further, $F_i(\theta) = B_d^{-1}(\mathbb{I} - \hat{A}_i(\theta))$ holds if $F_{il} = B_d^{-1}(\mathbb{I} - \hat{A}_{il}), l = 1, \ldots, 4$.

By considering that, for each mode i, $z_j \in [a_j; b_j], j = 1, 2$, the four following rules are taken into account:

IF $z_1(k)$ is about a_1 and $z_2(k)$ is about a_2, THEN, $A_i(z_1,z_2) = A_i(a_1,a_2) = A_{i1}$.
IF $z_1(k)$ is about a_1 and $z_2(k)$ is about b_2, THEN, $A_i(z_1,z_2) = A_i(a_1,b_2) = A_{i2}$.
IF $z_1(k)$ is about b_1 and $z_2(k)$ is about a_2, THEN,$A_i(z_1,z_2) = A_i(b_1,a_2) = A_{i3}$,
IF $z_1(k)$ is about b_1 and $z_2(k)$ is about b_2, THEN, $A_i(z_1,z_2) = A_i(b_1,b_2) = A_{i4}$.

The membership functions are given by:

$$\begin{aligned}
h_1(t) &= f_{11}(k)f_{21}(t), h_2(t) = f_{11}(t)f_{22}(t), \\
h_3(t) &= f_{12}(t)f_{21}(t), h_4(t) = f_{12}(t)f_{22}(t), \\
f_{j1}(t) &= \frac{z_j(t) - b_j}{a_j - b_j} \\
f_{j2}(t) &= 1 - f_{j1}(t) = \frac{a_j - z_j(t)}{a_j - b_j}, j = 1, 2.
\end{aligned}$$

For this real system, matrix B_d is common for all the subsystems. Note that $h_{il} = h_l, l = 1, \ldots, 4$. Parameters R_1, R_2, R_{12} are experimentally estimated as $R_1 = R_2 = 0.95, R_{12} = 0.52$. While $a_1 = 0.2236, b_1 = 0.4472$ (volume of tank 1 between 5 and 20 liters), $a_2 = 0.2582, b_2 = 0.4082$ (volume of tank 2 between 6 and 15 liters). The desired reference is $y_r = [18; 15]^T$ while the sampling time is taken as $T = 0.02mn$.

The different system matrices are given by:

$$A_{11} = \begin{bmatrix} 0.9911 & 0.0047 \\ 0.0047 & 0.9904 \end{bmatrix}, A_{12} = \begin{bmatrix} 0.9882 & 0.0033 \\ 0.0033 & 0.9918 \end{bmatrix},$$

$$A_{13} = \begin{bmatrix} 0.9930 & 0.0028 \\ 0.0028 & 0.9895 \end{bmatrix}, A_{14} = \begin{bmatrix} 0.9811 & 0.0104 \\ 0.0104 & 0.9818 \end{bmatrix}$$

$$A_{21} = \begin{bmatrix} 0.9958 & 0 \\ 0 & 0.9951 \end{bmatrix}, A_{22} = \begin{bmatrix} 0.9915 & 0 \\ 0 & 0.9951 \end{bmatrix},$$
$$A_{23} = \begin{bmatrix} 0.9958 & 0 \\ 0 & 0.9922 \end{bmatrix}, A_{24} \begin{bmatrix} 0.9915 & 0 \\ 0 & 0.9922 \end{bmatrix}.$$

In order to stabilize the T-S fuzzy system while imposing positivity in closed-loop, we solve the LMIs of Corollary 10.3. If these LMIs are feasible, one can compute the required controllers K_i and the corresponding Lyapunov function given by P_{il}.

The consequence of matrix B_d common, is that the number of LMIs (10.73)-(10.75) to be used is considerably reduced by letting $i = l$ in these LMIS.

In order to show the interest of taking into account the saturation in the design phase, the LMIs of Corollary 10.3 are used in two steps: firstly without and secondly with LMI (10.74).

Using LMI Toolbox in Matlab software, LMIs (10.73) and (10.75) are feasible for $\rho_i = 1$. The evolution of the component state and the control are plotted in Figure 10.5.

The first figure of Figure 10.5 plots the evolution of states x_1 and x_2 in liters starting at $x(t) = [5;7]^T$. The desired reference is reached in less than $0.1mn$ while the state remains always positive. The second figure plots the evolution of the two pump flows in $liter/mn$. The third one plots the used sequence of switching. One can notice that the used control for tracking the reference is very excessive and is not admissible.

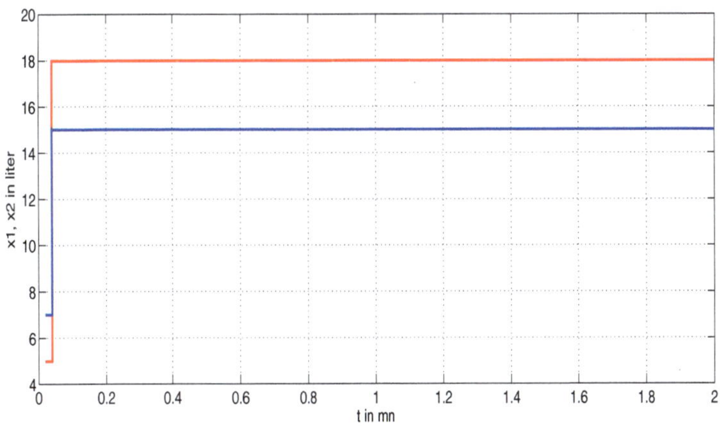

Fig. 10.3 This figure plots the evolution of the states x_1 and x_2 in liter

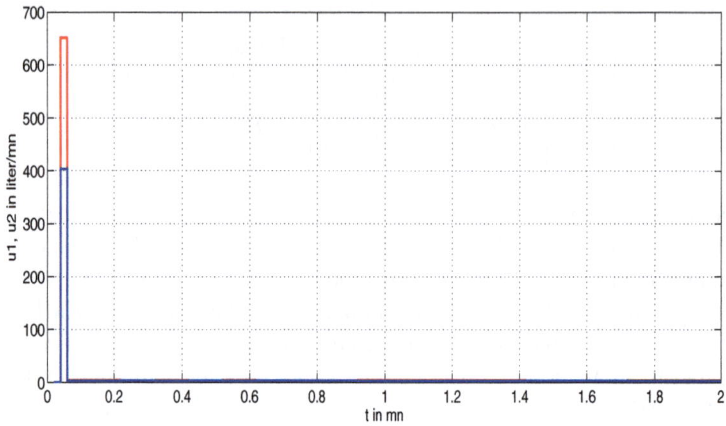

Fig. 10.4 Evolution of the two pump flows in liter/mn

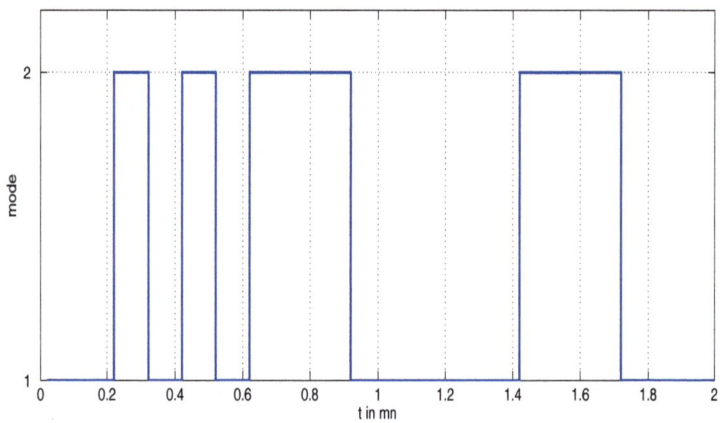

Fig. 10.5 Evolution of sequence of switching

We use now LMIs (10.73)-(10.75) to take account of saturation on the control by using the following system:

$$\begin{aligned} x(t+1) &= Ad(z_1, z_2)x(t) + \tilde{B}d\tilde{u}(t), \\ y(t) &= Cx(t), \end{aligned} \tag{10.79}$$

with $\tilde{B}_d = 5B_d$ and $\tilde{u}(t) = u(t)/5$.

The objective is to stabilize the saturated switching T-S fuzzy system while imposing positivity in closed-loop by choosing unsaturating controllers. LMIs (10.73)-(10.75) are feasible and give the following results:

$$
\begin{aligned}
P_{11} &= 0.2350\mathbb{I}_2, P_{12} = 0.2438\mathbb{I}_2, P_{13} = 0.4608\mathbb{I}_2, P_{14} = 0.4564\mathbb{I}_2,\\
P_{21} &= 0.2446\mathbb{I}_2, P_{22} = 0.2530\mathbb{I}_2, P_{23} = 0.7095\mathbb{I}_2, P_{24} = 0.7052\mathbb{I}_2,\\
K_{11} &= \begin{bmatrix} -0.4409 & -0.0252 \\ -0.0253 & -0.4396 \end{bmatrix}, K_{12} = \begin{bmatrix} -0.4015 & -0.0141 \\ -0.0139 & -0.4080 \end{bmatrix},\\
K_{13} &= \begin{bmatrix} -0.1748 & -0.0015 \\ -0.0015 & -0.1732 \end{bmatrix}, K_{14} = \begin{bmatrix} -0.1707 & -0.0054 \\ -0.0054 & -0.1710 \end{bmatrix},
\end{aligned}
$$

$$
\begin{aligned}
K_{21} &= \begin{bmatrix} -0.4133 & 0 \\ 0 & -0.4121 \end{bmatrix}, K_{22} = \begin{bmatrix} -0.3805 & 0 \\ 0 & -0.3866 \end{bmatrix},\\
K_{23} &= \begin{bmatrix} -0.1377 & 0 \\ 0 & -0.1368 \end{bmatrix}, K_{24} = \begin{bmatrix} -0.1369 & 0 \\ 0 & -0.1371 \end{bmatrix}.
\end{aligned}
$$

One can notice that all matrices P_{il} are diagonal. The obtained matrices in closed-loop, which are omitted, are all positive as required by Corollary 10.3 despite matrices $K_{[} = il$ are negative. Matrices F_{il} can be deduced by using $F_{il} = B_d^{-1}(\mathbb{I} - \hat{A}d_{il}), l = 1,\ldots,4$ and are also omitted. It is worth noting that in practical, the effective control to be applied to the real system is $u(t) = 5(K_i(\theta) + F_i(\theta))$.

The first figure of Figure 10.7 plots the evolution of states x_1 and x_2 in liters starting at $x(t) = [5;7]^T$. The desired reference is reached in $2mn$ while the state remains always positive. The second figure plots the evolution of the two pump flows in $liter/mn$. The sequence of switching is taken as in Figure 10.5. One can notice that the used control for tracking the reference is now admissible and does not exceed $5liter/mn$.

Remark 10.6. *It is important to note that the control of this model, which guarantees asymptotic stability and tracking of a reference, is saturated. However, the obtained results of LMIs (10.73)-(10.75) ensure only that the control $u(t) = 5(K_i(\theta)$ is admissible. Nevertheless, the used control is given by $u(t) = 5(K_i(\theta) + F_i(\theta))$. To ensure that this global control is admissible is until now an open question.*

In this section, a real plant model is studied. The obtained non linear model is directly used under T-S fuzzy form, without linearizing around a predetermined state. In this case, the system can be driven to any point of the state space, that is, to any levels x_1, x_2 without modifying the model. Nevertheless, a consequence of this, is that the states have to remain always positive in addition of the control, which is naturally positive. This additional constraint is also taken into account. Besides, it is shown that saturation on the control constitutes an inherent problem in practice. In this section, the interest of designing controllers for saturated systems is also shown in Figure 10.5 where the obtained control is not adequate at all for implementation in practice.

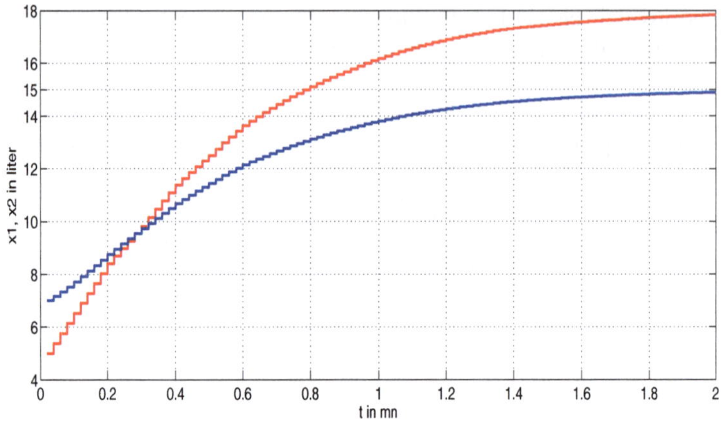

Fig. 10.6 Evolution of the states x_1 and x_2 in liter

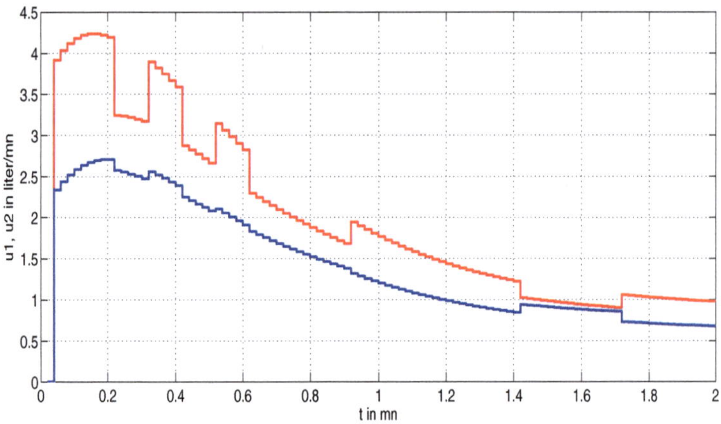

Fig. 10.7 Evolution of the two pump flows in liter/mn

10.7 Conclusion

In this chapter, sufficient conditions for the asymptotic stability of switching discrete-time T-S fuzzy systems are developed by using a state feedback control and two types of multiple Lyapunov functions. The first one is a function which is common for all the T-S fuzzy subsystems while multiple for the modes of the switching system; the second is a switched Lyapunov function which is multiple for both the T-S fuzzy subsystems and all the modes of the switching system. The main results of this chapter are given under LMI formulation, leading to a methodology for designing stabilizing state feedback controllers. The applicability of the proposed approaches is shown by using numerical examples, including a comparison with previous approaches. The extensions to the class of saturated switching T-S fuzzy systems and positive saturated switching T-S fuzzy systems are also presented. A real plant model is studied to show the interest and the applicability of the obtained results on saturated positive switching discrete-time systems.

10.8 General Conclusion

In this book, the background of the approach called positive invariance which was developed for linear systems to deal with saturations on the control of the system is recalled. Its extension to an important class of dynamical systems called singular linear systems is developed in this book. The approach of positive invariance is known by designing controllers working inside a region of linear behavior and does not allow saturation of the control. These results are then extended to switching systems allowing necessary and sufficient conditions of positive invariance to be obtained. A design method based on the resolution of equation $XA+XBX=HX$ is proposed. A main result obtained for the first time with this approach is also presented. This result consists in proving that the union of all the sets of local positive invariance constitutes a set of positive invariance of the switching system. The developed controllers with this technique work only in regions of linear behavior. Besides, the resolution of the algebraic equation $XA+XBXE=HXE$ is studied for its practical use for the controller design of saturated singular linear systems. These results are then applied to solve the problem of saturated switching şingular linear systems presented for the first time in this book.

A particular class of switching systems, where the switching is random and follows a Markovian jumping law, is studied in this book. This class is called Systems with Markovian Jumping Parameters. For this class of systems, the approach of positive invariance is again applied. Necessary and sufficient condition of positive invariance of a common set to all the subsystems is proposed. The cases of discrete-time and continuous-time systems are studied.

A second technique which allows the control to be saturated and known by technique of saturation is also recalled in this book. The main idea of the technique of saturation is to write the nonlinear saturation function under a combination of 2^m linear functions, where m stands for the size of the control. This technique presents conditions of building stabilizing controllers, for linear systems, allowing saturation to take effect on the control by using LMIs. This technique, is also applied to switching systems allowing the design of stabilizing controllers tolerating saturations to take effect on the control. The obtained results are presented in LMI form. The objective of using this technique is to obtain larger domains of initial values. However, this objective is not theoretically guaranteed at all. As particular results, one find those obtained with positive invariance approach. A comparison between the two techniques is presented for some examples showing the advantage and the drawback of each technique. The problem of stabilizing switching systems by using output feedback is also studied.

The extension of the obtained results for saturated switching systems to uncertain switching systems is presented. Two types of uncertainties are considered: polytopic uncertainties and parametric uncertainties. Stabilizing controllers are developed while tolerating saturations. Illustrative examples are presented for each type.

The stability problem for positive switching systems are then investigated with multiple Lyapunov functions. A necessary and sufficient condition for the positive switching system to admit a multiple Lyapunov function is proposed. A linear programming is presented to select positive diagonal matrices to construct this Lyapunov function. The obtained results are used to study the problem of stabilization of positive switching systems. Stabilizing controllers are designed by using state feedback and output feedback control. All the results are presented in LMIs form making their application easy. Illustrative examples are also studied. The extension to saturated positive switching systems is also developed.

The stabilization problem by state feedback and output feedback control for linear two-dimensional (2D) switching systems is studied in this book. Conditions of stabilizability are obtained by using common and multiple Lyapunov functions for both state feedback and output feedback controls. The case of saturated switching 2D systems is also studied. The unsaturating controllers ensuring asymptotic stability for this class of systems is then presented.

The last class of switching systems studied in this book is the one of switching Takagi-Sugeno fuzzy systems. Sufficient conditions of stabilizability are obtained for switching nonlinear systems described under Takagi-Sugeno form. A switched Lyapunov function and state feedback control are used. The proposed Lyapunov function is multiple for both the T-S fuzzy subsystems and the switching modes. This choice allows a general approach which may be less conservative. The obtained conditions are then worked out to be presented in LMI form. The extensions to the class of saturated switching T-S fuzzy systems and positive saturated switching T-S fuzzy systems are also presented. A real plant model composed of two linked tanks, where the link pipe is equiped with an on/off valve, is studied to show the interest and the applicability of the obtained results on saturated positive switching T-S fuzzy discrete-time systems.

References

1. Ait Rami, M., Tadeo, F., Benzaouia, A.: Control of constrained positive discrete systems. In: Proceedings of the American Control Conference, New York City, USA, July 11- 13 (2007)
2. Ait Rami, M., Tadeo, F.: Controller synthesis for positive linear systems with bounded controls. IEEE Trans. Circuits Sys. II 54, 151–155 (2007)
3. Ait Rami, M., Elfaiz, S., Benzaouia, A., Tadeo, F.: Robust Exact Pole Placement Via an LMI-Based Algorithm. IEEE Trans. Aut. Control 54, 394–398 (2009)
4. Anderson, B.O., Agathoklis, P., Jury, E.I., Mansour, M.: Stability and the Matrix Lyapunov Equation for Discrete 2-Dimensional Systems. IEEE Trans. Circuits Sys. CAS-33, 261–266 (1986)
5. Araki, M., Kondo, B.: Stability and Transient Behavior of Composite Nonlinear Systems. IEEE Trans. Aut. Control 17, 537–541 (1972)
6. Araki, M.: Application of M-matrices to the stability problems of composite dynamical systems. J. Math. Analy. Appl. 52, 309–321 (1975)
7. Baddou, A., Maarouf, H., Benzaouia, A.: Partial eigenstructure assignment problem and its application to the constrained linear problem. Int. J. Systems Sci. (to appear, 2012)
8. Bara, G.I., Boutayeb, M.: Switched output feed back stabilization of discrete-time switched systems. In: 45th Conference on Decision and Control, San Diego, December 13-15, pp. 2667–2672 (2006)
9. Benmessaouda, O., Benzaouia, A., Tadeo, F.: Stabilization of Uncertain State Constrained Discrete-time Switched Systems. In: Preprints of the 18th IFAC World Congress, Milano (Italy), August 28 - September 21 (2011)
10. Benzaouia, A., Burgat, C.: Regulator problem for linear discrete-time systems with non-symmetrical constrained control. Int. J. Control 48, 2441–2451 (1988)
11. Benzaouia, A., Burgat, C.: Existence of nonsymmetrical stability domains for linear systems. Linear Algebra Appl. 121, 217–231 (1989)
12. Benzaouia, A., Burgat, C.: Existence of nonsymmetrical Lyapunov functions for linear systems. Int. J. Systems Sci. 20, 597–607 (1989)
13. Benzaouia, A.: The regulator problem for linear discrete-time systems with non symmetrical constrained control. In: Proc. of the 30th Conference on Decision and Control, Brighton, December 13-15, pp. 1742–1743 (1991)
14. Benzaouia, A., Hmamed, A.: Regulator problem for linear continuous time systems with nonsymmetrical constrained control using nonsymmetrical Lyapunov functions. In: 31st Conference on Decision and Control, Arizona, December 16-18 (1992)

15. Benzaouia, A., Hmamed, A.: Regulator problem for linear continuous systems with nonsymmetrical constrained control. IEEE Trans. Aut. Control 38, 1556–1560 (1993)
16. Benzaouia, A.: Resolution of equation XA+XBX=HX and the pole assignment problem. IEEE Trans. Aut. Control 40, 2091–2095 (1994)
17. Benzaouia, A., Baddou, A.: Piecewise linear constrained control for continuous time systems. IEEE Trans. Aut. Control 44, 1477–1481 (1999)
18. Benzaouia, A., Mehdi, D.: The output feedback saturated controller design for linear systems. In: 11th Mediterranean Conference, Portugal (2002)
19. Benzaouia, A., Baddou, A., Elfaiz, S.: Piecewise linear constrained control for continuous-time systems: The maximal admissible domain. In: 15th World Congress IFAC, Barcelona (2002)
20. Benzaouia, A., Boukas, E.L., Daraoui, N.: Stability of continuous-time linear systems with Markovian jumping parameters and constrained control. Int. J. Hybrid Systems 3, 19–32 (2003)
21. Benzouia, A., Ait Rami, M., El Faiz, S.: Stabilization of linear systems with saturation: A Sylvester equation approach. IMA Journal of Mathematical Control and Information 21, 247–259 (2004)
22. Benzaouia, A., Saydy, L., Akhrif, O.: Stability and control synthesis of switched systems subject to actuator saturation. In: American Control Conference, Boston, June 30-July 2 (2004)
23. Benzaouia, A.: Constrained stabilization: An enlargement technique of positively invariant sets. IMA Journal of Mathematical Control and Information 22, 109–118 (2005)
24. Benzaouia, A., El Faiz, S.: The Regulator problem for Linear Systems with Constrained Control: An LMI Approach. IMA Journal of Mathematical Control and Information 23, 335–345 (2006)
25. Benzaouia, A., Saydy, L., Akhrif, O.: Stabilization of switched systems subject to actuator saturation by output feedback. In: 45th IEEE Conference on Control and Decision, San Diego, December 13-15 (2006)
26. Benzaouia, A., Tadeo, F., Mesquine, F.: The regulator problem for linear systems with saturations on the control and its increments or rate: An LMI approach. IEEE Trans. Circuits and Syst. Part I 53, 2681–2691 (2006)
27. Benzaouia, A., DeSantis, E., Caravani, P., Daraoui, N.: Constrained control of switching systems: A positive invariance approach. Int. J. of Control 80, 1379–1387 (2007)
28. Benzaouia, A., Mehdi, D., El Hajjaji, A., Nachidi, M.: Piecewise quadratic Lyapunov function for nonlinear systems with fuzzy static output feedback control. In: European Control Conference, Kos, Greece, July 2-5 (2007)
29. Benzaouia, A., Tadeo, F.: Output feedback stabilization of positive switching linear discrete-time systems. In: 16th Mediterranean Conference, Ajaccio, France, June 25-27 (2008)
30. Benzaouia, A., Benmesaouda, O., Shi, Y.: Output feedback Stabilization of uncertain saturated discrete-time switching systems. Int. J. Innovative Computing, Information Control 5, 1735–1745 (2009)
31. Benzaouia, A., Benmesaouda, O., Tadeo, F.: Stabilization of uncertain saturated discrete-time switching systems. Int. J. Control Aut. Sys. 7, 835–840 (2009)
32. Benzaouia, A.: Book Chapter: Stabilization of Saturated Switched Systems in Switched Systems. In: Kleban, J. (ed.) INTEC, Croatia (2009)
33. Benzaouia, A., Hmamed, A., Tadeo, F.: Stability conditions for discrete 2D switching systems, based on a multiple Lyapunov function. In: European Control Conference 2009 - ECC 2009, Budapest, Hungary, August 23-26, pp. 23–26 (2009)

34. Benzaouia, A., Tadeo, F.: Stabilization of positive switching linear discrete-time systems. Int. J. Innovative Computing, Information Control 6, 2427–2437 (2010)
35. Benzaouia, A., Akhrif, O., Saydy, L.: Stabilitzation and control synthesis of switching systems subject to actuator saturation. Int. J. Systems Sci. 41, 397–409 (2010)
36. Benzaouia, A., Hmamed, Tadeo, F., El Hajjaji, A.: Stabilization of discrete 2 D time switching systems by state Feedback Control. Int. J. Systems Sci. 42, 479–487 (2011)
37. Benzaouia, A., Hmamed, A., El Hajjaji, A.: Stabilization of controlled positive discrete-time T-S fuzzy systems by state feedback control. Int. J. Adaptive Control Signal Processing 24, 1091–1106 (2010)
38. Benzaouia, A., El Hajjaji, A.: Delay-dependant stabilization conditions of controlled positive T-S fuzzy systems with time varying delay. Int. J. Innovative Computing, Information Control 7, 1533–1548 (2011)
39. Benzaouia, A., El Hajjaji, A.: Delay-dependant stabilization conditions of controlled positive T-S fuzzy systems with time varying delay. Int. J. Innovative Computing, Information Control 7, 1533–1547 (2011)
40. Benzaouia, A., El Hajjaji, A., Tadeo, F.: Stabilization of switching Takagi-Sugeno systems by switched Lyapunov function. Int. J. Adaptive Control Signal Processing 25(12), 1039–1049 (2012)
41. Benzaouia, A., Gounane, S., Tadeo, F., El Hajjaji, A.: Stabilization of saturated discrete-time fuzzy systems. Int. J. Contr. Aut. Sys. 9, 1–7 (2011)
42. Benzaouia, A., Darouach, M., Hmamed, A.: Solution of equation XA+XBXE=HXE and the pole assignment in linear singular systems: Application to constrained control. IMA Journal of Mathematical Control and Information (to appear, 2012)
43. Berardi, L., De Santis, E., Di Benedetto, M.D.: Invariant Sets and Control Synthesis for Switching Systems with Safety Specifications. In: Lynch, N.A., Krogh, B.H. (eds.) HSCC 2000. LNCS, vol. 1790, pp. 59–72. Springer, Heidelberg (2000)
44. Blanchini, F.: Set invariance in control - a survey. Automatica 35, 1747–1768 (1999)
45. Blanchini, F., Savorgnan, C.: Stabilizability of switched linear systems does not imply the existence of convex Lyapunov functions. In: 45th Conference on Decision and Control, San Diego, December 13-15, pp. 119–124 (2006)
46. Blanchini, F., Miani, S., Mesquine, F.: A Separation Principle for Linear Switching systems and parametrization of all stabilizing controllers. IEEE Trans. Aut. Control 54, 279–292 (2009)
47. Bitsoris, G.: Positively invariant polyhedral sets of discrete-time linear systems. Int. J. Contr. 47, 1713–1726 (1988)
48. Bitsoris, G.: Existence of positively invariant polyhedral sets for continuous time linear systems. Control Theory and Advanced Technology 7, 407–427 (1991)
49. Bitsoris, G., Vassilaki, M.: Constrained regulation of linear systems. Automatica 31, 223–227 (1995)
50. Bochniak, J., Galkowski, K., Rogers, E., Kummert, A.: Control law design for switched repetitive processes with a Metal Rolling Example. In: Proc. of the IEEE International Conference on Control Applications, Singapore, October 1-3, pp. 1–3 (2007)
51. Boukas, E.L., Benzaouia, A.: Stability of discrete-time linear systems with Markovian jumping parameters and constrained Control. IEEE Trans. Aut. Control 47, 516–520 (2002)
52. Boukas, E.-K., Liu, Z.: Deterministic and Stochastic Time Delay Systems. Birkhasuer, Boston (2002)
53. Boyd, S.P., El Ghaoui, Feron, E., Balakrishnan, V.: Linear Matrix Inequalities in System and Control Theory. SIAM, Philadelphia (1994)

54. Branicky, M.S.: Multiple Lyapunov functions and other analysis tools for switched and hybrid systems. IEEE Trans. Aut. Control 43, 475–482 (1998)
55. Burgat, C., Tarbouriech, S., Klai, M.: Continuous time saturated state feedback regulators theory and design. Int. J. Systems Sci 25, 315–336 (1994)
56. Camacho, E.F., Bordons, C.: Model Predictive Control. Springer, London (2004)
57. Cao, Y.-Y., Lin, Z., Shamash, Y.: Set invariance analysis and gainscheduling control for LPV systems subject to actuator saturation. Syst. Control Lett. 46, 137–151 (2002)
58. Caravani, P., De Santis, E.: Observer-based stabilization of linear switching systems. Int. J. Robust Nonlinear Control (2008); published online
59. Castelan, E.B., Hennet, J.C.: On invariant polyhedra of continuous time linear systems. IEEE Trans. Aut. Control 38, 1680–1685 (1993)
60. Castelan, E.B., Gomes da Silva, J.M., Cury, J.E.R.: A reduced order framework applied to linearsystems with constrained controls. In: Proceeding of the 33rd IEEE Conference on Decision and Control, Florida, pp. 1036–1041 (1994)
61. Chadli, M., Maquin, D., Ragot, J.: An LMI formulation for output feedback stabilization in multiple model approach. In: Proc. of the 41st Conference on Decision and Control, Las Vegas, Nevada (2002)
62. Chegancas, J.: Sur le concept d'invariance positive appliquée à l'étude de la commande contrainte des systèmes dynamiques. U.P.S (Toulouse) Thesis. LAAS no. 85325 (1985)
63. Cobb, J.D.: Feedback and pole placement in descriptor variable systems. Int. J. Control 33, 1135–1146 (1981)
64. Daafouz, J., Bernussou, J.: Parameter dependent Lyapunov functions for discrete-time systems with time varying parametric uncertainties. Systems and Control Letters 43, 355–359 (2001)
65. Daafouz, J., Riedinger, P., Iung, C.: Static output feedback control for switched systems. In: Procceding of the 40th IEEE Conference on Decision and Control, Orlando, USA (2001)
66. Daafouz, J., Riedinger, P., Iung, C.: Stability analysis and control synthesis for switched systems: a switched Lyapunov function approach. IEEE Trans. Aut. Control 47, 1883–1887 (2002)
67. Dai, L.: Singular control systems. LNCIS, vol. 118. Springer, Heidelberg (1989)
68. Darouach, M.: Solution to Sylvester equation associated to linear descriptor systems. Systems and Control Letters 55, 835–838 (2006)
69. Darouach, M., Boutayeb, M.: Design of observers for descriptor systems. IEEE Trans. Aut. Control 40, 1323–1327 (1995)
70. Darouach, M., Benzaouia, A.: Constrained observer based control for linear singular systems. In: 18th Mediterranean Conference on Control and Automation, Congress Palace Hotel, Marrakech, Morocco, June 23-25 (2010)
71. Daraoui, N., Benzaouia, A., Boukas, E.K.: Regulator problem for linear discrete-time delay systems with Markovian jumping parameters and symmetrical constraints. In: The Fourth International Conference on Control and Automation, Montral, Canada, June 9-12 (2003)
72. Daraoui, N., Benzaouia, A., Boukas, E.K.: Regulator problem for discrete-time delay systems with Markovian jumping parameters and constrained control. In: 42nd IEEE Conference on Decision and Control, Hawaii, December 9-12 (2003)
73. Daraoui, N.: Commande des systèmes hybrides avec contraintes. Thesis of University Cadi Ayyad (2004)
74. Daraoui, N., Benzaouia, A.: Regulator problem for linear discrete-time delay systems with Markovian jumping parameters and symmetrical constraints. Int. J. of Systems Science 37, 179–285 (2006)

75. Decarlo, R.A., Branicky, M.S., Petterson, S., Lennartson, B.: Multiple Lyapunov functions and other Perspective and results on the stability and stabilizability of hybrid systems. Proceeding of the IEEE 88, 1069–1082 (2000)
76. De Santis, E., Di Benedetto, M.D., Berardi, L.: Computation of maximal safe sets for switching systems. IEEE Trans. Aut. Control 40, 184–195 (2004)
77. De Santis, E., Di Benedetto, M.D., Pola, G.: Can linear stabilizability analysis be generalized to switching systems? In: Proc. of Mathematical Theory of Networks and Systems (MTNS 2004), Leuven (Belgium), July 5-9 (2004)
78. De Santis, E., Giordano, P.: Positive switching systems. In: Proceedings of the Multidisciplinary International Symposium on Positive Systems, Grenoble, France (2006)
79. Duan, G.R.: Eigenstructure assignment and response analysis in descriptor systems with state feedback control. Int. J. Control 69, 663–694 (1998)
80. El Ghaoui, L., Ait Rami, M.: Robust State Feedback Stabilization of Jump Linear Systems Via LMIs. Int. J. Robust Control 6, 1015–1022 (1996)
81. El Hajjaji, A.: Benzaouia and M. Naib, Stabilization of fuzzy systems with constrained controls by using positively invariant sets. Mathematical Problems for Engineering Article ID13832, 1–17 (2006)
82. El Ghaoui, L., Oustry, F., Ait Rami, M.: A Cone Complementarity Linearization Algorithm for Static Output-Feedback and Related Problems. IEEE Trans. Aut. Control 42, 1171–1176 (1997)
83. Espinosa, J., Vandewalle, J., Wertz, V.: Fuzzy logic idenification and predictive control. AIC. Springer, Heidelberg (2005)
84. Farina, L., Rinaldi, S.: Positive Linear Systems: Theory and Applications. Wiley, New York (2000)
85. Ferrari-Trecate, G., Cuzzola, F.A., Mignone, D., Morari, M.: Analysis and control with performanc of piecewise affine and hybrid systems. In: Procceding of the American Control Conference, Arlington, USA (2001)
86. Fletcher, L.R., Kaustky, J., Nichols, N.K.: Eigenstructure assignment in descriptor systems. IEEE Trans. Aut. Control. 31, 1138–1141 (1986)
87. Fornasini, E., Marchesini, G.: State-space realization theory of two-dimentional filters. IEEE Trans. Aut. Control 21, 484–492 (1976)
88. Fornasini, E., Marchesini, G.: Doubly-indexed dynamical systems: state-space models and structural properties. Math. Syst. Theory 12, 59–72 (1978)
89. Georgiou, C., Krikelis, N.J.: A design approach for constrained regulation in discrete singular systems. Systems and Control Letters 17, 97–304 (1991)
90. Galkowski, K.: LMI based stability analysis for 2D contiuous systems. International Confonference On Electronics, Circuits and Systems 3, 923–926 (2002)
91. Galkowski, K., Rogers, E., Xu, S., Lam, J., Owens, D.H.: LMIs-A fundamental tool in analysis and controller design for discrete linear repetitive process. IEEE Trans. Circuits Syst. 49, 768–778 (2002)
92. Geromel, J.C., Colaneri, P.: Stability and sabilization of discrete-time switched systems. Int. J. Control 79, 719–728 (2006)
93. Givone, D.D., Roesser, R.P.: Multidimentional linear iterative circuits-General properties. IEEE Trans. Comp. 21, 1067–1073 (1972)
94. Gilbert, E.G., Tan, K.T.: Linear Systems with State and Control Constraints: The Theory and Application of Maximal Output Admissible sets. IEEE Trans. Aut. Control 36, 1008–1020 (1991)
95. Guerra, T.M., Vermeiren, L.: LMI-based relaxed nonquadratic stabilization conditions for nonlinear systems in the Takagi-Sugeno's form. Automatica 40, 823–829 (2004)

96. Gurvits, L., Shorten, R.N., Mason, O.: On the stability of switched positive linear systems. IEEE Trans. Aut. Control 52, 1099–1103 (2007)
97. Gutman, P., Hagandar, P.: A new design of constrained controllers for linear systems. IEEE Trans. Aut. Cont. 30, 22–33 (1985)
98. Hetel, L., Daafouz, J., Iung, C.: Stabilization of arbitrary switched linear systems with unknown time-varying delays. IEEE Trans. Aut. Control 51, 1668–1674 (2006)
99. Hmamed, A., Benzaouia, A., Ait Rami, M., Tadeo, F.: Positive stabilization of discrete-time systems with unknown delay and bounded controls. In: Proceedings of the European Control Conference 2007, Kos, Greece (2007)
100. Hmamed, A., Benzaouia, A., Ait Rami, M., Tadeo, F.: Memoryless control to drive states of delayed continuous-time systems within the nonnegative orthant. In: 17th World Congress IFAC, Seoul, Korea, July 6-11 (2008)
101. Hmamed, A., Alfidi, M., Benzaouia, A., Tadeo, F.: LMI Conditions for Robust Stability of 2D Linear Discrete-Time Systems. Mathematical Problems in Engineering Article ID 356124, 11 pages (2008)
102. Hmamed, A., Mesquine, F., Tadeo, F., Benhayoun, M., Benzaouia, A.: Stabilization of 2D Saturated Systems by State Feedback Control. Multidim. Syst. Sign. Process. 21, 277–292 (2010)
103. Hu, T., Lin, Z., Chen, B.M.: An analysis and design method for linear systems subject to actuator saturation and disturbance. Automatica 38, 351–359 (2002)
104. Hu, T., Lin, Z.: Control Systems with Actuator Saturation: Analysis and Design. Birkhäuser, Boston (2001)
105. Hu, T., Ma, L., Lin, Z.: On several composite quadratic Lyapunov functions for switched systems. In: Procceding of the 45th IEEE Conference on Decision and Control, San Diego, USA, pp. 113–118 (2006)
106. Hu, T.: Switching law construction for discrete-time systems via composite quadratic functions. In: American Control Conference ACC 2007, New York, July 11-13 (2007)
107. Igic, P.M., Towers, M.S., Mawby, P.A.: A 2D physically based compact model for advanced power bipolar devices. Microelectronics Journal 35, 591–594 (2004)
108. Kaustky, J., Nichols, N.K., Chu, E.K.W.: Robust pole assignment in singular control systems. Linear Algebra Appl. 121, 9–37 (1989)
109. Kaczorek, T.: Two-dimensional linear systems. LNCIS, vol. 68. Springer, Germany (1985)
110. Kaczorek, T.: Realization problem, reachability and minimum energy control of positive 2D Roesser model. In: Proc. 6th Ann. Int. Conf. Advances in Communication and Control, Hungary, August 11-17, pp. 765–776 (1997)
111. Kaczorek, T.: Two dimensional linear systems. Springer, Heidelberg (1985)
112. Kalpakjian, S., Schmid, S.R.: Manufacturing engineering and technology, 5th edn. Pearson, New Jersey (1992)
113. Kimura, H.: Pole assignment by gain output feedback. IEEE Trans. Aut. Control 20, 509–517 (1975)
114. Kimura, H.: A further result on the problem of pole assignment by output feedback. IEEE Trans. Aut. Control 22, 458–463 (1977)
115. Lam, H.K.: Stability analysis of sampled-data fuzzy controller for nonlinear systems based on switching T-S fuzzy model. Nonlinear Analysis: Hybrid Systems 3, 418–432 (2009)
116. Lee, E.B., Lu, W.-S.: Stabilization of Two-Dimensional Systems. IEEE Trans. Aut. Control 30, 409–411 (1985)
117. Lian, J., Zhao, J., Dimirovski, G.M.: Robust H_∞ sliding mode control for a class of uncertain switched delay systems. Int. J. Syst. Sci. 40, 855–866 (2009)

118. Liberzon, D., Morse, A.S.: Basic problems in stability and design of switched systems. IEEE Control Systems Magazine 19, 59–70 (1999)
119. Liberzon, D.: Switching in systems and control. Springer, Heidelberg (2003)
120. Lin, H., Antsaklis, P.J.: Switching stabilizability for continuous-time uncertain switched linear systems. IEEE Trans. Aut. Control 52, 633–646 (2007)
121. Liu, Y., Dimirovski, G.M., Zhao, J.: Robust output feedback control for a class of uncertain switching fuzzy systems. In: Proceedings of the 17th IFAC World Congress, Seoul, Korea, July 6-11 (2008)
122. Liu, J., Liu, X., Xie, W.C.: Exponential stability of switched stochastic delay systems with non-linear uncertainties. Int. J. Sys. Sc. 40, 637–648 (2009)
123. Liu, X.: Stability analysis of switched positive systems: A switched linear copositive Lyapunov function method. IEEE Trans. Circuits Sys. II 56, 414–418 (2009)
124. Lo, W.C., Wang, L., Li, B.: Thermal transistor: heat flux switching and modulating. Journal of the physical Society of Japan 77, 1–4 (2008)
125. Lu, W.-S.: Some new results on stability robustness of two-dimensional discrete systems. Multidimensional Syst. and Signal Proc. 5, 345–361 (1994)
126. Lygeros, J., Tomlin, C., Sastry, S.: Controllers for reachability specifications for hybrid systems. Automatica 35, 349–370 (1999)
127. Marszalek, W.: Two dimensional state-space discrete models for hyperbolic partial differential equations. Appl. Math. Models 8, 11–14 (1984)
128. Mignone, D., Ferrari-Trecate, G., Morari, M.: Stability and stabilization of Piecwise affine and hybrid systems: an LMI approach. In: Procceding of the 39th IEEE Conference on Decision and Control, Sydney, Australia (2000)
129. Mesquine, F., Tadeo, F., Benzaouia, A.: Regulator problem for linear systems with constraints on the control and its increments or rate. Automatica 40, 1378–1395 (2004)
130. Mesquine, F., Mehdi, D.: Constrained observer based-controller for linear continuous-time systems. Int. J. Sys. Sciences 27, 1361–1367 (1996)
131. Mulder, E.F., Kothare, M.V., Morari, M.: Multivariable anti-windup controller synthesis using linear matrix inequalities. Automatica 37, 1407–1416 (2001)
132. Nachidi, M., Tadeo, F., Hmamed, A., Benzaouia, A.: Static output-feedback stabilization for time-delay Takagi-Sugeno fuzzy systems. In: 46th IEEE Conference on Control and Decision, New Orleans, December 12-14 (2007)
133. Nachidi, M., Tadeo, F., Hmamed, A., Benzaouia, A.: Static output-feedback stabilization for time-delay Takagi-Sugeno fuzzy systems. In: 46th IEEE Conference on Control and Decision, New Orleans, December 12-14 (2007)
134. Nachidi, M., Benzaouia, A., Tadeo, F.: Based approach for output-feedback stabilization for discrete-time Takagi-Sugeno systems. IEEE Trans. Fuzzy Systems 16, 1188–1196 (2008)
135. Nachidi, M., Tadeo, F., Benzaouia, A., Ait Rami, M.: Static output-feedback design for Takagi-Sugeno systems with delays. Int. J. Adaptive Control Signal Processing 25(4), 295–312 (2011)
136. Naib, M., Benzaouia, A., Tadeo, F.: l_1-control using linear programming for systems with asymmetric bounds. Int. J. Control 78, 1459–1465 (2005)
137. Ozcaldiran, K., Lewis, F.: A geometric approach to eigenstructure assignment for descriptor systems. IEEE Trans. Aut. Control. 32, 629–632 (1987)
138. Petterson, S.: Synthesis of switched linear systems. In: 42nd Proceeding of the Conference on Decision and Control, Hawaii, December 9-12 (2003)
139. Porter, B.: Eigenvalue assignment in linear multivariable systems by output feedback. Int. J. Control 25, 483–490 (1977)

140. Radhy, N.E., Benzaouia, A., Boughari, H.: Constrained state regulation of linear continuous-time singular systems. Systems Analysis Modelling Simulation 42, 677–694 (2002)
141. Roesser, R.: A discrete state space model for linear image processing. IEEE Trans. Aut. Control 20, 1–10 (1975)
142. Sun, Z., Ge, S.S.: Switched linear systems: control and deign. Springer, Heidelberg (2005)
143. Santesso, P., Valcher, M.E.: Reachability properties of discrete-time positive switched systems. In: Proceedings of the 45th IEEE Conference on Decision and Control, San Diego, USA, December 13-15, pp. 4087–4092 (2006)
144. Seborg, D.E., Edgar, T.F., Mellichamp, D.A.: Process dynamics and control. John Wiley and Sons, New York (1989)
145. Shen, C., Ban, Y., Dimirovski, G.M., Jing, Y.: Robust delay-dependent stability and stabilization of polytopic systems with time delay and its application to flight control. In: Proc. 27th American Control Conference, Seattle, WA, June 11-13, pp. 1624–1629 (2008)
146. Shorten, R.N., Narendra, K.S.: On the existence of a commun Lyapunov function for linear stable switching systems. In: Shorten, R.N. (ed.) Proc. 10th Yale Workshop on Adaptive and Learning Systems, pp. 130–140 (1998); A sufficient condition for the existence of a commuon Lyapunov function for two second-order liinear systems. In: Proc. 36th Conference on Decision and Control, pp. 3521–3522, San Diego, December 10-12 (1997)
147. Takagi, T., Sugeno, M.: Fuzzy identification of systems and applications to modeling and control. IEEE Transactions on Systems, Man and Cybernetics 15, 116–132 (1985)
148. Tanaka, K., Masaaki, I., Wang, H.O.: Stability and smoothness conditions for switching fuzzy systems. In: Proceedings of the 19th American Conference on Control, Piscataway, NJ, pp. 2474–2478 (2004)
149. Tarbouriech, S., Burgat, C.: Positively invariant sets for constrained continuous-time systems with cone properties. In: Proc. of the 30th Conference on Decision and Control, Brighton, December 13-15 (1991)
150. Tarbouriech, S., Prieur, C., Gomes da Silva, J.M.: Stability analysis and stabilization of systems presenting nested saturations. IEEE Trans. Aut. Control 51, 1364–1371 (2006)
151. Vidyasagar, M.: On matrix measures and convex Liapunov functions. J. Math. Anal. Appl. 62, 90–103 (1978)
152. Wang, R., Zhao, J.: Exponential stability analysis for discrete-time switched linear systems with time-delay. Int. J. Innovative Computing, Information Control 3, 1557–1564 (2007)
153. Wang, M., Zhao, J., Dimirovski, G.M.: Stabilisation and L2-gain analysis for a class of uncertain switched non-linear systems. Int. J. Sys. Sci. 40, 1077–1085 (2009)
154. Wang, H.O., Tanaka, K., Griffin, M.F.: An approach to fuzzy control of nonlinear systems: Stability and design issues. IEEE Trans. Fuzzy Systems 4, 14–23 (1996)
155. Wicks, M.A., Peleties, P., Decarlo, R.A.: Construction of piecewise Lyapunov functions for stabilizing switched systems. In: Procceding of the 33rd IEEE Conference on Decision and Control, Lake Buena Vista, December 14-16 (1994)
156. Wonham, W.M.: On pole assignment in multi-input controllable linear systems. IEEE Trans. Aut. Control 12, 660–665 (1967)
157. Wu-sheng, L., Lee, E.B.: Stability analysis for two-dimensional systems via a Lyapunov Approach. IEEE Trans. Circuits Sys. 32, 61–68 (1985)
158. Wu, L., Shi, P., Gao, H., Wang, C.: Hinfinity filtering for 2D Markovian jump systems. Automatica 44, 1849–1858 (2008)

159. Xiang, Z.R., Xiang, W.M.: Stability analysis of switched systems under dynamical dwell time control approach. Int. J. Sys. Sci. 40, 347–355 (2009)
160. Xie, D., Wang, L., Hao, F., Xie, G.: Robust Stability Analysis and Control Synthesis for Discret-time Uncertain switched Systems. Appl. Math. Comput. In: 42nd IEEE Conference on Decision and Control, Hawaii, December 9-12 (2003)
161. Xu, S., Lam, J.: Robust control and filtering of singular systems. LNCIS. Springer, Heidelberg (2005)
162. Yang, H., Liu, H., Dimirovski, G.M., Zhao, J.: Stabilization control of a class of switched fuzzy discrete-time systems. In: Proc. IEEE International Conference on Fuzzy Systems, London, UK, July 23-28, pp. 1345–1350 (2007)
163. Yang, H., Dimirovski, G.M., Zhao, J.: Switched Fuzzy Systems: Representation Modelling, Stability Analysis, and Control Design. SCI, vol. 109. Springer, Heidelberg (2008)
164. Yaz, E.: On state-feedback stabilization of two-dimensional digital systems. IEEE Trans. Circuits Sys. CAS-32, 1069–1070 (1985)
165. Yu, J., Xie, G., Wang, L.: Robust Stabilization of discrete-time switched uncertain systems subject to actuator saturation. In: American Control Conference, New York, July 11-13 (2007)
166. Zhao, J., Dimirovski, G.M.: Quadratic stability of switched nonlinear systems. IEEE Trans. Aut. Control 49, 574–578 (2004)
167. Zhou, B., Duan, G.R.: A new solution to the generalized Sylvester matrix equation $AV - EVF = BW$. Systems Control Letters 55, 193–198 (2006)

159. Xiang, Z.R., Wang, W.M.: Stability analysis of switched systems under dynamical dwell time control approach. Int. J. Syst. Sci. 40, 347–355 (2009)

[illegible]

Index

MIX
Papier aus verantwortungsvollen Quellen
Paper from responsible sources
FSC® C105338

If you have any concerns about our products,
you can contact us on
ProductSafety@springernature.com

In case Publisher is established outside the EU,
the EU authorized representative is:
Springer Nature Customer Service Center GmbH
Europaplatz 3, 69115 Heidelberg, Germany

Printed by Libri Plureos GmbH
in Hamburg, Germany